D1085157

ENERGY

THE SOLAR-HYDROGEN ALTERNATIVE

ENERGY
THE SOLAR-HYDROGEN ALTERNATIVE

J. O'M. BOCKRIS

Institute of Solar and
Electrochemical Energy Conversion
The Flinders University of South Australia
Adelaide, Australia

A HALSTED PRESS BOOK

JOHN WILEY & SONS
New York — Toronto

First published in 1975 by
Australia & New Zealand Book Co Pty Ltd
23 Cross Street Brookvale NSW 2100 Australia

Library of Congress Cataloging in Publication Data
Bockris, John O'M.
Energy, the solar hydrogen alternative.
"A Halsted Press book."
1. Solar energy. 2. Hydrogen as fuel.
I. Title.
TJ810.B58 1976 333.7 75-19125
ISBN 0-470-08429-4

Published in the U.S.A. by
Halsted Press, a division of John Wiley & Sons, Inc., New York.

Wholly set up and printed in Australia by
Hogbin, Poole (Printers) Pty Ltd, Redfern NSW Australia
Printed in Australia

Contents

Preface

This book originated in 1973 and 1974, during which time several international meetings were held to discuss the consequences of using hydrogen as a medium of energy. Because many of the considerations depend upon the characteristics of the original energy source, I have coupled a presentation of the Hydrogen Economy with that of various possible energy sources, discussing largely solar, and related, sources, because information on the nuclear ones is easily available in other books.

The aim of the book is to give to energy planners and policy makers a number of relevant and related facts under one cover; and to give to scientists and engineers, the background of considerations which suggest a relation between the economic availability of solar energy at industrially active areas, and the cheap transport of energy over long distances in the form of hydrogen. The book should be of value not only to professionals as above, but also to students of energy conversion, environmentalists, and all those concerned in the effort to prepare for the coming era of difficulty in which a change in energy source has to be made from the diminishing fossil fuels to inexhaustable clean ones.

J. O'M. Bockris
1975

Dunk Island,
North Queensland

Flinders University of South Australia,
Adelaide

Acknowledgements

My acknowledgements are firstly due to Mrs Pam Paddick. They go firstly for her expertise in the typing of four to five drafts of the present work. But Mrs Paddick has excelled all former help which I have had in book writing, by invaluable information gathering in the library; in collating and investigating information; and in arranging of references, and carrying out many other tasks, including aiding me with her excellent memory.

I acknowledge with great appreciation also the work of Mrs Jenny Pugatschew, who has liaised with others in obtaining information, and particularly looking after the voluminous correspondence associated with the writing of this book, liaising with the draftswoman and acting in a minor capacity as draftswoman herself.

I thank Mrs Janet Crooks, the principal draftswoman, for carrying out her job with skill and accuracy.

Thereafter, I must thank many colleagues for their help to me in reviewing chapters (each chapter has been reviewed by several specialists in its area), and in giving their opinion on many points. Only the main contributions can be acknowledged.

Francis Bacon has helped me in many letters, particularly with the history of the Hydrogen Economy, as also matters concerning the storage of hydrogen. I have corresponded extensively with many other people about the history of the Hydrogen Economy, and Olle Lindström and Eduard Justi have been helpful in getting me to a maximum of appreciation of the pre-war work. R.J. Schoeppel of the Oklahoma State University has corresponded with me about the history of the Hydrogen Economy, and sent me early papers.

M. King Hubbert of the United States Department of the Interior, whose work is so basic to our evaluation of the exhaustion time of fossil fuels, has had my statements evaluated by David E. Root of his office and, apart from correspondence which he has had with me, Root's detailed evaluation of my chapter 3 has been used in later drafts.

I have been in correspondence several times with Martin A. Elliott of the Texas Eastern Transmission Corporation regarding his valued papers upon the lastingness of fossil fuels, and he has made excellent criticisms of my understanding of the situation. Lawrence Lessing, formerly of the Fortune Magazine, has helped me with discussion on the future of coal. W. Häfele has given me his views about how the coal supply might be extended. John E. Johnson of Union Carbide has evaluated my work and given criticisms of my various priorities. M. Steinberg of Brookhaven National Laboratory has also made an evaluation of my writing on coal, and I have used his criticisms in revision. I have gathered much from P.N. Thompson of the U.K. Operational Research Executive, and his appraisal of my writing on underground gasification has been valuable. L. Grainger of the U.K. National Coal Board has also given me his appraisal of the underground gasification.

Philip Hammond, formerly of the Atomic Energy Commission at

Oakridge, was one of those who helped me get into the Energy business, and he has helped me further in several conversations and with letters, papers and comments of my writing. Edward Teller has helped me in one conversation concerning the future of breeder reactors, and his office has been helpful in other matters with regards to underground gasification. Correspondence with R.E. Powe of the Montana State University concerning various methods of wind power has been most helpful. Perhaps one of my greatest degrees of thanks must go to L.F. Mullett, formerly of the South Australian Electricity Trust, who introduced me to concepts of wind power in its modern form, and taught me how to estimate mean annual wind energy from the various types of data which exist. J.V. McAllan of the Australian C.S.I.R.O. has been helpful in respect to various pieces of information on Australia, for example, about tidal power.

I have had many letters and much correspondence, and a visit, with David Jewett of the TYCO laboratories, who is directing work upon the new edge-defined growth method for photovoltaic silicon. A.F. Hildebrandt of the University of Houston has written me many extra notes and pieces of information about the work of him and his team on the mirror concentrator method for solar energy. Peter Glaser of Arthur D. Little has corresponded with me in several letter-discussions concerning the value of the orbiting satellite method of collecting solar energy compared with the earth-based method. Bill Escher has kept me informed of his latest developments in respect to considerations of floating platforms for ocean-based systems.

Kenneth E. Cox of the University of Kentucky has given me comments on prices of thermo-chemical hydrogen generation. James E. Funk, also of the University of Kentucky, has sent criticisms of my chapter on methods for obtaining hydrogen fuel. G. deBeni of Euratom in Italy has forwarded me several reports and chapters, plus some criticisms of my evaluation of the work on chemical methods of obtaining hydrogen.

J. Schröder of Philips in Aachen has sent criticisms of the thermal energy storage chapter. Krafft Ehricke has given me help in evaluating the distance at which a space-based satellite would be a preferable economic proposition to that of a pipe containing hydrogen.

Lawrence W. Jones of the University of Michigan has exchanged several informative letters with me on hydrogen safety, and sent me a table on estimates of the price of hydrogen.

R.J. Schoeppel has also helped me with regards to transportation matters, in particular the present state of the hydrogen engine.

Andre Sjoerdsma of the Future Shape of Technology Foundation of Amsterdam, has helped me with some erudite correspondence concerning predictions of technological progress, including many criticisms and evaluations of early drafts of my proposals and statements in chapter 17.

Nejat Veziroglu of the University of Miami, whose considerable efforts for a Hydrogen Economy are well known, has been good enough to look through and comment upon some eight chapters in the book.

Lastly, I have gained tremendously by the study of 'the blue book' by Derek Gregory (assisted by P.J. Anderson, R.J. Dufour, R.H. Elkins, W.J.D. Escher, R.B. Foster, G.M. Long, J. Wurm and G.G. Yie), *A Hydrogen Energy System,* and correspondence with Derek Gregory has enlightened some points.

CHAPTER 1

Introduction: Some Questions

IN THIS INTRODUCTORY chapter, themes for the other chapters of the book will be surveyed and the inter-relatedness of some of them shown.

ENERGY BASES TO THE ECONOMY

Technology since its inception has been associated with coal; then with oil and coal; and more recently with oil, natural gas and coal. The first two of these fuels are now undergoing price rises presaging their exhaustion. They will be replaced by atomic resources and perhaps by solar sources which will produce heat and electricity directly.

There arise from this situation of a coming change in energy base, the many questions which underlie this book.

How long will the present fuels last, and at what price? Will there be a substantial intermediate use of coal to replace segments of the economy now fueled by oil and natural gas? What of the rate of development of atomic and solar sources? What intermediate fuels will be used to convey the energy of the central sources to their use in industry, housing and transportation? In particular, what of the balance between an all-electrical economy, an economy in which methanol is the chief medium, and one in which this role is played by hydrogen?

Most of the considerations of this book concern the future; and the scientific knowledge available at present must be the basis of the *projections* which are made.

The title of the book, and its emphasis, concern that option which seems at present to be the one most advantageous to the longer-term future, that beyond 2000. The origin of the energy will be from solar and atomic sources. The transmission of energy will be in hydrogen, and its storage will be in hydrogen, liquid or gaseous. The role of hydrogen is connected particularly to the solar part of the energy supply; but it would also couple well with nuclear sources. Indeed, during the decades in which fossil fuels will still be the main source of energy (and particularly if coal is used to replace oil and natural gas), their gasification to hydrogen, and not to methane, would resolve the mounting pollutional problems, which are inextricably connected with the available options.

HOW MUCH TIME IS AVAILABLE FOR A CHANGE IN THE ENERGY BASE TO THE ECONOMY?

Change in the energy base from fossil fuels to atomic and solar sources is not a matter of choice which can be carried out according to the economic and social pressures of the time. It needs preparation over many decades. What time is there until the new sources must be producing and must have taken over from coal, natural gas and oil? For what time would the coal resources last? What time will the new energy system take to build, after the research and development has been decided upon – and carried out? How long will that research and development take? To what extent can conservation measures extend the time in which the fuels can be used? How much will the fall in living standards be in the Western countries before Abundant, Clean Energy sources are available?

TO WHAT EXTENT CAN COAL BE RELIED UPON AS A MASSIVE ENERGY SOURCE?

It is generally assumed that coal is plentifully available in most countries, and that a return to coal could be made for 'hundreds of years'. This concept is an opiate for the energy difficulties of the times, for it implies that the price rises of natural gas and oil would only lead to the difficulties associated with returning to coal, and not to the disastrous implication of energy exhaustion. Do the predictions upon which this confidence in coal is based depend upon implicit assumptions which may be untrue? How much will the growth of the population, and the living standards which it will expect, change these estimates? Do the estimates of the number of years upon which coal can be relied assume the energy demands of the present time, or the more appropriate ones, of, say 2000 and after?

Estimates of the lastingness of a resource depend upon an extrapolation of previous use, together with some estimate of the total amount available. Do estimates made in respect to coal involve the tacit assumption that energy demands made upon it will simply continue the previous course? Have the facts that oil and natural gas have to be replaced with synthetic gas from coal been taken into account in the estimates? To what extent do these estimates take into account the fact that, now and in the fairly recent past, coal has been mainly used for the production of electricity, which at present is only a small part of the total energy budget? What would be the predicted extinction times of coal were economies to run largely on coal from 2000?

Further, suppose that a decision is made to rely on coal in massive amounts, is it possible to get it out of the ground in the time necessary? What will be the environmental effects if all the sulphur cannot be completely removed from coal, as is the case at present?

ABUNDANT, CLEAN ENERGY

There is no doubt that Abundant, Clean Energy *can* be obtained for a population up to about three times the present one. The doubts are in respect to the *time* at which this new energy will be available.

The sources from which it can come are atomic, solar, wind, perhaps geothermal, and with small additions from hydro-electric and tidal sources. With present research plans, when will atomic energy be available to the extent of, say, 50 per cent of the demands of the U.S. economy at 2000? What could the other energy resources be at that time, when oil and natural gas will be very short? What are the difficulties of greatly accelerating the building of atomic reactors? What of the various schemes for clean atomic waste disposal? Why have none of them as yet been implemented? What of the dangers of atmospheric pollution from atomic reactors?

A widespread view about solar energy is that the conversion apparatus for it is so expensive that the availability of it in an electrical form would give rise to energy at least ten times higher in cost than previous energy. To what extent have post-1970 research developments changed the validity of this view? Is there a prospect that, within the next three or four decades, solar energy could become a main source of energy? What of the possible part which could be played by thin film photovoltaics?

What are the other paths to solar energy, apart from photovoltaics? The use of the temperature difference in the oceans was explored in the 1920s. Could machinery be set up to use such gradients and produce power perhaps as early as one to two decades from the present? What kind of cost projections do researchers in these fields make at the present time? What of the secondary uses of solar energy in households? The energy budget of households is some 25 per cent of the economy. To what extent could a substantial amount of this household heat and light be supplied by solar energy? Is it true that this secondary solar energy stands before us as an immediately realisable resource? Is there potentially a new industry which could be launched in this decade in respect to the manufacture of commercial conversion kits for household solar energy?

From these matters arises the relationship between money and time in energy research. Is distribution of money among the various energy sources an appropriate one, suitable for needs foreseen fifteen to twenty five years ahead? Again, what time is there to produce the new energy sources which will be clean, and will prevent a reduction of living standards in Western countries? How much research money is going into the various sections of this research? The concentration in the U.S. is largely upon the gasification of coal, and breeder reactors (hundreds of millions of dollars per year), whilst research into other sources, particularly solar ones, is at tens of millions of dollars per year? Should these priorities be changed?

METHODS OF TRANSMISSION OF ENERGY

In previous times 'power stations' were near to the sites of use, but already in the late 1960s citizen groups made it difficult to accept atomic reactors close to cities. What about the advantages and fall in the price of fuel produced from very large reactors? How can heat pollution (and perhaps other kinds of pollution) from these reactors be dealt with? Could reactors be advantageously set up hundreds of miles from the site of use, where their pollutional problems would be reduced? What of the cost of transmission of energy in the form of electricity from them? Would it be cheaper to electrolyse water at source and send the hydrogen from pipes to distant sites of use? Is the cheap transport of hydrogen through pipes, a method whereby energy sources thousands of miles distant from the sites can be economically used? What implications do such properties of hydrogen have for the development of solar energy at distant sources? Would it be possible for solar sources, for example in the Sahara, to supply Europe with energy? What of the possibilities of the use of Australia as an energy depot for the supply of hydrogen from solar energy to South-East Asia?

Very large scale transportation of energy might, of course, be carried out in other ways. What of the feasibility and economics of radiation beams directed to satellites, and their transmission thereby to distant towns? How would the cost of such schemes, relevant perhaps as soon as 2000, be comparable with the costs of long distance transmission by hydrogen?

Are these advantages of hydrogen as a medium of long distance transmission in the use of future energy resources taken into account in the estimation of the relative advantages of a hydrogen economy compared with those of other economies?

WHAT METHODS CAN BE USED FOR THE CHEAP PRODUCTION OF HYDROGEN?

The traditional methods for obtaining hydrogen, such as those from natural gas, are not relevant in the time scale of decades which must be considered in respect to New Energy. Water will be the only source of hydrogen in this time scale.

Traditional electrolysis is regarded as the method by which hydrogen can be obtained from water. However, electrolysis has a very big point against it, because it must suffer from any of the inefficiencies of the conversion of a basic energy source to electricity. Conversely, there is no chance of a direct chemical production of hydrogen from water, because of the temperatures involved. Can hydrogen be produced indirectly by chemical reaction of a substance M with water, for example how can it combine with water to give $MO + H_2$; and the MO thermally decomposed to oxygen and M? Would it be possible to get sufficient thermal efficiency in a series of cyclical chemical reactions to give hydrogen more cheaply that it can be obtained from electrolysis? Will the degree of completeness of the cyclical reactions be satisfactory so that no build-up of the intermediate can occur?

Electrolysis is usually rejected as the method to obtain hydrogen because it would be too expensive. Is this true in the time scale being considered? How much does electrode overpotential play a part in the efficiency of the production of hydrogen by electrolysis? What of the relevance of the fact that the electrolysers upon which the pessimistic cost estimates are built, represent a technology little different from that of the 19th century? How much funding is being put into research for new kinds of electrolysers?

What of hybrid methods in which both thermal and electrolytic methods are combined? The energy needed to decompose water decreases with increase in temperature. What of the electrolysis of steam at high temperatures? Is there a technology of solid electrolytes sufficient to give low resistances and make the driving of such electrolysis by heat an economic proposition?

Electrolysis of water has large overall advantages; a known process, a single process, a simple process, and two gases are produced separately.

What priorities of research are given to the technology of water electrolysis compared with other methods of producing hydrogen?

In the analysis of both thermochemical and electrolytic means of producing hydrogen, the principal cost is the cost of the fuel, heat or electricity, to break the bonding in the water. What of the possibilities of photochemical decomposition? Is photoelectrochemical decomposition, producing hydrogen and electricity at the same time, a possibility? What of photosynthesis?

QUESTIONS CONNECTED WITH THE STORAGE OF ENERGY

Coal, oil and natural gas have been used directly, and as they exist in large quantities in the ground, questions of their manufacture and storage have not arisen. The new sources will be nuclear and solar. Nuclear sources are capital intensive: it is wasteful not to use them steadily all the time. Storage of the energy is necessary because of the irregular use pattern, and solar energy must be stored because of its sporadic nature. In what form will the energy be stored? Is this a principal advantage of hydrogen as a medium of energy? Could hydrogen be stored on a large scale underground? What of the costs of liquefaction? Is there a future in the storage of hydrogen in hydrides, particularly for hydrogen as a transportation fuel?

ARE SAFETY MATTERS A SUBSTANTIAL DIFFICULTY IN RESPECT TO THE USE OF HYDROGEN?

Safety difficulties have been raised with every fuel; they were raised in the early days of the use of oil, and are raised now as a principal objection to the development of breeder reactors. In the same way, questions of safety are the first ones which arise in respect to the massive use of hydrogen as an energy medium. The reference substance must be methane. Is hydrogen less safe than natural gas? What are the properties which would lead to this conclusion? Are there counter factors, some of which may make hydrogen safer than the use of natural gas?

In particular, what about the experience obtained by NASA during the decade in which it has used hydrogen as a rocket fuel? Hydrogen certainly has an image problem, to what extent is it a real problem?

WHAT WOULD BE THE PROBLEMS WITH MATERIALS?

Hydrogen embrittlement is a well-known phenomenon. Would it perhaps cause the type of steels used in piping hydrogen to be too expensive; so that the costs of piping this medium of energy – based upon known costs of piping natural gas – could be greatly increased? Are there any special kinds of embrittlement associated with the contact of coal-hydrogen with steel undergoing changes in pressure twice daily and therefore repeated changes in stress levels.

WHAT OF THE REDEVELOPMENT OF ENERGY FROM HYDROGEN?

Eventually, fossil fuels are a medium of solar energy. However, their presence in the ground makes us regard them as primary fuels. This is not so with hydrogen. It may become an intermediate of atomic or solar energy. Thus, it has to be reconverted to mechanical and electrical energy to go into immediate use. How will the alternatives for this conversion compare in cost and feasibility? Will a path *via* fuel cells and electric motors be cheaper for the same performance than a path *via* direct chemical conversion in a combustion engine? Will combustion engines, with their Carnot efficiencies, working conventionally with generators, be cheaper than large fuel cells? What are the lifetime difficulties of fuel cells and how may these affect their economics?

WOULD THE LARGE SCALE AVAILABILITY OF HYDROGEN AND OXYGEN HAVE CONSEQUENCES IN TECHNOLOGY?

Hitherto concentration has been on raising questions concerning the energy future in terms of time at which the fossil fuels will be largely replaced by atomic or solar sources, and the comparative costs of various pathways. But one of the main accompanying problems, which if not considered will make the evaluation of the options incomplete, is the pollutional problem. This problem comes not only in the way in which energy is produced, but in the way it is used; in manufacture, synthesis, extraction and recycling. In the coming decades, the development of an energy system must take place with considerations of the polluting nature of the manufacturing process well in mind. Could hydrogen be used massively in the production of steel? Of copper? Of nickel? Of aluminum? What about the availability of cheap oxygen? Could it be used to facilitate sewage disposal? What of the extensive replacement of air by oxygen in manufacture? Would the economic gains in the reduction of plant size be sufficient to pay for the piping of the oxygen from electrolysis plants, or for its extraction from air? To what extent are long-term ecological and environmental effects considered when methanol and methane are considered as future fuels, in comparison with hydrogen?

IS HYDROGEN-BASED TRANSPORTATION A PRACTICAL OPTION?

The tranquil picture of energy development, so broken at present, first began to fray in the 1960s with the realisation that the smog prevalent in Los Angeles, and increasing in many other cities, arose mainly from automobiles. The car remains the principal polluter. What are the alternatives? Electric cars are the main ones considered, but their research and development problems come largely down to the one of developing economic high energy density batteries. However, little research has been put into this area since the cut-back in space research, in which such batteries underwent rapid development during the '60s.

What of the advantages of hydrogen cars over electrically powered cars? What performance possibilities do they have? What degree of change in present car engines would be necessary to convert them to hydrogen? Would hydrogen driven transports have the performance of gasoline driven ones? What about the expense of cryogenic condensers? What of the cost of storage in hydrides which easily release hydrogen upon moderate heating? Supersonic and hypersonic transports have difficulties in respect to range when run on contemporary fuel. Is liquid hydrogen the answer for the development of future concepts in air transportation? What about materials problems arising during the heating of the body of supersonic transports? Would the presence of hydrogen cooling change prospects by giving lower weight by the use of light construction materials which would normally lose strength at higher temperatures?

It is important when considering options available in the development of energy not to compare hydrogen upon the simple basis of cost, for instance with methanol. The simple criterion of cost must be accompanied by the evaluation of long-term environmental considerations. The use of hydrogen to provide clean industrial processes, and to give clean land and air transportation, are points which must be weighed when hydrogen is compared as a medium with the costs and properties of other media.

ENVIRONMENTAL QUESTIONS

The strong and obvious claim of hydrogen to primacy in respect to clearing up air pollution would seem to leave few questions concerning the environment. But this is so only if environmental questions are limited to the consideration of pollution. It is not so if the environmental factors include the use-up of resources. In this respect, how does a Hydrogen Economy link with solar energy from photovoltaics compare, say, with a Hydrogen Economy link with nuclear sources? How does a Nuclear Economy linked with a methanol medium compare with one in which hydrogen is the medium? What of the Solar-Electrochemical Economy? Can these environmental damage weights be given a quantitative measure?

COMPETING ENERGY ECONOMIES

The choice and needed avenues of research for the development of a new energy economy depend upon many aspects. They depend firstly, on the technical feasibility of the schemes, and secondly on the cost of the final product. They depend on the time scale concerned, and this time scale is at least one and perhaps as many as six decades into the future. They also depend on the long-term ecological aspects, the fittingness of the energy base to industry and the connection with the underdeveloped countries, leading eventually to the evolution of a world economy.

What are the competing energy systems for consideration in research funding? What of synthetic natural gas from coal? Of methanol from coal? Of methanol from atmospheric CO_2, and nuclearly derived hydrogen? How much will the disadvantages of hydrogen (gaseous fuel and hence image of danger) be able to compete with the advantages of methanol (liquid, safe and like gasoline)? How will this depend upon the energy source; coal, atomic or solar? What will be the effect of the distance between the source and the site of use? What about the possibility of hydrogen being converted to more convenient liquids, such as the boranes? Will the extra weight per unit of energy in the fuel be compensated by a lighter storage system?

Availability, price, pollution and long-term aspects of ecology are some of the factors which will be taken into account in deciding the merits of this or that kind of combination of energy source and energy intermediate. But it is no use considering what this is at the present time.

The projections have to be taken into the future, the ten and twenty year future, even the seventy five year future. For this reason, the cost and even the feasibility of the estimates will be associated with considerable uncertainties. However, the basis of the development of new energy economies have to be laid out in this decade for the economies of four to five decades ahead. The research and development will be a two or three decade matter, but the building and spread of the new energy economies would be over a further two to five decades. Are the uncertainties of attempting to project fifteen, twenty five or fifty years ahead so great that they invalidate any attempt to distinguish between possibilities? Do we have other ways in deciding which research lines shall be built up and funded now and, where the concentration of the funds should be?

Does the degree of futurity in the necessary thinking make actual price estimates meaningless? What do we know of inflation and economic changes even five years from now, let alone twenty five years, or even fifty years ahead? What have these limitations and difficulties in economic projections to do with numerical results stated in this book?

PROSPECTS

The energy supply and demand, and the options for a new energy basis to the economy in the next fifty years, roughly the area of investigation of this book, are the province of science and engineering, firstly, and economics secondly. The subject is one in which the separation of science and engineering from economics, and then from politics and sociology, is not possible. The main rate-determining aspects are more likely in the latter aspects than in science and engineering. Indeed, if the research and development could go forward 'freely' (that is with research budgets which are large enough to cope with the ideas available), there would be little doubt of the available outcome, as indeed dramatically illustrated in NASA's moon programme. But the extent of the R & D funding depends upon sociological and political factors. Politics is the practice of the distribution of wealth. Per capita income is proportional to per capita energy. Energy, income and politics, are all closely connected, and differently so for each country.

The questions to which this gives rise are very wide indeed. The International Corporations are the essence and the bastions of capitalism. But they are geared to short-term considerations of profitability. Will it be possible for them to work to develop an energy economy with considerations fifteen, twenty five or fifty years ahead? Is it possible for a corporation to hold back the maximisation of yearly profit so that it can exist healthily one to two generations ahead? Would any shareholders ever vote to reduce their income, or even to avoid it, for a future generation? Capital is invested in the present technology. Will its owners bring obsolescence to that technology, and so devalue their capital for the sake of long term environmental considerations? What of the effect of these concepts on the likelihood of the funding of new bases to the energy economy which do not use the machinery in which the capital is invested? Is massive research funding of new energy sources in the U.S. economy, which is so accountable to, and so affected by, corporate thinking, conceivable? But what if such funding is not given in time?

How is research funding decided? Are U.S. Government committees often peopled by those who – however solid their personal integrity – are immersed in their company's thinking? Would they be likely to recommend the necessary funding of the development of energy sources which will decrease the long-term viability of the companies in which they have evolved?

As far as can be seen with the data available, there seems to be a period of danger in the U.S. as early as the mid-1980s when fossil fuel imports will have to be so great that their interruption would cause disaster within months. Is such an interruption so unlikely that the acceptance of the large-scale imports necessary in the 1980s is tolerable? What are the alternatives which can be mobilised in the U.S. between the late-80s and the coming of the Abundant, Clean Energy sources after 2000? If the price rises in energy give rise to a reduction of living standards, would that reduction be sufficient to have revolutionary implications? And is there any new energy source which could be built up to significant proportions at a time scale of as little as fifteen years?

The material in this book ranges from stable, definite and scientific engineering matters of the present, through fairly well based projections in the several-decade scale in which only inflational changes will make important distortions to the figures estimated. But it also includes material which is *speculative* and which concerns the *possibilities* which are foreseeable at this time.

CHAPTER 2

The Hydrogen Economy

COMPONENTS OF DOOM

THE WELL-KNOWN graphs of Meadows and his collaborators[1] clearly indicate the occurrence of a Doom Decade* somewhere between 2020 and 2070. Thus, by such a time, pollutional levels from fossil fuels will become unacceptable, even if massive research funds are deployed to reduce them. The growing CO_2 concentration in the atmosphere from fossil fuel combustion and/or aerosol concentrations from such combustions will have adverse climatic effects.[2] Pollutants, such as NO, CO and SO_2, will provide substantial health hazards.

Important resources, particularly some metals, will exhaust before 2000 (Table 2.1).[1] A fossil fuel economy, even if gasified coal is included, may not pass 2000, because of the exhaustion of economically minable coals[3] (see Chapter 3).†

Meadows' work has been criticised.[4,5] For example, a detailed examination of the computer programs used reveals an error in one so that a modification of the dates for doom are set. But Meadows' work seems to contain an essential (and indeed self-obvious) truth: sufficient growth of production would have a back reaction which would poison the surface on which it lived. Meadows *et al.*[1] have quantified this qualitatively obvious situation.

Progress from the present to the situation where there is independence from fossil fuels will be bumpy, because the period of plenty (and hence cheapness) in these fuels has gone, and political influence, amounting to a degree of control over some policies in their client states, will increasingly be executed by the possessors of the remaining bank of fuel. Correspondingly, an attempt to save the situation by a changeover to a Latter Day Fossil Fuel Age using coal to struggle through two or three decades after 2000, would give many problems (Chapter 3).

Meadows' work proves that *continuation of the present kind of fossil fuel-based and resource-consuming economy* will bring difficulties impossible to withstand in affluent countries before mid-century. However, Meadows' work implies that *sufficiently nimble* changes in technology

---

* Doom decade means a decade when a negative trend (for example, lessening food per capita) undergoes an average lowering of ten per cent or more per year over the decade.

† 'Exhaust' means here 'pass through the maximum on its amount of production-time curve'.

TABLE 2.1[1]
NATURAL RESOURCES WHICH WILL EXHAUST BEFORE 2000

Resource	Known Global Reserves[a]	Static Index (yrs)[b]	Projected Rate of Growth (% per year)[a] High	Av.	Low	Exponential Index (yrs)[c]	Exponential Index Calculated Using Five Times Known Reserves (yrs)	Countires or Areas with Highest Reserves (% of World Tota!)[a]	Prime Producers (% of World Total)[e]	Prime Consumers (% of World Total)[f]	US Consumption as % of World Total[a]
Aluminum	1.17×10^9 tons[f]	100	7.7	6.4	5.1	31	55	Australia(33) Guinea(20) Jamaica(10)	Jamaica(19) Surinam(12)	US (42) USSR (12)	42
Copper	308×10^6 tons	36	5.8	4.6	3.4	21	48	US (28) Chile(19)	US (20) USSR (15) Zambia(13)	US (33) USSR(13) Japan(11)	33
Gold	353×10^6 troy oz	11	4.8	4.1	3.4	9	29	Rep. of S.Africa(40)	Rep. of S.Africa(77) Canada(6)		26
Lead	91×10^6 tons	26	2.4	2.0	1.7	21	64	US(39)	USSR(13) Aust.(13) Canada(11)	US(25) USSR(13) W.Germany (11)	25
Mercury	3.34×10^6 flasks	13	3.1	2.6	2.2	13	41	Spain(30) Italy(21)	Spain(22) Italy(21) USSR(18)		24
Natural Gas	1.14×10^{15} cu.ft.	38	5.5	4.7	3.9	22	49	US(25) USSR(13)	US(58) USSR(18)		63
Petroleum	455×10^9 bbls	31	4.9	3.9	2.9	20	50	Saudi-Arabia(17) Kuwait(15)	US(23) USSR(16)	US(23) USSR(12) Japan(6)	33
Silver	5.5×10^9 troy oz	16	4.0	2.7	1.5	13	42	Communist Countries (36) US(24)	Canada(20) Mexico(17) Peru(16)	US(26) W.Germany (11)	26
Tin	4.3×10^6 lg tons	17	2.3	1.1	0	15	61	Thailand (33) Malaysia (14)	Malaysia (41) Bolivia(16) Thailand(13)	US(24) Japan(14)	24
Tungsten	2.9×10^9 lbs	40	2.9	2.5	2.1	28	72	China(73)	China(25) USSR(19) US(14)		22
Zinc	123×10^6 tons	23	3.3	2.9	2.5	18	50	US(27) Canada(20)	Canada(23) USSR(11) US(8)	US(26) Japan(13) USSR(11)	26

[a] Source: US Bureau of Mines, *Mineral Facts and Problems*, 1970, Government Printing Office, Washington, D.C., 1970.

[b] The number of years known global reserves will last at current global consumption. Calculated by dividing known reserves (column 2) by the current annual consumption (US Bureau of Mines, *Mineral Facts and Problems*, 1970).

[c] The number of years known global reserves will last with consumption growing exponentially at the average annual rate of growth. Calculated by the formula

and the engineering of available abundant clean energy could avoid Catastrophe. If the number of people on the planet can be brought under the control of ecological demands, if re-cycling of materials can be universal and near to 100% efficient – Catastrophe *will* be avoided. There is even a prospect for the spread of high living standards to those in less developed countries. High energy per capita and degree of completeness of recycling are the key material determinants.

THE MEDIUM OF FUTURE ENERGY

The likely sources of energy for the future are atomic and solar. Atomic reactors can provide electricity which would become cheaper as the reactors increase in size, but with size there comes the difficulty of thermal pollution, so that large atomic reactors, which would give relatively cheap electricity at source, would have to be placed either on the ocean, far from population centres, or in remote areas, such as Northern Canada, Siberia or Central Australia.

Correspondingly, massive solar collectors are likely to be far from the population centres which need them, for they would be most advantageously situated in North Africa, Saudi-Arabia and Australia (see Chapters 6 and 7). Hence, the electricity to which they would give rise is liable to have to travel at least 1,000 miles, and, in some situations, as much as 4,000 miles, to go from the site of production to the site of use (Fig. 2.1).[6]

The likelihood of this situation, and the energy loss in conduction, gave rise to the concept of a 'Hydrogen Economy'.[7] Thus, it could be cheaper to convert electrical energy, which will be a product of solar and atomic reactors, to hydrogen at the energy source.[8] Thereafter, the hydrogen would be transmitted through pipes – the pumping energy being relatively small – and converted back to electricity at the site of use (fuel cells) or used in combustion to provide mechanical power.

The distance at which it becomes cheaper to transmit energy by hydrogen rather than in the original electrical form, depends upon the voltage of transmission and the dependence as shown in Fig. 2.2.[9] For situations in which there are distant energy sources (the typical energy source

---

$$\text{exponential index} = \frac{ln(\text{r.s}) + 1}{r}$$

where r = average rate of growth from column 4
s = static index from column 3

[d] The number of years that five times known global reserves will last with consumption growing exponentially at the average annual rate of growth. Calculated from the above formula with $5s$ in place of s.

[e] Source: UN Department of Economic and Social Affairs, *Statistical Yearbook 1969*, United Nations, New York, 1970.

[f] Sources: *Yearbook of the American Bureau of Metal Statistics 1970*, Maple Press, York, Pa., 1970.

World Petroleum Report, Mona Palmer Publishing, New York, 1968.
UN Economic Commission for Europe, 'The World Market for Iron Ore', United Nations, New York, 1968.
US Bureau of Mines, *Mineral Facts and Problems*, 1970.

of the future)[10,11], therefore, hydrogen is likely to be the medium of energy.

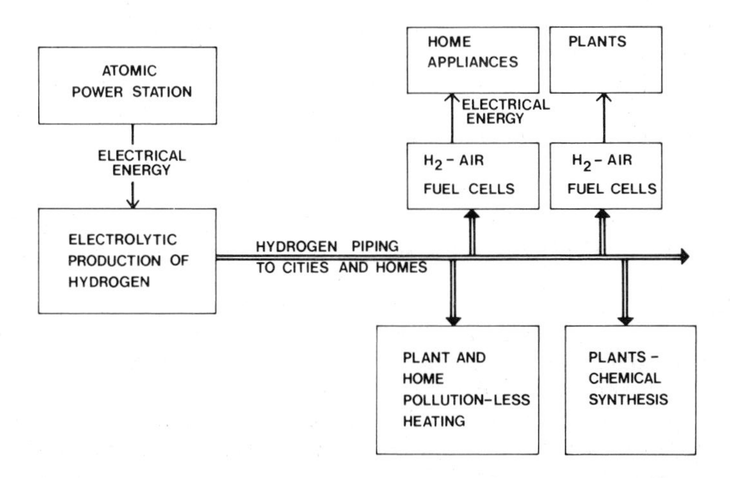

Fig. 2.1 A Hydrogen Economy.[6]

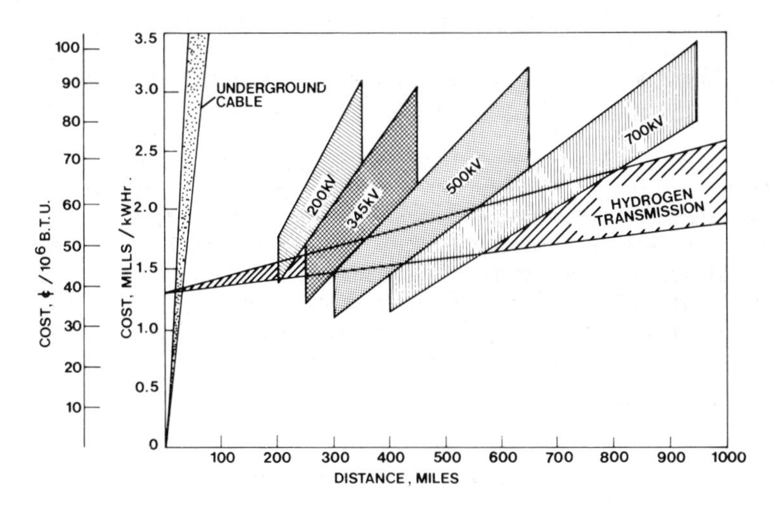

Fig. 2.2 Relative costs of energy transmission by electricity cables and by hydrogen pipeline.[9]

THE HYDROGEN ECONOMY

Hydrogen as the medium of energy has the economic basis projected for a future energy situation above, but the Hydrogen Economy is a concept with a broader meaning.[12,13] It arises from the desirability of utilising the coming sources of abundant clean energy with maximum economy, but it also strives to remove the pollutional aspects of the present technology. The Hydrogen Economy can be described thus: it is the use of hydrogen to transfer energy from large remote sources (Fig.

2.3)[14]; combined with the massive use of hydrogen as a chemical in technology and for transportation and household energy; with the conversion of it to fresh water (and its subsequent rejection, after use, to the sea) as an essential part of the cyclical concept involved.

Some of the goals of a Hydrogen Economy are[13,15–18]:

Fig. 2.3 Floating energy platform.[14]

(1) To increase the energy per capita available to all groups of peoples without running into an unacceptable pollutional situation. (Were this to be attempted *via* fossil fuels, there would be unacceptable increase in pollution, even if there were enough coal for a number of decades which would justify the building of a coal-based system.)

(2) To increase the efficiency of energy produced from atomic or solar sources, by eliminating Carnot cycle losses between the energy at source and mechanical work.

(3) To remove pollution associated with the availability of extensive cheap transportation, and with the increase in extent of chemical and metallurgical industry which would be associated with increasing the living standard of the rest of the world to some 10 kW per person.

HYDROGEN VERSUS NATURAL GAS

Natural gas contains about three times more energy per unit volume than does hydrogen, though about one-third as much per unit weight. Natural gas is easier to store and handle.

Stress should be on hydrogen, for the availability of natural gas has (in Western countries) only a decade or two more of life; and the potential availability of synthetic natural gas from coal some three to six decades.

There are reasons for beginning to go to hydrogen as a fuel as soon as it is economic, and not to wait until natural gas has been exhausted. One of these reasons is to reduce air pollution and to reduce the greenhouse effect which will tend to come by the use of coal (see Chapter 4).

CONSEQUENCES OF A HYDROGEN ECONOMY[10,11,13,18–21]

At present, some 10% of energy is provided by the transmission of electricity through wires. Hydrogen would not only fulfil this requirement, but the other major energy requirements.

(1) *Chemical Technology:* Sufficiently cheap hydrogen could be used in a large number of reduction reactions, with cheapened costs and pollution greatly reduced, or negligible.

(2) *Metallurgy and Refining:* A number of processes may be carried out more cheaply with hydrogen than by using reducing agents such as carbon. Another reason for increasing the use of hydrogen is to reduce air pollution at present associated with metallurgy and refining.

(3) *Effluents:* The upgrading of effluents could occur more easily with cheap electricity and/or oxygen available via a Hydrogen Economy. An in-house treatment of sewage through its electrolytic oxidation to carbon dioxide, or its chemical treatment in a molten salt at high temperature (producing CO_2) may become practical.

(4) *Water:* A fraction of drinking water needs could be met by the use of hydrogen as the medium of energy. The end reaction involved is the production of liquid water. For a community at 10 kW per person, there could result about 14 gallons per person per day of fresh water, of which about 3 would be produced in the house.

(5) *Transportation:* Hydrogen could be used in internal combustion engines, which run well on hydrogen after only small modification (Chapter 15). Fuel cells would probably eventually be used because of their greater efficiency. Trains, likewise, could be driven from fuel cells running on gaseous or liquid hydrogen.

Sea transportation: Large freighters will be necessary in the foreseeable future. They could run on hydrogen-oxygen fuel cells with liquid hydrogen storage. They would use energy islands, at which the hydrogen could be produced, as refuelling centres.

Air transportation: Independently of the more general concept of the Hydrogen Economy, subsonic, supersonic and hypersonic aircraft would all run with advantage on hydrogen.

The use of hydrogen as a fuel in transportation would allow attractive concepts to be realised. One of these is economic hypersonic flight (Chapter 15). Single seat individual helicopters would become feasible and perhaps economic.

IMPLEMENTATION

A recent study by Veziroglu[11] has concentrated on a world model in which there is calculated the pollutional rise with time with hydrogen being introduced at various times from the present. If no hydrogen is introduced, fossil fuels will run out about 2035 and the pollution at that time will be some twenty times that of the present. The introduction of hydrogen in place of fossil fuels meets fuel needs, reduces pollution and improves the quality of life. At present rates of fuel consumption,

Veziroglu calculates that hydrogen introduction doubling times of 1 and 2 years could save 84% and 21% of the fossil fuel reserves. Fig. 2.4[11] illustrates the effects.

Williams has suggested an implementation schedule as shown in Fig. 2.5.[17] (See also Chapter 18.)

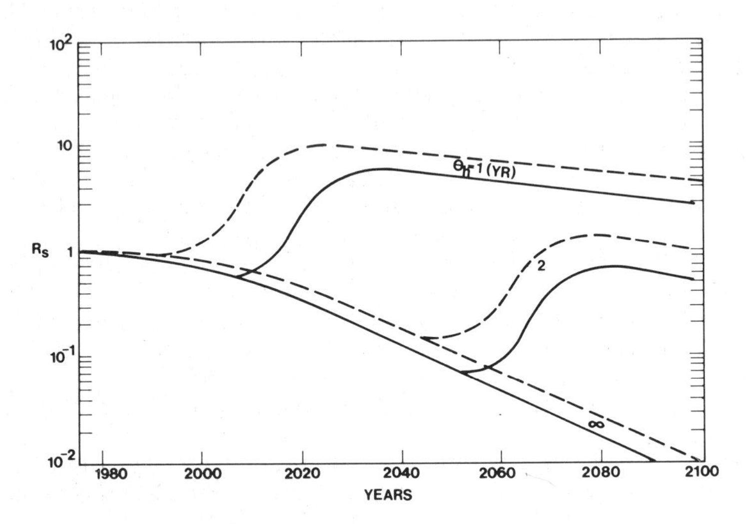

Fig. 2.4 Quality of life *vs* time.[11]
ϴ = the doubling time of the hydrogen production rate (= 1 or 2).
– – – = constant growth rate.
——— = variable growth rate.

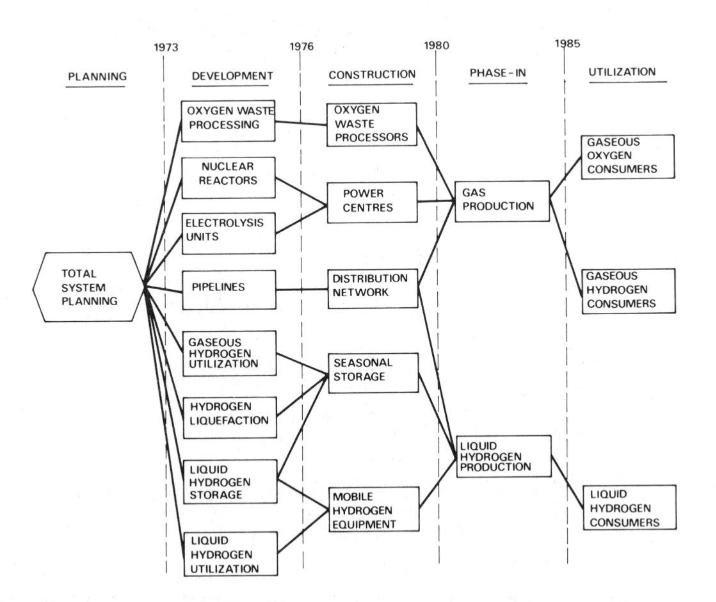

Fig. 2.5 Implementation steps.[17]

THE ORIGIN OF THE CONCEPT OF A HYDROGEN ECONOMY

A distinction must be made between a Hydrogen Economy – the energetic, ecological and economic exploitation of the availability of cheap hydrogen[12], to be the underlay of the entire Society – and suggestions that hydrogen might be used as a fuel. Jules Verne[22] had Captain Nemo say that water would be the origin of the fuel of the future. Errin and Hastings-Campbell[23], and also King[24], were advocates of the use of hydrogen in internal combustion engines, in 1933. They, in turn, refer to a prior general suggestion by Haldane[25] that hydrogen from wind-power via electrolysis, liquefied and stored, would be the fuel of the future. Sikorski[26] suggested hydrogen as an aviation fuel in his Steinmetz lecture of 1938. Bacon and Watson[27] put forward again the Haldane suggestions and developed the first practical hydrogen-air fuel cell (cf. Bacon[28 29]). In a fine example of prophetic journalism, Lessing called hydrogen 'the Master Fuel of a New Age' as early as 1961, although he seemed to be referring largely to the fusion programme.[29a]

In 1962, Bockris[7] proposed a general plan of supplying American cities with solar-based energy *via* hydrogen. He suggested the use of floating platforms containing photogalvanic devices producing hydrogen by the electrolysis of sea-water. This was to be piped to land and reconverted to electricity in fuel cells.

Lindstrom[30], in 1963, attributed the first suggestion that transport of energy in pipes would be cheaper than in wires, to the turbine constructor,

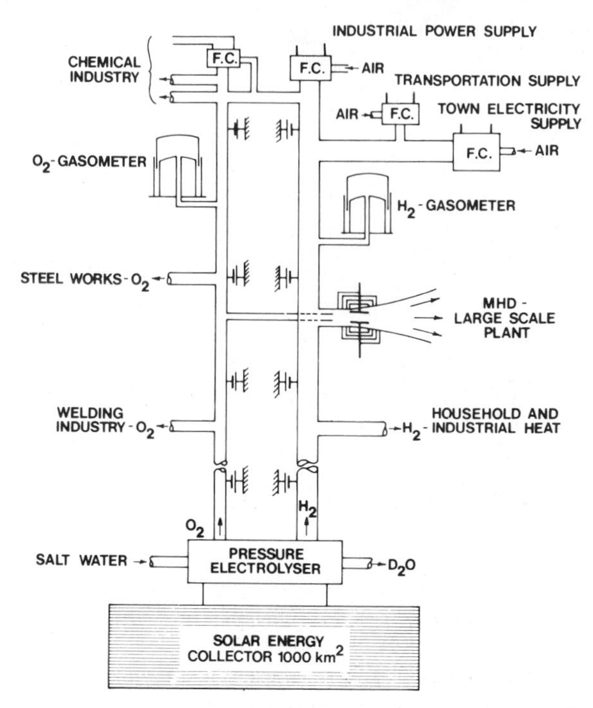

Fig. 2.6 Schematic presentation of a solar plant working in North Africa and having an area of about 1,000 kilometres square. Hydrogen would be transmitted at 50 atmospheres.[31]

Lawaczeck[8], in the 30s, and advocates long distance passage through pipes. Strong advocacy of hydrogen as a fuel was made by A.T. Stuart* in a little-known paper[8a] of 1927. Niederreither suggested the use of electricity to produce hydrogen from electrolysis; the hydrogen to be stored and distributed, and the electricity to be re-obtained from gas cells.[31a] A similar scheme was proposed independently in Australia by J. S. Just in 1944, with reference to motor fuel.[31b] Lindstrom[30] stated that the costs of distribution as a gas are only 10 to 20% of those through wires.

Justi[31], in 1965, published the suggestion that solar energy should be used to originate hydrogen by electrolysis (Fig. 2.6[31]). The hydrogen and oxygen would be piped and provide electricity via fuel cells and MHD engines, or used directly in industry.

In the Fall of 1969, Bockris, seeking to develop the use of hydrogen widely as a fuel, contacted H.R. Linden of the Institute of Gas Technology, to solicit an article from members of the staff of this Institute, which would involve the carrying out of calculations to examine a general concept. This involved sea-based atomic reactors, or solar collectors, and piped hydrogen, with the hydrogen piped to land and used in industry, transport and housing.[7] Linden expressed enthusiasm for the use of hydrogen as a fuel and it was agreed that he would request some of his staff to make the investigations suggested.[31c]

The necessary work was carried out by Gregory, Ng and Long[9] in 1972. Their article contained the quantifications requested. Gregory and his co-workers, with the resources of the Institute of Gas Technology in Chicago, have been from that time a primary source of quantitative knowledge in the field of hydrogen energy systems.[32]

The phrase 'A Hydrogen Economy' arose for the first time in a discussion between Bockris and Triner of the General-Motors Technical Centre, 3 February 1970. They had been discussing (along with others in a Group) the various fuels which could replace polluting gasoline in transportation and had come to the conclusion that hydrogen would be the eventual fuel for all types of transports. The discussion went to other applications of hydrogen in providing energy to households and industry, and it was suggested that we might live finally in what could be called 'A Hydrogen Society'. The phrase 'A Hydrogen Economy' was then used later in the same conversation. The suggestion of a link between the cheap transmission characteristics of hydrogen as a medium of energy, and its general use as a basis to most fuel-using functions in the Economy, as a replacement for natural gas and gasoline, was published by Bockris in 1971. About the same time, Dean and Schoeppel presented a paper[36] in which they referred to the earlier suggestions of Weinberg[37] that hydrogen be used to power vehicles. Dean and Schoeppel independently referred to a Hydrogen Economy and lauded hydrogen as a general fuel, particularly in transportation.

Thus, the concept of electrolytically evolved hydrogen as a special purpose fuel for occasional use was suggested some 40 years ago[25], and the use of hydrogen for certain special purposes mentioned in the 1930s.[26] The suggestion that cheaper energy could be available a long way from

* Internal company memoranda by A.T. Stuart of 1934 suggests a widespread use of hydrogen in industry, with hydrogen storage.

the original source by the use of hydrogen as a transmission medium was made in 1933 by Lawaczek. The use of this hydrogen (solar-origined) as the medium of a non-polluting technology for towns was first set down by Bockris (1962)[7]; referred to independently by Lindstrom (1963)[30]; developed and diagramated by Justi (1965)[31]; named a Hydrogen Economy by Bockris and Triner in 1970[33]; formulated in terms of a non-polluting economy by Bockris (1971)[12], and Bockris and Appleby (1972)[13]; and quantified in a published document for the first time by Gregory, Ng and Long (1972)[9] (cf. also Marchetti, 1972[34] and Lessing, 1972[35]).

REFERENCES

[1] D.H. Meadows, D.L. Meadows, J. Randers & W.W. Behrens III, *The Limits to Growth,* a report for the Club of Rome's project on the Predicament of Mankind: Potomac Associates, Universe Books, New York, 1972.

[2] G.N. Plass, in *The Electrochemistry of Cleaner Environments,* ed. J.O.'M. Bockris, Plenum Press, New York, 1972.

[3] M.A. Elliott & N.C. Turner, presented at the American Chemical Society Meeting, Boston, Massachusetts, April 1972.

[4] J. Boyle, *Nature,* **245,** 127 (1973).

[5] D.H. Meadows & D.L. Meadows, *Nature,* **247,** 97 (1974).

[6] J.O.'M. Bockris & D. Drazic, *Electrochemical Science,* Taylor and Francis, London, 1972.

[7] J.O.'M. Bockris, Memorandum to Westinghouse Company (C. Zener), 1962.

[8] F. Lawaczeck (1930), quoted by R.O. Lindstrom in ref. 30, and also quoted by E.W. Justi in ref. 31; also in priv. comm., 6 May 1974.

[8a] A.T. Stuart, *Industrial and Engineering Chemistry,* **19,** 1321 (1927).

[9] D.P. Gregory, D.Y.C. Ng & G.M. Long in *The Electrochemistry of Cleaner Environments,* ed. J.O'M. Bockris, Plenum Press, New York, 1972.

[10] W. Hausz, G. Leeth & C. Meyer, 'Eco-Energy', Proceedings of the 7th IECEC, San Diego, 1972.

[11] T.N. Veziroglu & O. Basar, *Hydrogen as a Fuel,* University of Miami, Coral Gables, Florida, 1974.

[12] J. O'M. Bockris, *Environment,* **13,** 51 (December 1971).

[13] J.O'M. Bockris & J. Appleby, *Environment This Month,* 29 (July 1972).

[14] R.P. Hammond, in *The Electrochemistry of Cleaner Environments,* ed. J.O'M. Bockris, Plenum Press, New York, 1972.

[15] J.W. Michel, presented at the American Chemical Society Symposium, Chicago, Illinois, August 1973.

[16] W.E. Winsche, K.C. Hoffman & F.J. Salzano, *Science,* **180,** 1325 (1973).

[17] L.O. Williams, presented at Cryogenic Engineering Conference, Colorado University, Boulder, August 1972.

[18] *Chemical and Engineering News,* p. 14 (June 1972).

[19] R.L. Savage, L. Blank, T. Cady, K. Cox, R. Murray & R.D. Williams, edd., *A Hydrogen Energy Carrier,* NASA-ASEE, 1973.

[20] L.W. Jones, *Journal of Environmental Planning and Pollution Controls,* **1,** 12 (1973).

[21] D.P. Gregory & J. Wurm, presented at IGT/AGA Conference on National Gas Research and Technology, Atlanta, Georgia, June 1972.

[22] J. Verne, *Twenty Thousand Leagues Under the Sea,* Paris, 1869.

[23] R.A. Erren & W.A. Hastings-Campbell, *J. Inst. Fuel,* **VI,** 277 (1933); see also *Chem. Trade J.,* **92,** 239 (1933).

[24] R.O. King, W.A. Wallace & B. Mahapatra, *Can. J. Res.,* **26F,** 264 (1948).

[25] J.B.S. Haldane, in a lecture, 'Daedulus, or Science of the Future', at Cambridge, 4 February 1923.

[26] I.I. Sikorski, Steinmitz Lecture on Aviation (1938).

[27] F.T. Bacon and R.G.H. Watson, priv. comm., 1950.

[28] F.T. Bacon, *BEAMA Journal,* p. 2 (January 1954).

[29] F.T. Bacon, *World Science Review*, p. 21 (April 1959).
[29a] L. Lessing, *Fortune*, p. 152, (May 1961).
[30] R.O. Lindstrom, priv. comm. to F.T. Bacon (1963); also R.O. Lindstrom, *ASEA Journal*, **37**, No. 1 (1964).
[31] E.W. Justi, *Leitungsmechanismus und Energieumwandlung in Festkorpen*, Vandehoeck & Ruprecht, Gottingen, 1965.
[31a] H. Neiderreither, German Patent 21b, 648941, 1937.
[31b] J.S. Just, Gas and Oil Power, *Annual Technical Review, 326 (1944).*
[31c] H.R. Linden, 'Gas Scope', Institute of Gas Technology, publication 19, 1971.
[32] D.P. Gregory, assisted by P.J. Anderson, R.J. Dufour, R.H. Elkins, W.J.D. Escher, R.B. Foster, G.M. Long, J. Wurm & G.G. Yie, 'A Hydrogen-Energy System', prepared for the American Gas Association by the Institute of Gas Technology, Chicago, August 1972, p. X-22.
[33] J.O'M. Bockris, *Chemical & Engineering News* (October 1972).
[34] C. Marchetti, *Chemical Economic Engineering Review*, **5**, 7 (1963).
[35] L. Lessing, *Fortune*, p. 138 (November 1972).
[36] J.L. Dean & R.J. Schoeppel, paper presented to the 1971 Frontiers of Technology Conference, Oklahoma State University, 30 September-1 October 1971.
[37] A.M. Weinberg, *Physics Today*, p. 18 (1959).

CHAPTER 3

The Time Available for the Research, Development and Building of a New Energy Base

INTRODUCTION

LOOKED AT FROM the viewpoint of the earth's people, hydrogen as a fuel should be introduced as soon as it is feasible, because it would slow, and finally solve, the growing air pollution problem.[1] But air pollution which increases imperceptibly year by year is a small irritant compared with the diminution of living standard caused by price rises in fuels. This factor, more than pollution, will create the public pressure for the development of new energy sources, if the energy from them can be cheaper than energy from coal.

THE EXHAUSTION OF FOSSIL FUELS

A threat of imminent exhaustion of oil has been made regularly over decades, but earlier, each decade saw discoveries which increased oil reserves. This fact was reinforced by statements of some geologists[2], to the effect that much oil remained to be discovered.

As to coal, that exhaustion could threaten within times of interest to living people, would not have been regarded as likely, by any but a few specialists[3], even in the early 70s. 'The U.S. has at least 1,500 years of coal', was a statement at that time.[4]

Why, then, have ideas on the availability of fossil fuels changed so suddenly? The answer originates from three sources:

(1) In resource exhaustion, the suddenness with which onset symptoms (the main one is steep extra-inflational price rise) arrive is a characteristic symptom of the disease. Thus, Glaser[5] refers to a riddle set to French school children. A farmer has a pond on which lilies have begun to grow. The amount of the lilies covering the pond doubles each day, and the pond will be completely covered within one month. The farmer decides that he need not start to deal with the nuisance until the pond is half covered. On what day will the farmer start to bother? The answer, of course, is upon the 29th day, when he at last recognises that the situation has become serious – but won't now have the time to take effective action.

(2) A reason why statements[4] such as 'the U.S. has 1,500 years of

coal' turn out to need further elucidation, depends upon the extrapolation law used for the prediction of future use rates. It used to be assumed that a resource would last that time given by dividing the total amount of the resource by the amount used in the year of the estimate. Latterly, the assumption seems to have been made that if an Economy were expanding at a certain rate per year, resource exhaustion could be calculated by *linear* extrapolation of the use rate. In fact, use rates have been increasing exponentially for years (Fig. 3.1[6]), and it is this factor, more than others, which has made earlier estimates of how long energy resources would last grossly optimistic.

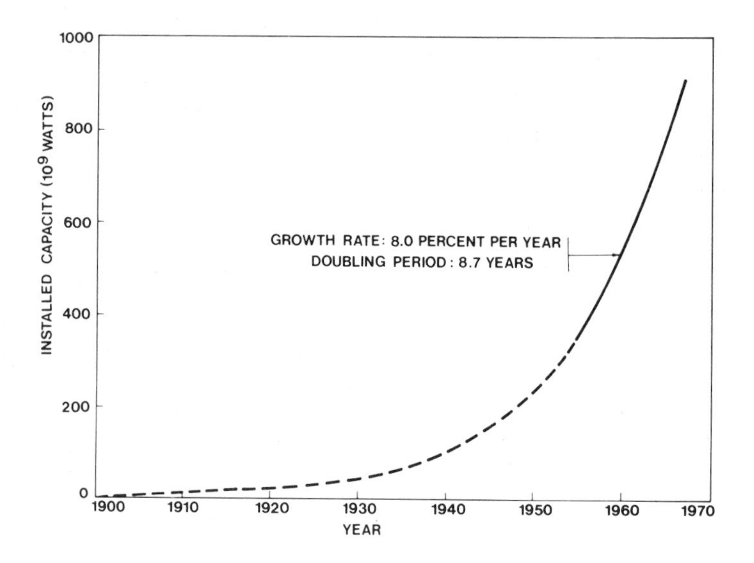

Fig. 3.1 World electrical generating capacity.[6]

(3) The concept of extrapolation following a mode of previous behaviour (even the true exponential one) implies continuation of the same areas of use of the resource; and the same use of the resource in the economy. Thus, for many years, the principal use of coal has been in the production of electricity. Exponential calculations of the time at which coal will exhaust, have neglected the fact that the uses to which coal would be put may change with the years. Thus, as liquid and gaseous fossil fuels exhaust, one possibility is that gasified coal may be applied to fuel that part of the economy (in the U.S.A., still at least 87% by 2000) not fueled by atomic energy. *But the use of electricity is a small fraction of the economy's energy*. Its use fraction is about 10% (c. 30% of fuel used, because the efficiency factor of coal to electricity conversion is some 0.3). Thus if an extrapolation, which was based on coal consumption in past years when its uses were limited to the production of electricity and some metallurgical applications (i.e. *traditional* uses), showed that a country had coal for, say, 150 years,[6]

such an extrapolation would be grossly optimistic. It would be optimistic because if the coal were used to make synthetic natural gas, and other fuels, to replace the exhausting natural gas and oil, the time of exhaustion in the example taken would be as little as 50 years.

Conversely, advances in mining techniques can act in the opposite direction, and extend the supply. Thus, statements of resource availability should have added to them a reservation of the type: 'if no more advanced means are found for recovering the resources more economically than those of the present technology'. Thus, as exhaustion continues, technology may be developed for recovering resources at greater depth.*

An example of statements which are true but which may easily mislead is the Chase-Manhattan Bank's statement[4] on the availability of fossil fuels, which refers to 1,500 years of coal in the U.S.A. 'at the present (1972) rate of use'. By 2000, the energy need per year will be 3-4 times the present (thus, at 2000, there would be some 400 years supply, if need stopped growing at 2000). Only about one-quarter of the coal is recoverable by present technology, so that, with this technology, and the 2000 A.D. use-rate, coal would last 100 years. But, even this estimate assumes coal is being used all this time in the traditional ways. If coal were used as the general base to energy, its rate of use would increase about three times over that assumed in the projections made above. Thus, the '1,500 years of coal' is reduced to some 33 years from *c.* 2000.

In spite of this evidence, it is still sometimes asked – even by scientists[7] – if it is true that we are *really* running out of fossil fuels, or is it a cry of 'wolf', with the economics of scarcity as the objective on the side of commercial interests, and the desire for funds for research as an influence on the scientist. This chapter contains the evidence which the present author selected (in the presence of his own biases). The reader must weigh the conflicting statements of the companies ('centuries of coal') and scientists ('three or four decades past 2000') for himself. A statement as: 'fossil fuels (effectively coal) will exhaust in the third decade of the 21st century' must be made with reservations. Its validity depends upon the absence of the development of much improved technologies of recovery; upon roughly constant price (except for inflational changes); upon no radical change in the degree of inflation; *upon no technological breakthroughs of competing resources*; and upon political

---

* An example is given by the South African goldfields. The reef which emerges to the surface around Johannesburg, upon which much of South African wealth has been based, is saucer-like in shape. The gold band is a few feet in thickness. Thus, in the past, mines have been founded at various parts of the territory surrounding Johannesburg, near where the reef emerges to the surface. They have depths up to several thousand feet. South African goldmines are beginning to exhaust, but there is probably a large supply of gold at lower levels. With increase of depth, the difficulty of providing miners with cooled air begins to make the cost of mining uneconomic. Two factors influence this. The first is the price of gold, but the second would be the development of an automated 'moon-buggy' which carried out the miners' work at temperatures up to, say, 100°C. The availability of such technology would increase the availability of gold very greatly. The necessary technology is developable, but those who could finance it will not do so until the economics are right for them. Owners of resources do not want a resource to seem to be abundant.

interaction (that silent giant) between the images thrown up by corporate interests and their credibility among the consumers. The public will exert very strong pressure for a continuation of the living standards which are based upon cheap energy. But the drive for the continuation of an affluent life style has to be compounded with the tendency to the alternate *low* energy life style among those, such as some University students, which gradually makes itself visible in the affluent world.

A factor likely to be over-estimated is the possibility of the discovery of large new resources of fossil fuels available through present technology at, say, four times the present prices (or their inflational equivalents). Thus, predictions of exhaustion times of fossil fuels (with reservations as above) are based not only upon *known* availability, but upon a resource base which discounts the discoveries projected according to extrapolation of earlier rates of discovery. Thus, although discoveries of oil and natural gas deposits will certainly continue to be made in the coming decades, it is unlikely that they will be outside that allowed for in the estimates described below, because such estimates discount the fuel likely to become available on the basis of known data of amount discovered per year.

This factor, the likelihood of discoveries of great new fossil fuel caches, is diminishing (Fig. 3.2[12]).*

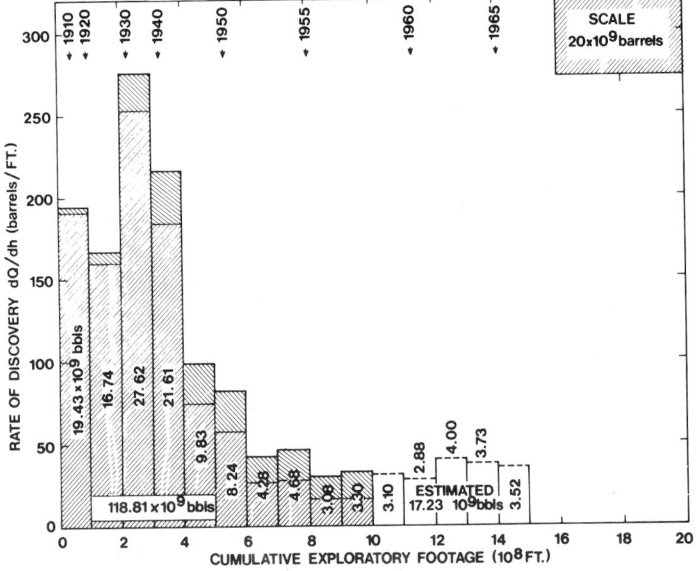

Fig. 3.2 Average U.S.A. discoveries of crude oil per foot of exploratory drilling for each 10^8 ft interval of exploratory drilling.[12]

---

* However, exploration of Antarctica could lead to significant discoveries. Such discoveries must be viewed in respect to ease of recovery and transportation. Discovery of Middle East-sized oil fields beneath some thousand feet of ice would not have the same impact as their discovery under, say, the flat deserts of Australia. Correspondingly, large oil fields might exist beneath the seas, but only those on continental shelves are of interest because of the increase in expense of recovery with depth. Further, pre-occupation with continued discoveries of fossil fuels which would last past about 2050 is pointless because of climatic changes which continued injection of CO_2 into the atmosphere would bring about (see Chapter 4).

Other factors which bear upon the desirability of future discoveries of fossil fuels – in particular coal – arise from the economies of conversion[8] (see Chapter 4). Thus, a popular view is that coal can be relied upon for at least some decades. There is only one oil from coal plant remaining in the world from the many functioning, particularly in Germany, in earlier times. It is at Sasolburg in South Africa, but the costs of the production are such that it can only be sold if supported by a 50% Government subsidy.[9] This situation comes about in spite of the fact that South African coal is cheaper than that of any Western country, because it is surface-mined with cheap labour, and the plant uses it at the mine area, avoiding transport costs.

Artificial barriers inhibit knowledge of the short term future. There is even difficulty in discovering the availability of oil in the U.S. in the 1970s. This knowledge is not available to U.S. Government institutions, but is retained by the companies. The media present the subject in colours, using as a criterion of what is presented that which will excite, in other words, selecting the evidence available.

ENERGY AND LIVING STANDARDS

Faced by a rise in price of a commodity, the normal reaction is to seek a substitute. However, Energy is *the* basic commodity and affects the price of all but the most primitive activity. A gross cut-back in

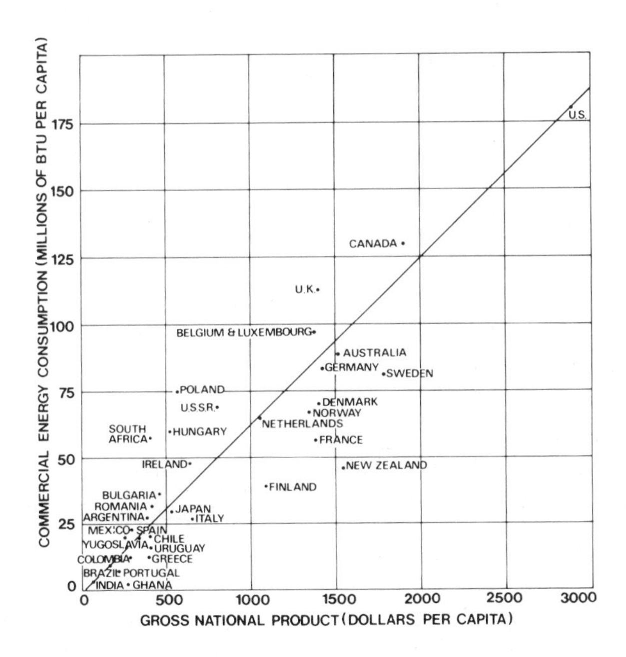

Fig. 3.3 Rough correlation between per capita consumption of energy and gross national product is seen when the two are plotted together.[11]

the availability of energy within *urban* communities would not merely be unpleasant, but fatal. 10% cut-backs in energy per capita might be acceptable, perhaps even cathartic. The affluent society involves a population concentrated in cities which is viable only if certain mechanisms (distribution of goods, lighting and sewage disposal) work.[10] All of them depend upon the supply of oil, natural gas and coal or oil-based electricity. During the strike of coal miners in England in 1974, it was stated that one of the early effects of a sufficient rundown of coal supplies would be the failure of the sewage system in London, which is operated electrically. Thus, the cities would become suddenly uninhabitable at some critical value of the energy per capita.* In England, in 1974, the Government fell, the miners' claims were met, and inflation continued upwards in excess of historic norms.

The dependence of living standards upon the supply of energy per capita is given in Fig. 3.3.[11] One has to visit Ghana, Columbia, and particularly Indian cities, to understand the effect of a sufficiently lessened per capita supply of Energy in an urban situation in terms of squalor, starvation, and the trucks sent daily to remove corpses from the streets.

THE AVAILABILITY OF VARIOUS ENERGY RESOURCES

The most comprehensive review of this has been given by Hubbert† (1970)[6], and other authors use Hubbert's work as a basis to develop commentaries and representations. There is a need for a repetition and extension of this work by University and Government scientists who would be free of conflict of interest difficulties in making interpretations of facts.

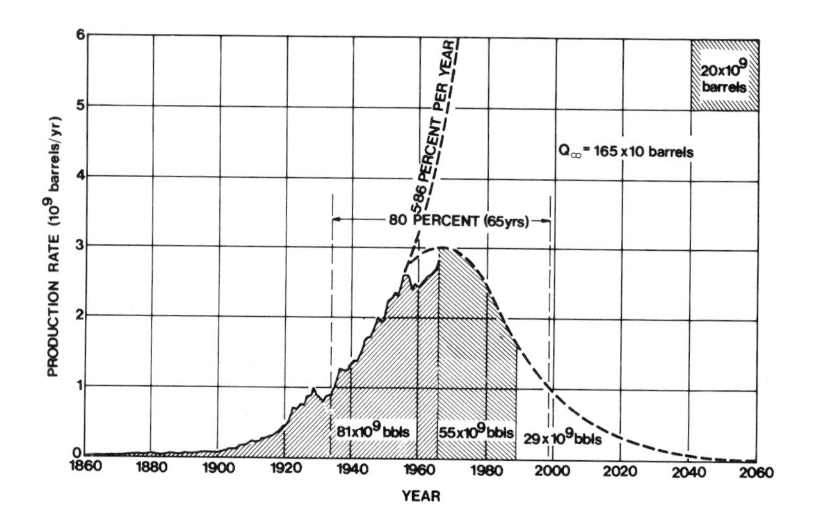

Fig. 3.4 Complete cycle of U.S.A. crude-oil production.[12]

---

* Of course, were the cities to become uninhabitable, starvation, disease and massive number of deaths would follow within weeks.

† M. King Hubbert is a research geophysicist who works for the U.S. Department of the Interior, on the Geological Survey. His address is: National Centre, 12201 Sun Rise Valley Drive, Reston, Virginia, 22092.

FOSSIL FUELS

Hubbert found the U.S. crude oil production maximised about 1968 (Fig. 3.4[12]). Petroleum liquids in the U.S. (crude oil plus natural gas liquids) maximised about the same time (Fig. 3.5[12]). Natural gas is available

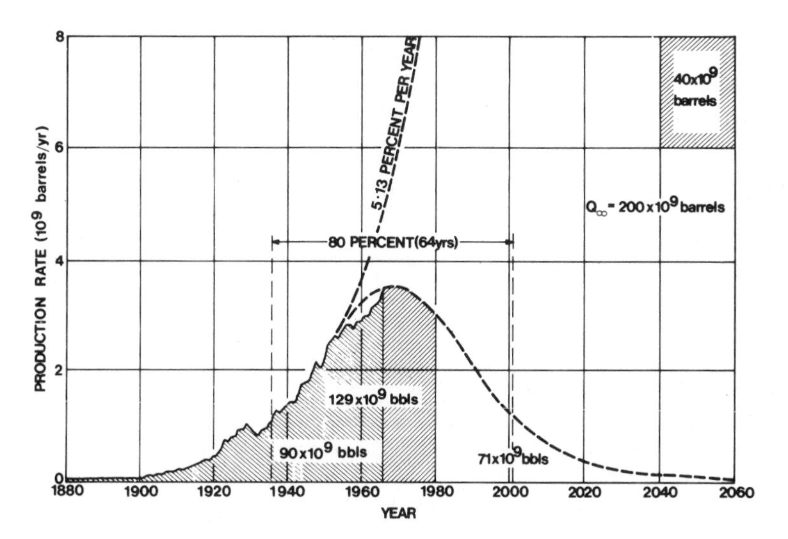

Fig. 3.5 Complete cycle of U.S.A. production of petroleum liquids.[12]

TABLE 3.1[6]

ENERGY CONTENTS OF THE WORLD'S INITIAL SUPPLY OF RECOVERABLE FOSSIL FUELS

Fuel	Quantity	Energy Content (10^{21} J(th))	Energy Content (10^{15} kWH(th))	Percent
Coal & lignite	7.6×10^{12} tonnes*	201	55.9	88.8
Petroleum liquids	2000×10^9 barrels (272×10^9 tonnes)	11.7	3.25	5.2
Natural gas	10000×10^{12} ft^3 (283×10^{12} m^3)	10.6	2.94	4.7
Tar-sand oil	300×10^9 barrels (41×10^9 tonnes)	1.8	0.51	0.8
Shale oil	190×10^9 barrels (26×10^9 tonnes)	1.2	0.32	0.5
TOTALS		226.3	62.9	100.0

* This is a high estimate. It includes coal beds as thin as 12 inches and as deep as 6000 feet. 2×10^{12} tonnes is an estimate of beds between 28 in and 1000 ft thick.

in the U.S. in the same amounts as petroleum liquids. Alaskan oil represents only a few years extra supply for the U.S. Tar sand oil and shale oil are some 25% of the total oil supply in the U.S.A., so that even if separation of oil from shale could be satisfactorily achieved it is unlikely to compete with coal. A summary is given in Table 3.1.[0]

WATER POWER

The world's hydro-electric power is given by Hubbert[6] as 3.10^{12} watts. 8% of it is developed. However, it would not be good to rely on water power for long-term development, because reservoirs will gradually silt up. If the entire water power resources of the world were developed, they would supply one-third of the present need. Hence, fuller development of water power cannot provide a general solution to the energy needs.[13] This does not imply a lack of interest in increasing development of some of the 92% of hydro-electric power which is undeveloped. A Hydrogen Economy would facilitate this. The undeveloped sources are in remote areas (for example, in Portugese East Africa) and the expense of taking energy from them to high use areas would be diminished by passage in hydrogen.

TIDAL POWER

Hubbert[6] considers that the developable tidal power would be 2% of that available from water power. Thus, conversion of tidal power could not provide a general solution. It could be a useful additive to the energy supply in special regions, where tides are particularly high.[6a] Such a region exists in the north-west of Australia. Its use as a supply for Western Australian industry would be made more economic through electrolysis at source and transmission of energy in hydrogen.

GEOTHERMAL POWER

Conventional geothermal power comprises utilisation of hot springs. The technology is developed in New Zealand. Were the hot (that is, more than 100°C) springs resources of the world to be fully developed, they would add about 2% to present energy.

However, a new concept is available in geothermal energy conversion, and is explained in Chapter 5. It is the recovery of energy from hot rocks at depths of several miles.[14] If this form of energy conversion proves to be both technically feasible and economically acceptable, it may make a contribution to the world energy supply. It should not be regarded as an inexhaustible source of energy, for as the artificially made cavities in the earth give out heat to water introduced into them, their surfaces cool down and a new cavity has to be made every few years. (The older ones would, of course, heat up again over the succeeding years and be re-usable.)

WIND POWER

This is an inexhaustible source, the possibilities of which are little realised. It is discussed in Chapter 5.

THE TIME OF EXHAUSTION OF THE WORLD'S FOSSIL FUELS

Fossil fuels are about 88% coal.[12] The coal is largely in the U.S.S.R., and in the U.S.A. Australia has about 10% of world coal reserves.

It is vital to estimate the exhaustion of this resource, because of the dependence of the time which we have to develop the inexhaustable clean energy supply. Two source-papers on the subject, those of Hubbert[6] and Elliott and Turner[3], differ significantly in prediction. Thus, Hubbert's conclusions are summarised in Fig. 3.6[12], and Elliot and Turner's latter conclusions are summarised in Fig. 3.7.[3] According to Elliott and Turner (1972), the maximum plot of production of world fossil fuels, including coal, may occur as early as 2030. Hubbert's figure originated in 1962, and, according to him, the maximisation of the production of this resource will occur between 2100 and 2150.

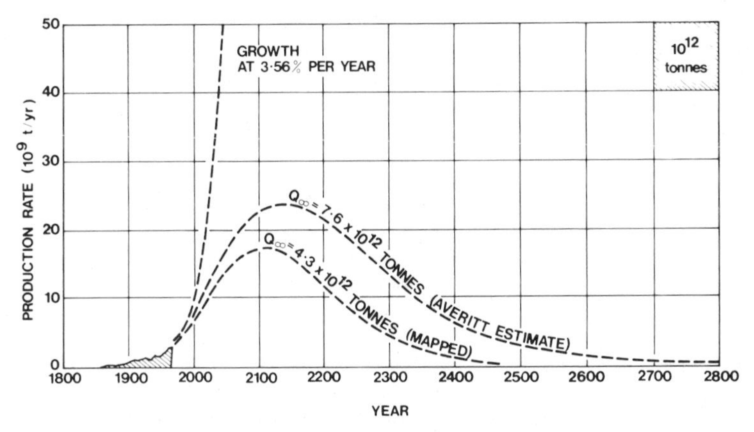

Fig. 3.6 Complete cycle of world coal production.[12]

The interpretation of this discrepancy[15] has an implication for other estimates. Hubbert's projection is based upon the use of coal in the past, which has been mainly for electricity production with some uses in metallurgy. Elliott and Turner's projection for 'world fossil fuels' takes into account the use which it is planned to make increasingly of coal, to take over the use now made of liquid and gaseous hydrocarbons. Thus, Elliott and Turner take into account the fact that coal will be gasified and liquefied. Their projection corresponds to what now seems to be the most probable short-term energy development.

Predictions made for the 'number of years left' of the fossil fuels have been decreasing.[16] 3,000 years of coal was the supply supposed to be available in the U.S. about a decade ago; but is now reduced to some 50 years. Elliott and Turner's prediction may, therefore, be optimistic. Conversely, new technology could increase the fraction of coal which can be recovered at a price which would make it economically competitive with the inexhaustible clean sources.

In Figs. 3.6 and 3.7 the implication is that a date at which a maximum resource use occurs is the end of the availability of the resource. However, oil and natural gas resource prediction passed its maximum (from U.S. sources) in the late 1960s (Fig. 3.4), though the majority of the U.S.

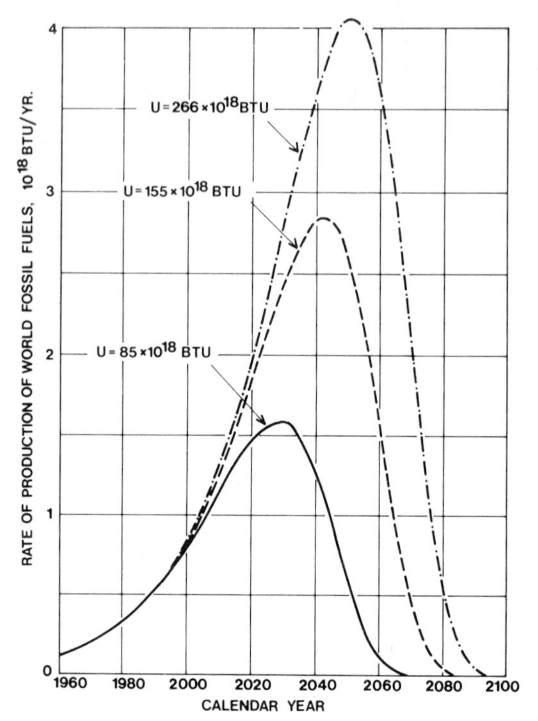

Fig. 3.7 Effect of the value of the resource base on the projected rate of world fossil fuel yields. Elliot and Turner's results.[3]

energy supply still comes from indigenous U.S. oil and natural gas. What the maximum means is that the end of the resource is near, in terms of a decade, or so, and that other sources must be increasingly introduced to supplement the dying resource. If, at the time of maximation, an alternative source is not in an advanced stage of development, the situation is that functions based on this resource will probably have to lapse.

MATHEMATICAL MODES FOR PREDICTING RESOURCE EXHAUSTION TIME

The rate of production of a resource reaches a maximum and then decays; and the area under the curve of the rate of production *versus* time is the total of the ultimate source.

According to Elliott and Turner (cf. Hubbert), the rate of production of natural resource as a function of time for the resource, and the quantity remaining, may be expressed as[3]:

$$\frac{dP}{dt} = f(D) \cdot f(S) \qquad (3.1)$$

in which P = cumulated production of the resource; t = time, calendar years; $\frac{dP}{dt}$ = annual rate of production; f (D) = function of demand; f (S) = function of remaining supply, but the supply is a function of the fraction of the resource remaining, so that one gets:

$$\frac{dP}{dt} = \frac{b'P}{U}\left\{1 - \frac{P}{U}\right\} \qquad (3.2)$$

in which U = ultimate resource and b' = a constant.

Integration of equation (3.2) gives:

$$-\ln\left\{\frac{U}{P} - 1\right\} = \frac{b't}{U} + n = bt + n \qquad (3.3)$$

in which n = constant of integration and $b = \frac{b'}{U}$

Thus:

$$P = \frac{U}{1 + \exp(-b-n)t} \qquad (3.4)$$

One may then take the second derivative of equation (3.4), equate it to zero, and let t_m be the time at which the maximum rate of production occurs. The solution is that:

$$n = -bt_m. \qquad (3.5)$$

Combining eqs. (3.4) and (3.5), we have that:

$$P = \frac{U}{1 + \exp[-b(t-t_m)].} \qquad (3.6)$$

This eq. (3.6) is that derived by Elliott and Turner, and is similar to that of Hubbert, in which:

$$P = \frac{U}{1 + A\exp[-b(t-t_r)]} \qquad (3.7)$$

where t_r = a reference calendar year and A = a constant, when the substitution, $A = \exp b(t_m - t_r)$ is made. It is $\frac{dP}{dt}$ which maximizes.

The Elliott and Turner equation has the limitation that 50% of the resource is produced at the time of maximum rate of production. Other more complex relations remove this limitation, which does not seem to be important for the present calculations.

THE ESTIMATION OF THE EXHAUSTION OF FUEL SUPPLY

The principal resource of fossil fuels is coal, and Elliott and Turner's analysis suggest exhaustion as early as 2030, if coal is used to supply the needs from oil and natural gas.

Brennan[18] has reviewed world resources, developing the work of Starr[19].

one uses the quantity Q. Q = 10^{18} Btu ~ 10^{21} joules

2[18] comes from Brennan's paper. World reserves of ces of power are about 40Q, whereas the estimated of the world energy between 1970 and 2000 is about ... rennan's analysis, world energy consumption maximises at 2000, an earlier estimate than that of Elliott and Turner with its maximisation between 2020 and 2050.

TABLE 3.2[18]

ANNUAL CONSUMPTION* OF ENERGY IN Q†

Year	World	U.S.	Australia
1800	0.006	0.001	
1900	0.03	0.01	
1970	0.3	0.1	0.002
2000	1	0.2	0.006

* The figures are drawn from a number of sources, but primarily from ref. 19.
† 1 Q = 10^{18} Btu ≃ 10^{21} joule ≃ 3×10^{14} kWH.

However, Brennan's estimate of passing through a maximum at 2000 depends upon the assumption that the world's consumption of energy stops increasing after 2000, an optimistic assumption. With this *optimistic* assumption, our present reserves would be doubly consumed by 2020.

Should estimates such as that of Brennan be pessimistic by a factor even as high as 5 (cf. Table 3.1), growth of world population, and its consumption, etc., will raise the annual consumption well above Q. Thus, from Brennan's viewpoint, the second decade of the next century – Turner and Elliott's time of turndown of the production rate of fossil fuels is 2030 – is an *upper limit* for the time at which fossil fuels could be the main source of energy. Brennan's analysis is consistent to a degree with the entirely independent analysis of Elliott and Turner.

Somewhat similar times have been calculated by Linden.[20] If fossil fuels are to carry the entire burden of providing us with energy (and the development of atomic power by 2000 will still supply less than 10% of the whole energy need in the U.S.), fossil fuels will exhaust by 2010. Extending the resource base to include not only that which seems recoverable now, but the total present, one gets (Linden) to 2040-2065 as the dates in which the production rate will begin to fall. Some arguments[20] would push the highest limit out to 2073. These estimates are similar to the '33 years from 2000' derived earlier from a crude dissection of the '1,500 years' of the Chase-Manhattan Bank's statement.[4]

Thus, there is a fair degree of agreement between some four authors. The average of their estimates is 2038 and the span 2000-2073. It is reasonable to conclude that society should be ready with the new energy base researched, developed and beginning to be built by 2000 (that is, extensively available by 2020). Later dates gamble on the availability of methods being developed for recovery which do not at present exist.

A brief summary, then, is: there are about 25 years for the research and development, and about the same time for the building of a new energy system.

TABLE 3.3

ESTIMATES OF THE YEAR IN WHICH A MAXIMUM PRODUCTION RATE OF COAL COULD OCCUR*

Source	Year	Comment
Chase Manhattan Bank, elucidated[4]	2033	Original report states 1,500 years at 1972 rate of consumption. Elucidation in text.
Brennan[18]	2000	Assumes energy demand stops increasing at 2000.
Elliott and Turner[3]	2040	Takes into account use of coal to replace oil.
Linden[20]	2010	Assumes mining technology not radically improved on 1970s.
Linden[20]	2052	Assumes all coal can be mined.
Linden[20]	2073	Highest credible limit for resource base assumed.

* In general, the assumption is made that the logistical difficulty of mining coal *at a sufficient rate* has been solved.

These conclusions (see Table 3.3), are those which spring from considerations of availability *below* the ground. However (cf. Chapter 4), there are doubts that coal could be extracted from the earth at sufficient speed to replace the exhausting oil and natural gas. If coal is not to become available in considerable amounts within a very short time, for example, 10 to 15 years, then the situation would indeed become less favourable than that discussed above.

THE TIME NEEDED TO REALISE A MAJOR TECHNOLOGICAL CHANGE

There is a long interval between the first laboratory demonstration of a new scientific device and its realisation in commercial practice. This time is sometimes assumed to be in the range 15 to 25 years. Closer study shows that the gap is often longer. An example is atomic energy. Fission was discovered in 1939 by Hahn and Strassmann[21] and the first practical realisation of the working reactor was in Fermi's Chicago Pile of 1942. However, atomic energy supplies less than 1% of the energy of the American economy, three decades after the initial discovery.

Space flight seems an example of a speedy development. However, the first equations which showed the possibilities of space flight were given in 1887 by Konstantin Tsiolkovskii.[22] Thus, if the moon landing is taken as an indication of the practical attainment of space flight, 82 years is the time of development from the first equations. Further, the development of space flight is a special case – free from the entropy of conflicts of interest, and influence of lack of political acceptability, which reduce the driving force of many projects which are in the long-term interest of the community.

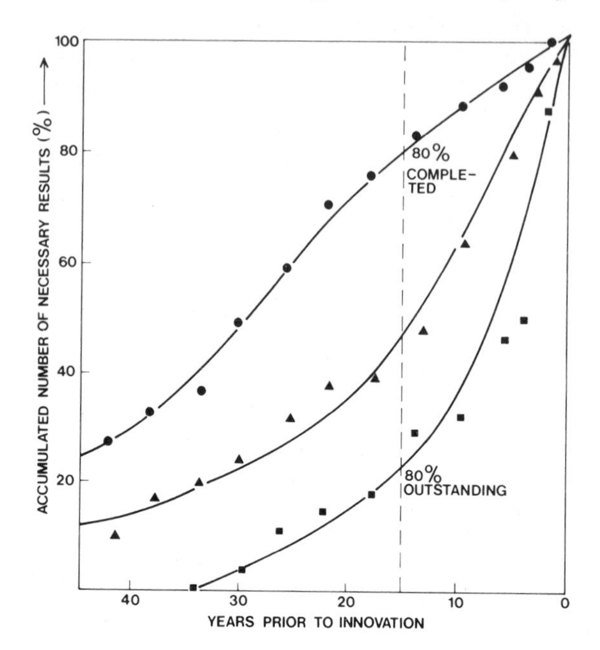

Fig. 3.8 Number of years of basic research (O), applied research (△) and development (□) necessary for the realisation of innovations.[23]

Friborg[23] has discussed the time of development of a technology from origination of ideas to its practical use by the public. His work is summarised in Fig. 3.8[23]. Thus, according to Friborg, the basic research is 80% complete after about 60 years, the applied research 40% complete after about 50 years; and the time which takes about 35 years is merely the development work needed to make the technology commercial.

The time for the development of massive solar energy may be examined in this context. The photovoltaic cell was discovered at the Bell Telephone Laboratory by Chapin, Fuller and Pearson[24] in 1954. The practical innova-

TABLE 3.4[25]

TIME DEVELOPMENT OF SOME ENERGY SOURCES

Source	Electricity	Steam Engine	Fission	Fast Breeder	Fusion	Solar
Scientific feasibility established	Faraday (1831)	Newcomen (1712) Watt (1765)	1942	1950	1975?	1954
Useful power	Sturgeon (1836)	Many developments	1955	1960	?	?
Economic power	Siemens (1856)	1785	1965	1980-90?	?	?
Time	25 yrs	20 yrs	23 yrs	35 yrs+		

tion should, therefore, be made some 60 to 70 years later, or between 2010 and 2020. Applied research should have begun in the 1960s and commercialisation in the 1980s. These figures seem to fit well what can now be judged of the likely course of events for solar energy.

Watson-Munro[25] has recently composed an interesting table (cf. Table 3.4), which lists the time from that at which scientific feasibility was established to that at which economic use could be made of the resource. 26 years is the average time for this situation to be achieved, and this would correspond somewhat to 60-35 $\simeq$ 25 in Friborg's scheme.

It is reasonable to conclude that about 25 years will be needed to research and develop a new energy source (except atomic, where much has been done). Conversion to the new system might be carried out over the next 25 to 50 years. It is interesting to compare these estimates with those of the time available (Table 3.3).

LIMITS TO GROWTH

Material living standards rise linearly with the energy available per capita (Fig. 3.3[11]). What are the limits to growth for the energy and population aspects of the future?[17] How many people, at what energy (that is, income) level, could the planet's resources maintain? Considerations will be limited to atomic and solar energy resources, for it has been seen that the fossil fuels may last for only quarter to half a century and that this is about the time needed to develop and build up alternative and inexhaustible energy sources.

LIMITS FOR THE SUPPLY OF ATOMIC ENERGY

(1) *Will there be enough fuel?*
Were fission reactors to be used as the major source of energy, the available U *ores* (see below) would be exhausted in 2 to 3 decades. Were Australia to rely entirely upon *fission,* without breeder reactors, from 2000 onwards, the energy supply would last only till about 2015. With the breeder reactor, these times could be expanded greatly, about 100 times. These statements, however, do not take into account radioactive material contained (10 to 30 ppm of thorium) in many granite rocks. Thus, the breeder process for thorium gives us:

$$\begin{array}{l} U^{235} + n \rightarrow 200\text{Mev} + \eta\, n^{1} \\ \underline{(\eta-1)n^{1} + (\eta-1)\text{Th}^{232} \rightarrow (\eta-1)U^{235}} \\ (\eta-1)\text{Th}^{232} \rightarrow 200\text{Mev} + (\eta-2)U^{235} \end{array} \qquad (3.8)$$

and, if $\eta = 2$, there is a catalytic burning of Th, so that:

$$\text{Th}^{232} \rightarrow 200\text{Mev}. \qquad (3.9)$$

Suppose that 20 kW per person is allowed and a world population of 10^{10} people, about 250 tons of Th or U per day would be needed. In certain granites there are about 30 . 10^6 tons of Th at about 30 ppm of radioactive material. Hence there should be breeder energy for an affluent world for 250 years. 7 . 10^6 tons of rock would have to be handled every day and this is the same order of material as the coal mined in the world at present.[26]

Including other rocks, and their uranium and thorium content, the world could exist upon breeder reactors for a thousand years[27], with the population limited to 10 billion.

Thus, there is no reason for changing from the development of an atomic technology, on the basis that there is not enough fuel, *so long as* satisfactory breeders can be developed; and *so long as* a solution to their pollutional problems can be worked out economically.[28]

(2) *Dissipating Heat*

Very large reactors (such as, 30,000 MW), which are a desirable aim of the future – because of the decreases in the price of electricity to which they will give rise – demand placement on water, which is 1 to 2 km in depth. Thus, the water below must be sufficiently cold, so that pumping it to the surface brings abundant cooling of the reactor environs.* Considerable distance from population centres would provide a diminution in pollutive difficulties and the corresponding disadvantage of the expense of lengthy transmission would be overcome if hydrogen were used as the medium of energy.

For 10^{10} people, about 2,000 reactors of 40,000 MW each would be needed.[26] How long would it take to build this number of reactors? To build sufficient reactors for $2 \cdot 10^8$ people by 2010, it would be necessary to set up about 40 giant reactors in 35 years, that is, one every 11 months until 2010. At $500 per kW (1974 dollars) each would cost about $2 billion, or there would have to be an investment rate of about $2 billion per year, an amount still small compared with defence commitments, but about the same as the U.S. space programme budget.†

The rate-determining step here is breeder technology – 40,000 MW breeder reactors cannot yet be built. Even if they could be, the U.S. public's negativity to the building of atomic reactors (because of fears concerning waste disposal) would stop their construction at sufficient rates to avoid a massive consumption of coal.‡

(3) *The Possibility of Explosion*

The explosion to which reference is made here is that which might occur should overheating of a reactor cause a melt-down. Reactors are

---

* If the water is from the sea-bottom, it will contain nutriments associated with shallow water, so that the surroundings of the reactor could become fish catchment areas.

† There is little doubt that atomic reactors, including separation plants and waste disposal *could* be made clean enough to avoid significantly increasing the death rate (e.g., they could be built far underground). The question is: could that be done *and* energy produced at an acceptable price, over a long term, in practical working conditions, using fallible technicians, and *in time*? The question is similar to that with automotive pollution. To build i.c. engines which pollute below certain limits in bench running at constant speed was relatively easy: it proves not possible to maintain such standards with acceptable prices for city driving above 10,000 miles.

‡ Pollution from coal-based stations can no longer be neglected. Thus, in 1974, the Victoria, Australia, EPA refused to license a large coal power station on grounds of insufficient protection against odours and toxic fumes.[29]

built to safeguard against this. Cooling water would spray on them if they overheated, and the steel top seen on reactor buildings is designed to withstand the pressure of the steam which would be produced. The difficulty is that experimentation to test the situation would be expensive and dangerous. Because no actual test has been hitherto made of this situation, insurance companies will not insure atomic reactors against this hazard.

(4) *Pollutional Hazards of Atomic Energy*
At present, radioactive wastes are held in steel canisters and this is unsafe in terms of decades because the steel will stress-corrosion crack in this time; and the radioactive materials would leak out and probably find ground water, with the possibility of contamination of municipal supplies.

The *intentions* of the Atomic Energy Commissions are to modify wastes and put them into salt mines. There is room in such mines and the radiation from the depth would be negligible. Alternatively, there is the possibility of allowing hot radioactive wastes to bury themselves in the antarctic ice.

In the meantime it is not unreasonable that some anxiety exists[8]:

(a) What is the damage from low intensity radiations? It used to be thought that they were non-damaging below a critical value. However, Sternglass has gathered information which suggests that the low intensity radiation which escapes from atomic reactors does have an effect upon health.[30]

Sternglass' work has been disputed. Could extra radiation below the background radiation level be a health hazard? It is claimed[31] that, although the steady radiation from atomic energy processing plants, and reactors, is small, there are times when gross amounts of radioactivity are rejected into the atmosphere (mainly from processing plants). Further, it is claimed[31] that the earlier view of a threshold in radiation absorbed per year in respect to health damage is incorrect. The damage in the community is simply proportional to the degree of radiation. To detect slight changes in death rates in communities surrounding reactors needs laborious examination of the records of municipal authorities and is hence easy to overlook.

A mechanism may be emerging whereby the effects of slight changes in background radiation can be a health hazard. Health is dependent upon the body's defence mechanisms, which depend partly on antibody activity. The latter depends on cell wall properties. These settle down to certain properties in natural background radiation, although it may be that their deterioration, and aging, are related to this background. Increase of the background increases cell wall damage and originates a decline in health.[32]

(b) The work of Stewart[33] upon the effect of radiation on the unborn child has shown that the foetus is more liable to radiation damage than the born human.

(c) Plans of the Atomic Energy Commissions are 'ideal cases'. Waste products can indeed be taken to salt mines. However, they have to be taken in something, and from time to time this transport will meet with an accident. Transports of liquid hydrogen on road or rail would certainly be involved in accidents. But, the damage would be small compared to that of the results of the spread over the country of the radiation from an atomic waste carrier involved in a serious collision (or attacked by saboteurs).

(d) Genetic damage: This is only, as yet, a fear, but the encouragement of such an anxiety by Nobel Laureates leads one to consider it. It will not be possible, scientifically, to measure it till many decades after a widespread atomic technology has been built up. Once the investment has been made, it will be as difficult to displace, as the radioactive contamination of the world it may produce.

Should the development of atomic technology be limited? Atomic technology *could* be made safe, but the costs of making it so may be too large. The evidence that there is no lower limit of radiation[30] which does not affect health has come at the time when society is beginning to become conscious of the imminent end of fossil fuels, that is, when there is increasing pressure to develop atomic resources more quickly. Conservationists will not be able to withstand public pressure which will arise if there is a significant diminution in living standards, through extra-inflational price rises of fossil fuels.

In summary, there is no limit *energy-wise* for an atomic technology for more than 10 billion people at 10 kW each for hundreds of years, so long as research development is funded with sufficient speed and at sufficient magnitude. It is the *pollutional hazards* from an atomic technology which may be substantial.

The limits to the growth of communities may come firstly from the absence of sufficiently efficient and widespread recycling, though this itself is much dependent in feasibility on the price of energy.

An Energy Exhaustion Disaster (an exhaustion of fossil fuels before the New Energy System has been built) is, however, conceivable as would be the case in the absence of a mechanism to build the reactors needed *in the time available*.

(5) *Time of Building of Atomic Power Stations*

Suppose the pollutional difficulties of atomic reactors (and the fact that the breeder reactor is still experimental) are neglected, then it can be asked for instance, at what rate could an entirely atomic supplied country be built up? Suppose a limit of 2000 A.D. is set as that at which the U.S.A. should be entirely (breeder) atomic, and thus independent of foreign sources. At that time – extrapolating from the present – the American need per person will be about 30 kW. Assuming a zero population growth in the U.S.A. at that time[34] (the population would be 225 million), about 7 million megawatts would be needed. Let us suppose all the stations are giant 10,000 MW stations. 700 such stations would have to be built in some 25 years, namely, a new one completed every 2 weeks from the mid 1970s until 2000. It is obvious that this cannot be done, that is; *the U.S.A. cannot build sufficient atomic power stations by the year 2000 for atomic energy to take over from exhausting fossil fuels.*

SOLAR

The total insolation of the world is $1.7 \cdot 10^{17}$ watts. The limit is how much can be in practice collected. If 10% of this radiation could be converted from 1% of the land area, sufficient energy could be given to an affluent world – everyone at the same standard as present Americans and 1.5 times more people than at present. Thus, solar energy is plentiful – abundant – and ecologically safe. Its most obvious point is its inexhaustability, so that it is not necessary to consider it further in this Chapter (see Chapters 6 and 7). This factor, the absence of pollution from it, and its abundance, makes it by far the best energy source.

SUMMARY ON ENERGY SUPPLY

In terms of the energy supply which *could* be researched, developed and finally built, it should be feasible to maintain 10 billion people at 10 kW per person. The energy could be supplied by atomic energy, but there seems to be pollutional difficulties of a massive atomic technology, based on breeder reactors, and the cost and time for the solution to these pollutional problems is not known. There is little time to evaluate it, before there will have to be very large national investment in new inexhaustible, clean energy sources. The uncertainty of the long-term effects on the population of massive atomic energy is part of the case for evolving away from atomic technology at this time to solar technology. There appears to be (see Chapters 6 and 7), with *present emerging technology*, a practical possibility of using solar energy as a clean, ecologically acceptable, and inexhaustible supply. But, in competition with the development of atomic energy, its difficulty is not scientific or technological, but socio-economic: atomic energy already represents a large investment, in money, and in men. Legislators listen to what they are told by those who have the money, time and interest to inform them of one part of the picture. There is no Solar Energy Commission and Industry, nor lobbyists, with the interest to present the solar case to legislators.

THE ORIGIN OF LIMITS TO GROWTH

As clarified above, there *need* not be limits to growth, if one looks at the total energy which could be made available if the collecting and conversion technology were developed. The limits arise from other directions. There *will* be limits to growth on the basis of energy shortages (lack of sufficiently rapid response to the need to develop and build converters of the abundant, clean energy sources [Chapter 5]).

Some limits are psychological. Thus, animals stop breeding when the space between each reaches a certain lower limit (and before the food runs out). For those species which do not stop breeding, an urge for death invades the population, when its numbers become too high, and then suddenly fall.[35] Chemical changes take place in the animals, which lead to a destructive urge. In humans, a similar mechanism may perhaps be weakly seen in respect to insurrectional trends associated with close city living.

A limit to growth which will set in before the energy limit needs to affect the situation is lack of equipment and organisation for speedy recycling of materials, most of which are rapidly exhausting.[36] The decade in which recycling will have to be extensive, differs for each material and country, but, all metals will have to be recycled within a generation, some within a few years; finally, everything will have to be recycled. One of the non-psychological limits to growth may be the population which a certain degree of recycling of a vital material can maintain. For, once society comes into a zero growth situation, there will be a definite – final – amount of each material available. (See Table 2.1.)

DOOM?

In the present chapter material has been presented from which the following conclusions can be made:

(1) Fossil fuels will be exhausted in less than six decades.

(2) It will probably not be possible to use all which seem available, because of the economics (and logistics) of their recovery. On this basis, exhaustion could occur within three decades and before an alternative supply can be built (effectively).

(3) A new energy system is likely (according to previous norms) to take some 25 years of research and development, and at least 25 to 50 years to build.

We could perhaps achieve the necessary changes – given funding priorities of the appropriate order, and an extensive special government organisation – and particularly the availability of the very large amounts of capital needed for building the new system.

However, there is a case for reading the evidence as indicating a Disaster and the Breakdown of an affluent Western Civilisation. The cost of fossil fuels has (1974) begun a rise which may be precipitous and continuous. Doubtless the rate of rise may vary over the next two or three decades – and perhaps there will be short periods when the oil supply will increase. But there are indications that this is the last decade of oil and natural gas at less than four times the 1973 prices, and is the last half century of fossil fuels within ten times the 1973 prices. Thus, as time advances, and energy costs rise, the possibility of adequate preparation for the very large research and development investments – and the much larger building investments – necessary for the new source, will decrease. What is needed, is an organisation similar to the Manhattan Project or the NASA moon project, but on a bigger scale. Investment has to be made with a similar *surge*, investment, in 1974 dollars, in the 10 to 100 billion dollars per year range, based on an Energy Research and Development Administration. But this, however, would only be 'acceptable' in a vaulting, booming economy. A booming economy depends, as a first factor, on the availability of cheap energy.

The more expensive fossil fuel energy becomes, the more corporate and government leaders will decide on a cheaper short-term course (for instance, a new method for extracting oil from shale or etc.) and thus the time left for development of inexhaustible clean energy sources will become shorter and the needed expenditure level to achieve a new supply in time still greater. But all will become more expensive as the cost of the underlying energy rises and the living standards fall. The disaster which threatens is that a start will be made too late on the conversion needed. The Western World may have become too poor – energy too expensive – to pay for the new system.

SPECULATIONS IN POSSIBLE COURSES OF LIVING STANDARDS

Some of these are shown in Figs. 3.9 and 3.10. (See also Fig. 3.30.)

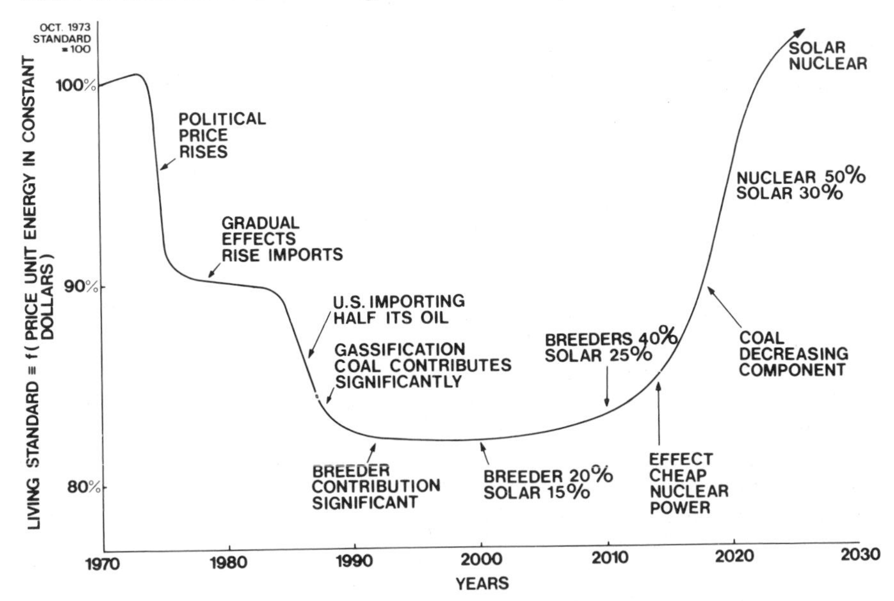

Fig. 3.9 Speculation on a positive solution for the near energy future.

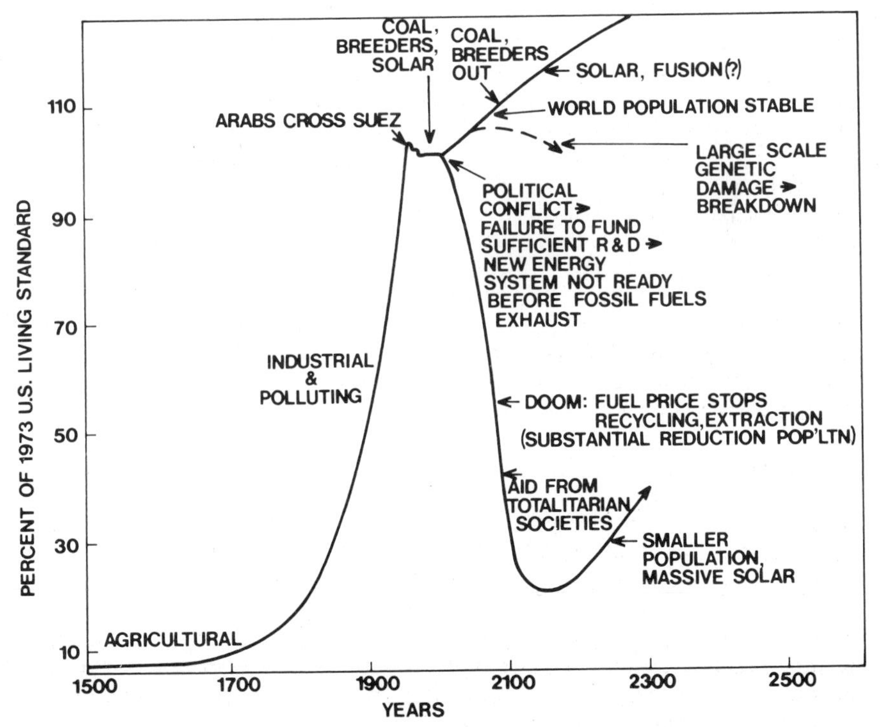

Fig. 3.10 Speculation on Triumph and Disaster (U.S.A.).

THE NEGATIVE SIDE OF THE SHORT-TERM (UP TO 2000 A.D.) SITUATION

The speculations of Fig. 3.9 assume a positive scenario: the possible will be done – dynamic and clear government leadership, an appropriate new and powerful Government Energy Department with a sufficient budget for a huge task and a planning range in the decades, which correspond to realistic time scales.

The Disaster possibilities arise in two ways. Firstly (see Chapter 4), it looks very unlikely that enough coal can be mined in time. The best hope would be in underground gasification, but the feasibility of this may be low. If the coal cannot be obtained from the ground at a suitable rate, then certainly the other energy producing techniques cannot be developed and built in sufficient amounts before the exhaustion of oil and natural gas. Economic breakdown in the Western World would reasonably have to be expected within one or more generations.

The other side of the Disaster aspect of the situation lies in the lack of ability to communicate the message of what is necessary to a sufficient number of political leaders; and the fact that resolve to spend billions and do much which will temporarily decrease living standards, will not be acceptable to the voting populace.

REFERENCES

[1] J.O'M. Bockris, 'On Methods for the Large Scale Production of Hydrogen from Water', presented at The Hydrogen Economy Miami Energy (THEME) Conference, March 1974.

[2] T.A. Hendricks, 'Sources of Oil, Gas and Natural Gas Liquids in the United States and the World', Geological Survey Circular 522, Washington, 1965; Energy Facts, Library of Congress, Serial H, November 1973; U.S. Energy Resources, Committee on Interior Affairs, U.S. Senate, Serial No. 93-40 (92-75), 1974.

[3] M.A. Elliott & N.C. Turner, presented at the American Chemical Society Meeting, Boston, Massachusetts, April 1972; M.A. Elliott & C.G. von Friedonsdorff, Natural Gas of North America, Memo. No. 9, 2171, **2**, 1968; M.A. Elliott, Proc. American Power Conference, **24**, 541, 1962.

[4] J.G. Winger, J.D. Emerson, G.D. Gunning, R.C. Sparling & A.J. Zraly, 'Outlook for Energy in the United States to 1985', Energy Economics Division, Chase Manhattan Bank, June 1972.

[5] P. Glaser, 'The Case for Solar Energy', presented at Conference on Energy and Humanity, Queen Mary College, London, 1972.

[6] M.K. Hubbert, 'Energy Resources for Power Production', in *Environmental Aspects of Nuclear Power Stations*, Vienna, International Atomic Energy Agency, p. 13-43 (1971).

[6a] T.J. Gray & O.K. Gashus, *Tidal Power*, Plenum Press, New York, 1971.

[7] 'Guide to National Petroleum Council Report on United States Energy Outlook', December 1972.

[8] A.L. Hammond, W.D. Metz & T.H. Maugh II, *Energy and the Future*, American Association for the Advancement of Science, Washington, 1973.

[8a] *Science*, 184, p. 247, (1974).

[9] P.E. Rousseau, presented at World Power Conference, Paper No. 158, Tokyo, October 1966.

[10] *Bulletin*, p. 23, 16 March 1974.

[11] E. Cook, *Scientific American*, September 1971.

[12] M.K. Hubbert, 'Energy Sources' in *Resources and Man*, A Study and Recommendations by the Committee on Resources and Man of the Division of Earth Sciences, National Academy of Sciences – National Research Council, Freeman, San Francisco, 1969,

pp. 157-242; 'Degree of Advancement of Petroleum in the United States', *American Assoc. Petroleum Geologists Bulletin,* vol. 51, pp. 2207-2227, 1957.
[13] Ref. 8, p. 48.
[14] Ref. 8, p. 55-60.
[15] M.A. Elliott, priv. comm., 1 February 1974.
[16] 'U.S. Energy Outlook, Coal Availability', National Petroleum Council, 1973.
[17] D.H. Meadows, D.L. Meadows, J. Randers & W.W. Behrens III, *The Limits to Growth*, Potomac Associates, Washington, 1972.
[18] M.H. Brennan, 'Energy and Fuels over the Next 50 Years', presented at the Australian National University, (February 1972), in 'Physics and the Energy Industry', Flinders University, pp. 1-14, (1974).
[19] C. Starr, *Scientific American*, September 1971.
[20] H.R. Linden, World Congress on Gas, Chicago, 1973. Analysis of world energy supplies, World Energy Conference, 1974, paper, 1.2.28; Energy Self Sufficiency, Institute of Gas Technology, 10/8/74.
[21] O. Hahn & F. Strassmann, *Naturwiss.*, **27**, 11 (1939).
[22] K. Tsiolkovskii, 'Free Space', 1883 (begun 28 February 1883, completed 13 April 1883), apparently first published in B. Vorob'yev's, *K.E. Tsiolkovskii*, Moscow, Molodaya Gvardiya, 1940, p. 45.
[23] G. Friborg, Forskning och Framsteg (Stockholm), No. 3, 2 (1969).
[24] D.M. Chapin, C.S. Fuller & G.L. Pearson, *J. Appl. Phys.*, **25**, 676 (1954).
[25] C.N. Watson-Munro, 'World Energy Resource for the Next Century', University of Sydney, 1974.
[26] R.P. Hammond in *The Electrochemistry of Cleaner Environments*, ed. J.O'M. Bockris, Plenum Press, New York, 1972.
[27] A.M. Weinberg & R.P. Hammond, *American Scientist*, **58**, 412 (1970).
[28] 'The Potential for Energy Conservation: Substitution for Scarce Fuels', A Staff Study, Office of Emergency Preparedness, Washington, January 1973, in 'Energy Resources of the United States', Geological Survey Circular, 650, 1972.
[29] *Machine Design*, p. 34 (1974).
[30] E.J. Sternglass, *Low Level Radiation*, Earth Island, 1973, p. 130-135.
[31] E.J. Sternglass, priv. comm., March 1974.
[32] E.J. Sternglass, priv. comm., April 1974.
[33] N.M.M. Stewart, *Proc. 9th Annual Hanford Biology Symposium*, p. 681-92 (1969); *Brit. Med. J.*, **1**, 1495 (1968).
[34] *Washington Post*, 27 January 1974: 1973-74 growth was 0.8%.
[35] V.C. Wynne-Edwards, 'Population Control in Animals', *Scientific American*, p. 68, 1964.
[36] Ref. 17, p. 170.

CHAPTER 4

Coal as a Source of Hydrogen (and Synthetic Carbonaceous Fuels)

INTRODUCTION

IN THE PREVIOUS chapter, it has been assumed that, in terms of decades, natural gas and oil are virtually exhausted (that is, they will be available in the U.S. predominantly from O.P.E.C. countries as from the mid-80s), and the major discussion concerned coal. It was learned that coal will *not* last in great abundance for hundreds of years, but that if its presence in the ground is the point, it could act as the sole energy source till the end of this century, and probably for a few decades more, variously estimated at one (if there is no new technology for mining coal) up to seven, if the total known and projected coal (as distinct from the *available* coal) is taken as an energy base.

However, the considerations of Chapter 3 neglected all those connected with the actual process of mining and extraction of energy from coal.

The conversion of coal to a form of natural gas and oil has been examined, and practised, for decades, and plants have been built to carry out such a conversion. Thus, coal gasification could be carried out without much development work, except for the sulphur removal which is not yet practical. That a process can be made to work economically is not necessarily an indication that it should be developed. The post-1970 sort of question must now be asked and answered – what are the long-term ecological consequences? What amount of sulphur containing compounds will be injected into the atmosphere in the gasification of coal? What of dust and aerosols similarly injected? What of the carbon dioxide content of the atmosphere and the climatic consequences of its steadily increasing concentration? There are a number of less obvious questions. They concern, for example, the rate at which coal mines could be built and that at which a great number of new miners could be trained. Considerations of these matters are relevant to a book on a Solar-Hydrogen Economy: they relate to estimates of the time at which achievement of the latter will be necessary.

WHAT IS COAL?

What is the composition of coal? The briefest answer is: $CH_{0.8}O_{0.1}$. Coal is not a simple substance: it resulted from complex organic processes, which took place on and in the ancient earth. There are many different coals, with a changing scale of C content. The main ones are called lignite, sub-bituminous, bituminous and anthracitic. Apart from being a dehydrogenated hydrocarbon, coal contains other elements in small concentrations, like chlorine, sulphur and traces of several metals.

Coal is not only chemically variable: it is a complex sort of solid. Thus a reason why oil is preferred to coal as a fuel – and natural gas preferred to oil – depends on the greater difficulty of handling and manipulating the complex solid.

Coal *is* the 'Ugly Duckling' of Fuels. Its products corrode and foul the plant. A plant built for use by one coal may be fouled by the use of another.[1] Coal is gritty. It must be mined and then pulverised. The prospect of its re-introduction as a main source of energy clashes with environmental considerations.

RESEARCH IN THE CONVERSION OF COAL TO GASEOUS AND LIQUID FUEL

Nearly all research in the field of coal conversion to oil and gas up to the 60s was carried out in Germany, where was produced in 1925, the Fischer-Tropsch method for converting coal to gasoline and the Lurgi process which produces 'town gas' from coal.

Only in 1960 did the U.S. Office of Coal Research start to work on coal conversion to liquid and gaseous fuels. It had negligible funding for a long time. Even in 1974, the government money being spent on research into coal to oil conversion in the U.S. is only about $50 million per year (breeder reactors $360 million, but solar energy, $20 million).* Lessing[1] suggests that expenditure on research and development of about $2 to 3 billion per year is needed, and that, then, commercial processes might be speeded up in development by 2 to 3 years.

One reason why so much money for development is needed is that the stage of cheap, fundamental work has long passed in coal research. It is Development Work, rather than Research. Pilot plants and small full-scale plants, must be built to find out the practical costs of obtaining energy from coal.[2]

THE BASIC CHEMICAL PROCESSES

There are several chemical processes by which carbon can be turned to useful fuels and they are summarised in Fig. 4.1[3].

(1) *Partial Combustion*

Steam is blown over hot coal and the product is CO and H_2. The steam cools the coal and this is brought back to about 1500°C by injecting air. The process is, thus, a basic process for making very impure hydrogen from coal.

Two changes in this process have been made in Germany. The intermittent process mentioned above is made continuous by substituting oxygen for air and feeding a mixture of oxygen and steam steadily to the reaction. If the pressure is raised to 300 lb per sq inch, the CO combines with the hydrogen to form methane (which has a greater

* However, the major oil corporations are spending several hundred million dollars per year in research and development on the conversion of coal to synthetic fuels.

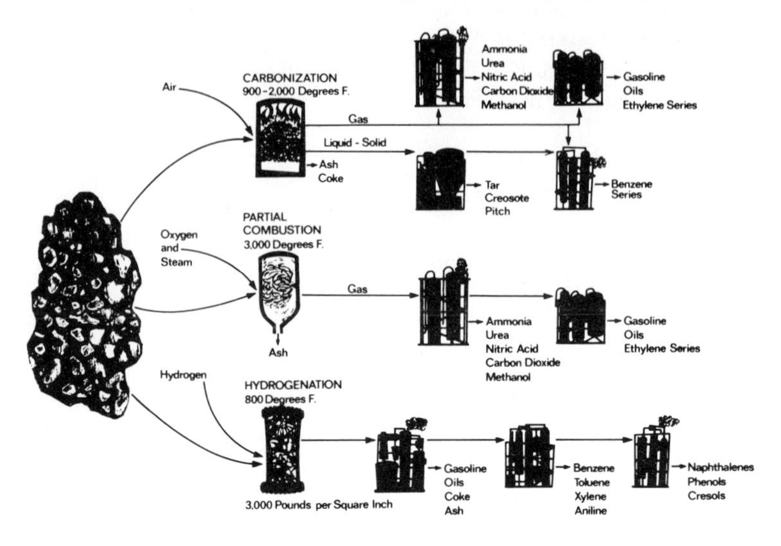

Fig. 4.1 Three basic processes for extracting various substances from coal are illustrated. In carbonization, coal is heated primarily to produce metallurgical coke. Partial combustion is based on the 'water gas' reaction. In hydrogenation coal is heated to 800° F at high pressure.[3]

calorific value per unit volume than does hydrogen). The calorific value is increased from 'low Btu gas' (a few hundreds) to 'high Btu gas' (1000 Btu).

(2) *Fischer-Tropsch*

The residual gas from coal combustion is fed across an iron catalyst at low temperature and pressure. Many hydrocarbons, oils and alcohols result. Some of them are gaseous at room temperature.

The most noteworthy of the application of the Fischer-Tropsch process is at Sasolburg, the large oil from coal plant in South Africa.[4] The plant is the only oil from coal plant still in existence in the world[5] and uses 6 million tons of coal per year to make South Africa 80% independent of imported fuel. It is also the centre of a large petrochemical complex.

(3) *Hydrogenation*

Hydrogen is injected into a paste of pulverised coal and of cobalt sulfomolybdate catalyst[6] at temperatures near 400° C, but at pressures of several hundred atmospheres. The coal reacts with the hydrogen to form many products: gasoline, diesel and heavy oils, benzene, phenols and nitrogen aromatics. Which of the compounds predominates depends upon the structure of the catalyst.

Various techniques of bringing coal into contact with hydrogen are being developed. One of these is called 'flash hydrogenation'. A rotating fluidised bed reactor (Fig. 4.2[7]) is used to bring hydrogen into brief contact with fluidised coal at high temperature and pressure.

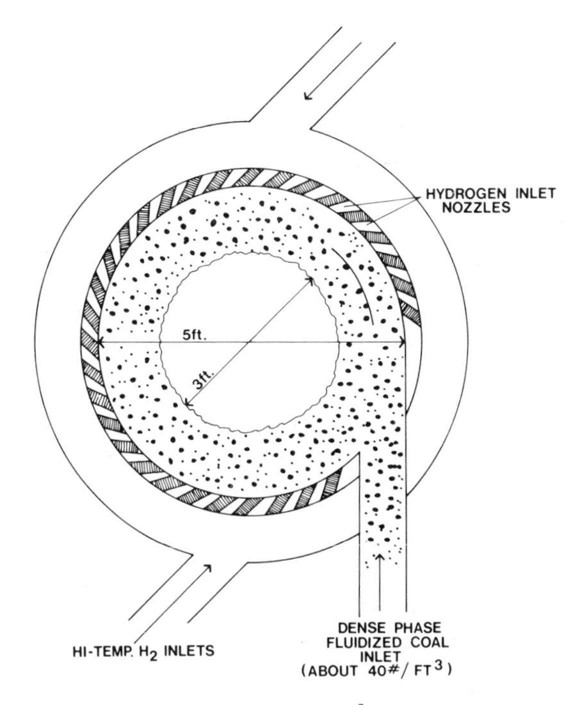

Fig. 4.2 Double wall cyclone concept plan view.[7]

THE HYGAS PROCESS[8 9]

This process (Fig. 4.3[10]) works by using internally generated hydrogen on coal in the following way:

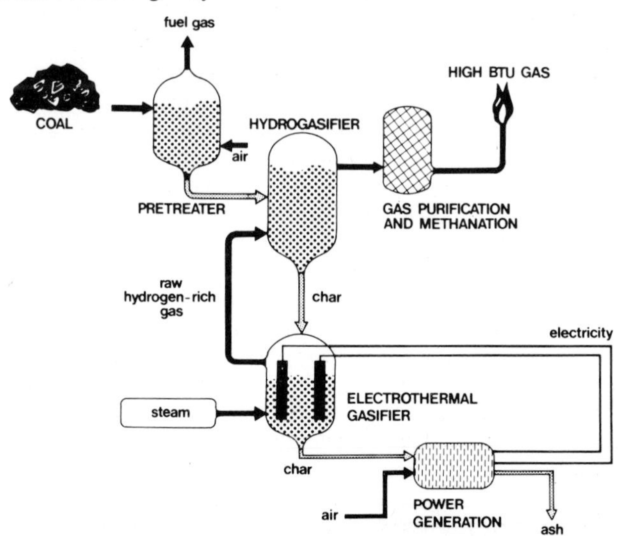

Fig. 4.3 Synthetic pipeline gas from coal by the HYGAS process, using electrochemical gasification for hydrogen generation.[10]

$$COAL + H_2O \rightarrow CO + H_2 \quad (4.1)$$

$$CO + 3H_2 \rightarrow CH_4 + H_2O \quad (4.2)$$

The thermal efficiency is increased by maximising direct methane formation. Therefore, the conditions for the HYGAS process are tailored towards this end. High temperature is needed to obtain reasonable reaction rates and high pressure is used to increase equilibrium methane yields. That fraction of the coal that is most active is hydrogasified to give methane. The less active fraction is used to generate hydrogen from hydrogasification. The raw gas from the hydrogasifier (Fig. 4.3) contains substantial amounts of CO and hydrogen. These are converted to methane *via* reaction (4.2) in a catalytic methanation unit. The object is to boost the heating value of the gas and reduce the CO content to less than 0.1%. About 65% of methane is formed by the direct route and 33-35% by the indirect route. The overall efficiency is 70% in the HYGAS process.

The key feature of the HYGAS process is the use of a mixture of hydrogen and steam. The coal-hydrogen reaction is exothermic, but the coal-steam reaction is endothermic. By using a mixture of hydrogen and steam instead of hydrogen alone, the heat released in the first process is absorbed by the second, so that there is a built-in temperature control and an internal hydrogen generator.

The coal is first crushed and dried (Fig. 4.3). It may be sent straight to the hydrogasifier, but, if it is an *agglomerating* coal – bituminous – it is pre-treated to reduce the agglomerating property.

Most of the sulphur in the coal is converted to hydrogen sulphide and removed. A high degree of sulphur removal is a necessary part of the process, because of the nickel catalyst, which has a low tolerance for sulphur. The H_2S could be converted to sulphur and this could perhaps be sold. The final products contain 95% methane. The process is not yet commercial.

RECENT IDEAS ON PROCESSES FOR TREATING COAL[1]

New development work for power generation from coal is being funded on a liberal scale in the U.S. There is solvent-refined coal, and a pilot plant has gone into operation at Fort Lewis, Washington. The process uses a heavy, recoverable solvent to dissolve coal. The solution is filtered to remove the impurities and resolidifies into a dense, refined coal. The process reduces sulphur and ash in the coal which has, then, a higher heat value than does raw coal. However, it does not yet give us a product which is liquid or gaseous, but only in a less dirty form for burning in coal burning furnaces for electricity.

Another process in pilot plant stage at Princeton, New Jersey, uses coal in four successive fluidised bed reactors to give synthetic crude oil, of medium Btu (which may be upgraded to pipeline gas), and char which could be used to fire utility boilers. The process, called COED, has been running on a variety of coals since 1970. One obtains one barrel of oil per ton of coal. This is the only coal to oil pilot plant process working (1974) in the U.S.

A negative view of coal as a source of energy has been recently taken by the National Petroleum Council[11,16], who published (March

1973) a report based on a two-year study of energy needs. The National Petroleum Council rejected coal completely, calling it a 'drop in the bucket resource' [sic].

THE SYNTHESIS OF METHANOL FROM COAL

This is possible by operating at 3000-4500 psig and using $ZnO-Cr_2O_3$ catalysts. Recently, the Lurgi company has been concerned with development of 700-1200 psig with a copper catalyst. The efficiency is about 59%.[12]

PREDICTED COSTS FOR SYNTHETIC FUELS FROM COAL

It is premature to give cost estimates. Lessing gave estimates in 1973 when coal averaged $8 per ton. These have been multiplied by 1.5 to give 1974 prices when coal is about $12 per ton.

A high Btu synthetic gas (effectively CH_4) would cost $1.65 to $2.17 per MBtu. Synthetic crude oil would cost $1.72 to $2.40 (equivalent to $10 to $15 per barrel). By comparison, high purity hydrogen from offpeak power using classical electrolysers could be as low as $1.60 per MBtu. With 7 mils power, and emerging technology electrolysis, it could be about $2.10. Various *hypothetical* photo-oriented methods give estimated H_2 costs at less than $1 per MBtu (see Chapter 9).

COSTS OF PRODUCTION OF OIL FROM SHALE

Proposals for producing oil from shale and tar sands present problems of material handling, which are greater than those for coal. They are now at a stage which warrants demonstration projects, but oil from shale is unlikely to compete with power gas or oil from coal.

DIFFICULTIES OF COAL GAS GASIFICATION PROCESSES

Of the classical processes developed in Germany, the Lurgi process does not fit big scale industry. Thermal efficiency is poor, as is the coal processing, and unit size is small. The ash must be continuously removed through a rotating grate. Tests run at $15 to $30 million per test for some experimental designs and about $200 million for a large demonstration plant.[1] Only when the pilot plant is built does coal show its enormous clogging power, and its strong tendency to corrode the machinery used. If these problems are solved, there would be doubt about dealing with larger volumes. Such developmental problems may need 10 years before solution.[1]

If coal gas is to be burnt once more in appliances in the U.S., it is desirable that it should have a high content of methane. The first step is in the gasifier, but it must operate at a higher temperature to maximise the methane. The gas moves on to be cleaned in the scrubber and then undergoes catalytic methanation.

HYDROGEN FROM COAL

It has been outlined above how power gas and oil could be obtained from coal at a price (very much dependent on that of coal of about $1.65 to $2.40 per MBtu). However, if use is continued of this type of fuel in an expanding economy beyond 2000, there may be a CO_2 greenhouse effect problem (see below). The building up of a coal-based fossil fuel industry would take 1 to 2 decades, and is unlikely to begin before the price of oil and natural gas has climbed considerably above their present values. Assuming that coal might substantially have replaced oil and natural gas by 2000, there would only be some 30 years before reaching Elliott and Turner's estimate[13] of the maximum production rate for coal (see Chapter 3). Conversely, coal as a source of hydrogen would give reduced pollutional problems, because it would be hydrogen gas which would be diffused over a large area in towns (and the product of its reaction in fuel cells, liquid water). The impurities generated from the chemical conversion of coal would be at central plants where they could be better handled.

Thus, the generation of hydrogen from coal could be an intermediate process which should be considered on the way to a Hydrogen Economy, where the hydrogen would arise from nuclear or solar plants. One could begin immediately to diminish air pollution problems.

Production of hydrogen from coal could be carried out by two methods.[14] There is the synthane process and the CO_2 acceptor process. When coal is reacted with steam at 450 psi and 800-900° C, the gaseous products are CO, CO_2 and H_2. Small amounts of methane are produced. The methane becomes a major product as the pressure is increased to 1000 psi. The CO_2 gas is removed from the final gas by washing by monoethanomine or potassium hydroxide. The final gas is only 97 to 98% pure. CO_2 is a by-product.

The alternative process, called the CO_2 acceptor process, involves lime which is introduced with coal when this is reacted with steam. The CO_2, the reaction product gas, is removed by the lime as calcium carbonate. With the CO_2 removed, the shift reaction occurs in the main reactor, and this eliminates the need for an external shift reactor and the washing to remove CO_2. Heat must now be applied to drive the coal-steam reaction, and this is given by the shift reaction, but also by the reaction of lime with the CO_2. In a separate reactor, the calcium carbonate can be heated to give off CO_2 to the atmosphere and reproduce lime. More coal is used for this process.

A flow sheet has been produced by Kincaid.[15] Thus, the fuel is ground up in both processes. Lignite (preferred for both processes) is dried. If bituminous coal is used, it has to be heated to remove volatile hydrocarbons. Gasification is done for both the steam-oxygen and the CO_2 acceptor processes in fluidised beds at 450 psi. If lignite is the fuel, 1600° F is acceptable, but bituminous coal has to go to 1800° F. In the steam-oxygen process, 12 lb of lignite, 8.8 lb of steam and 5.2 lb of oxygen produce 1 lb of hydrogen.

ESTIMATES OF COSTS OF HYDROGEN FROM COAL[15]

No plants exist to produce hydrogen from coal at present in the U.S. There is the Lurgi plant at Sasolburg, South Africa. The predictions of cost are therefore hypothetical.

Kincaid has produced a detailed cost estimate of hydrogen.[14] The major cost of producing hydrogen from coal is the fossil fuel and this has been treated as the variable. Plant capacity is the other variable.

TABLE 4.1[15]

CAPITAL COST OF COAL TO HYDROGEN PLANT

Process	Capital cost (10^6\$) for plant size of			
	250 tons/day	500 tons/day	1000 tons/day	2000 tons/day
Steam-methane reforming	10.7	17.4	28.3	53.7
Steam-oxygen process for lignite	20.5	33.2	54.0	102.5
Steam-oxygen process for bituminous coal	25.2	40.9	66.4	126.1
CO_2-acceptor process for lignite	25.8	41.9	68.9	129.1

The capital cost for the processes is as shown in Tables 4.1 and 4.2.[15]

Aside from fossil fuel costs a per centage breakdown given by Kincaid of a 1,000 tons per day is as shown in Table 4.3.[15] The cost estimate of about \$1.03 per MBtu from 1973 (coal at \$7 per ton) is probably low and is more likely in the range of \$2.00 to \$2.50.

Australian brown coal is available in large quantities in Victoria. Its 1974 cost was about one-third the cost per energy unit of the cost of oil.[17] Such coal should yield hydrogen at below \$1.50 per MBtu.

If coal, petroleum and natural gas were replaced by hydrogen from coal, immediately, 25 billion tons of hydrogen would be required by the end of the century and this would use up about 20% of U.S. coal reserves. After that, the remaining coal would be used up at an exponentially increasing rate. The time of exhaustion would follow considerations such as those of Elliott and Turner[13] (Chapter 3).*

THE EXPANSION OF COAL INDUSTRIES: PROBLEMS

Figures given in *Industrial Week*[19] bring up the question of whether a sufficient number of coal mines could be opened in time, so that we could start an all-coal economy by 2000. Thus, the investment costs per kW are stated to be:

> Capital costs for a mine producing 1 million tons per year: \$20 million
> Thus, the dollars per kW for plant is about \$200.

---

* In a pamphlet of 1974 ('Clean Energy from Coal Technology'[18]), the statement is made: 'It is reassuring to know, however, that proven reserves of coal in the U.S. would supply an expanding U.S. economy for hundreds of years'. This statement is in the category of 'true but apt to mislead'. Any resource can be made to last for any time and still 'supply' an economy if the fraction of need met by the resource is not specified. In the statement quoted, the fraction of energy need assumed to be supported is not stated, and hence the statement of itself remains true, although it then has little meaning. It has been seen, here, that, if the whole U.S. energy need is to be based on coal by 2000 A.D., the requirement in the opening of new mines, recruitment of miners, and the environmental effects, are impractically large. But oil and natural gas will be largely exhausted at that time; and atomic sources cannot be built up in sufficient size and number in the few remaining years.

TABLE 4.2[15]

OPERATING COST OF COAL TO HYDROGEN PLANT

Total operating costs, including 15% fixed charges, are provided in this tabulation for natural gas priced at 60¢/10^3 SCF and coal or lignite at $7 per short ton

Process	Total annual operating cost (10^6 $) for plant size of			
	250 tons/day	500 tons/day	1000 tons/day	2000 tons/day
Steam-methane reforming	11.7	22.4	42.9	102.7
Steam-oxygen process for lignite	16.1	29.1	53.5	121.0
Steam-oxygen process for bituminous coal	15.4	27.1	48.8	108.8
CO_2-acceptor process for lignite	15.8	28.4	51.6	117.0

TABLE 4.3[15]

ECONOMICS OF COAL TO HYDROGEN: KINCAID'S ESTIMATE (1973)*

	Steam-oxygen process Bit. coal (10^6 Btu)	CO_2 acceptor process (10^6 Btu)
Fixed charges (15%)	28.7	37.1
Oxygen	27.4	
Power	7.4	18.1
Direct labour	1.0	1.3
Materials	8.5	10.7
Maintenance	15.7	20.2
Other utilities	2.6	1.5
Oper. allocation	4.3	5.5
General & administrative	4.4	5.6
	$1.02	$1.03

* 1974 estimates suffer from a 50-100% increase in the cost of coal; and other cost increases. A corresponding estimate to this from Sasolburg gives $2.00-$2.50 per MBtu.

In the U.S. in 1974 there are some 5,600 mines producing 603 million tons of coal each year, giving an average output per mine of 0.1 million.[20] To be independent of foreign oil, according to Arthur[21], a production rate of about 3 billion tons of coal a year will be necessary by 1985. At present 0.6 billion are produced. Therefore coal mines will have to be increased by a ratio of 3:0.6 or about 5 times. Arthur also says the manpower increase would be from about 128,000 to about 1 million people, or about 7.8 times.

As the U.S. will have to import half the oil it consumes by 1985, it seems reasonable to take 10 times as the increase in the coal production needed to become independent of oil, through coal conversion. This makes it possible to calculate the rate of opening of new mines which would have to occur (on average) between now and 2000 to change over to coal as the main energy source,—if U.S. energy needs stopped growing in 1985.

At present, the U.S. has about 5,600 mines. If it is assumed that the future mines are on average the same size as the present ones, we need to produce 10 times more mines, or 56,000 new mines. Hence, the number of mines which have to be opened per day from now until 2000 will be 5. This seems an excessive expectation, but it is conservative, since it represents the 1985 U.S. need, fulfilled by 2000. Even if the mines were giants of 10 times the average size, they would have to be opened at the rate of one every two days until 2000.

Correspondingly, Arthur[21] has stated that 1 million more miners would be needed by 1985, and about 2 million by the year 2000. Thus, the number of new miners to be trained each year would be about 100,000. This seems unlikely short of drafting soldiers to become miners.

It takes some 2 to 3 years to put a strip mine into operation. It takes about a year to manufacture shovels for a mine and another year to assemble everything on site. Deep mines take 3 to 5 years to start up[2].

THE UNDERGROUND GASIFICATION OF COAL

Were it possible to carry out gasification underground, along the coal seam, the logistical difficulties of coal gasification would be greatly reduced.

However, there are difficulties in this process, which has been worked on extensively for decades in many countries. Supposing that it were indeed possible to find a way to react coal in the seam, for example, with high pressure steam, the 'coal gas' produced would then rise through the surrounding soil. It would spread and diffuse over a wide area and concepts for its efficient collection have not so far seemed feasible.

STRIP MINING

This process would relieve the logistical difficulties. However, only a few per cent of U.S. coal deposits can be strip mined.

THE GREENHOUSE EFFECT

An increase in the burning of fossil fuels brings with it the difficulty of climatic changes which would accompany a substantial increase of CO_2 injection into the atmosphere.[22] The mechanism rests on the change in shift in the intensity-wave length pattern as a function of the emitter temperature. The solar radiation originates from material at about 6000° C. The light radiated back into space has a distribution which is shifted towards much longer wave lengths because the emitter is much cooler. It is in the infra-red and that is where CO_2 absorbs. The more CO_2 builds up in the atmosphere – due to the fact that equilibrium with the sea takes about 1,000 years, and the photosynthetic back reaction to oxygen cannot keep pace with the extra, artifical CO_2 injection – the warmer will be the atmosphere.

Whilst it was thought that fossil fuels would be *decreasingly* emitted into the atmosphere after the end of the century, the CO_2 greenhouse effect did not seem to offer any danger. The fact that the mean temperature has been slightly *decreasing* in the last 20 years (due probably to increasing dust levels in the atmosphere) further reduced attention to the matter.

However, the prospect of coal being used as a main source of energy through the earlier decades of the next century – until all fossil fuels approach exhaustion – provides a serious prospect. Commentators have

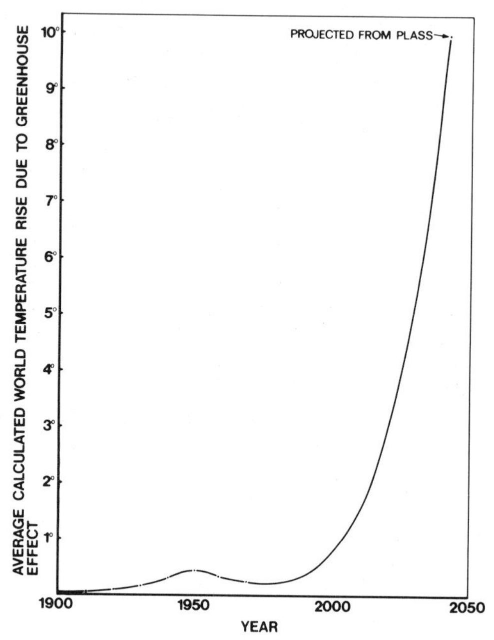

Fig. 4.4 A possible result of relying upon coal as the main source of energy until it is near exhaustion. The figure is derived as follows. Data through 1970 are from the experimental decade temperatures.[25] Because the dust level in the atmosphere increased some 5 times between 1940 and 1960, it seems likely that the dip in world temperatures exhibited in the last 20 years is due to this. To obtain the point at 2040, Elliott and Turner's estimate[13] of a maximum in production rate of world fossil fuels between 2030 and 2050 is accepted, as is Plass's estimate[24] that burning of all fossil fuels would increase the temperature of the atmosphere by 12°. The greenhouse effect factors are *assumed* to overcome the dust rise factors. The assumption of the absence of significant contributions from atomic and solar sources is an unlikely one. However, in the U.S. at 2000 A.D., atomic sources are projected as supplying only a few per cent of *total* energy. The indications of the figure would be realized if successful underground coal gasification allowed coal to replace oil and natural gas; the breeder reactor was not developed (pollution); and massive solar energy conversion is not built before 2030.

calculated[23 24] the rise in temperature which would occur if all fossil fuels were burned. Manabe and Wetherald[23] calculated a temperature rise of 24°C. Plass[24] thinks this is too great and calculates 12°C (see Fig. 4.4).

A 12°C temperature increase for overall world temperatures would be intolerable. Climatic effects would be severe and uncertain. One effect would be a melting of that ice now at temperatures above −12 (to −24)°C. The projected rise in world sea level would be significant.

Such climatic changes would be difficult to avoid were coal used as the main source of energy until its exhaustion in the third or fourth decade of the next century. They provide a strong reason for not pursuing large-scale coal gasification after about 2000.

CONCLUSION

Chemically, power gases, similar to natural gas and to hydrogen, can be made from coal; as can oil and petroleum. The economics have not been acceptable in the past, but have become so, because of the continually rising prices of oil and natural gas.

The reasons against the use of coal as a main source of energy are three in number, the last mentioned overwhelms the others.

Economical: Investment and amortisation problems have been looked on in the past on the basis that there were inexhaustible amounts of coal. Because of the environmental and exhaustion difficulties early in the next century, many new plants might have a life of less than 30 years, and hence not be attractive propositions without government subsidy.

Environmental: Removal of S is as yet uncertain. The greenhouse effect may become significant.

Logistical: The coal industry could not be expanded in the time necessary, and in the size necessary, to make coal the main source of energy by the end of the century.

Considerations such as these make reliance on coal for the post-2000 situation as a *main* source of hydrogen questionable, and potentially dangerous. But, until government funds* to produce inexhaustible clean energy sources have taken effect, coal may have to be regarded as the beginning source for a Hydrogen Economy.

Processes for obtaining hydrogen from coal are simpler than those for obtaining the high Btu SNG (synthetic natural gas). No polluting fuel would be diffused into the community. Pollutional products would be under control at the plant. Plants at the seaboard could reject CO_2 directly into the sea (no atmosphere build-up). The conversion could be gradual and the utensils, apparatus, etc., gradually changed: it would remain the same for H_2 from nuclear and solar plants. Amortisation problems cease to be important in the production of hydrogen from coal, because the plants would not become outdated before the end of their normal life: they could continue to operate as the Hydrogen Economy based on nuclear and solar sources was built up.

---

* A recent Annual Report of the U.S. Office of Coal Research[2] shows that a number of the individual research projects for coal conversion to polluting fuel have funding in the tens of millions, that is equal to the funding for the entire U.S. Government clean Solar Energy Research Programme.

REFERENCES

[1] L. Lessing, *Fortune*, November 1972, p. 210.
[2] 'Coal Technology: Key to Clean Energy', Office of Coal Research, U.S. Department of the Interior, Annual Report, 1973-74.
[3] L. Lessing in *Chemistry in the Environment*, Freeman, San Francisco, 1973, p. 159.
[4] P.E. Rousseau, World Power Conference, Tokyo Section of Meeting, 16-20 October 1966, p. 912.
[5] *South African Review*, 7, 2 (1973).
[6] E. Cohn, priv. comm., 17 June 1974.
[7] T.V. Sheean & M. Steinberg, *Comparison of Different Types of Fluidised Bed Operations for Hydrogenation and Liquefaction of Coal*, Brookhaven National Laboratory, Upton, New York, October 1973.
[8] C.W. Matthews, presented at the American Power Conference, Chicago, Illinois, April 1972.
[9] *The Oil and Gas Journal*, 23 July 1973, p. 27.
[10] B.S. Lee, presented at the American Power Conference, Chicago, Illinois, April 1970.
[11] 'U.S. Energy Outlook – Coal Availability', National Petroleum Council, 1973.
[12] P.F.H. Rudolph, American Lurgi Corporation, Publ. 24072, 1974.
[13] M.A. Elliott & N.C. Turner, presented at the American Chemical Society Meeting, Boston, Massachusetts, April 1972.
[14] W. Kincaid, priv. comm., 1973.
[15] W. Kincaid, paper presented at the Conference on the Hydrogen Economy, Cornell, 1973.
[16] *Science,* 184, 331 (1974).
[17] *Energy Profile*, February 1974, p. 23.
[18] 'Clean Energy from Coal Technology', U.S. Department of the Interior, Office of Coal Research, 1974, p. 28.
[19] D.B. Thompson, *Industrial Week*, 26 November 1973, p. 17.
[20] *Chemical and Engineering News*, 12 November 1973, p. 11.
[21] J. Arthur, Chairman, Duquesne Light Company, Pittsburg, Pennsylvania, quoted in ref. 19.
[22] G. Plass in *The Electrochemistry of Cleaner Environments*, ed. J.O'M. Bockris, Plenum, New York, 1972.
[23] S. Manabe & R.T. Wetherald, *J. Atm. Sci.*, **24**, 241 (1967).
[24] G. Plass, *Tellus*, **8**, 140 (1956).
[25] J.O'M. Bockris in *The Electrochemistry of Cleaner Environments*, ed. J.O'M. Bockris, Plenum, New York, 1972, p. 11.

CHAPTER 5

Sources of Abundant, Clean Energy

INTRODUCTION

FIG. 5.1 portrays some sources of inexhaustible energy and Fig. 5.2 gives methods of converting them to electricity. Gravitational energy is the 'primary' energy, because it triggers the release of nuclear energy as in the fusion which is the origin of solar energy. Thus, gravitation causes the cosmic dust to aggregate and form an increasing mass. Inside this, gravitational forces operate, increasing with the mass. When these are sufficiently large, the hydrogen undergoes nuclear fusion.

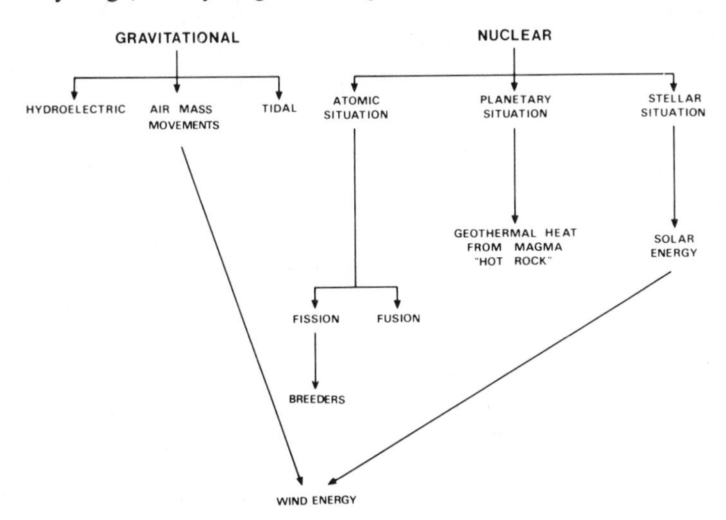

Fig. 5.1 Abundant, clean energy sources.

Geothermal sources, too, are secondary to the nuclear, because the heat of the earth's interior arises from the molten core, this heat arising from nuclear heat and radioactivity.

Few of the sources which arise from gravitational, nuclear and geothermal energy have been converted to useful work. This does not mean that there are large untapped energy sources available to the technology now in sight. Correspondingly, some of the methods of conversion, referred to in Fig. 5.2, though old in principle, are little researched and have not yet been brought to the development stage.

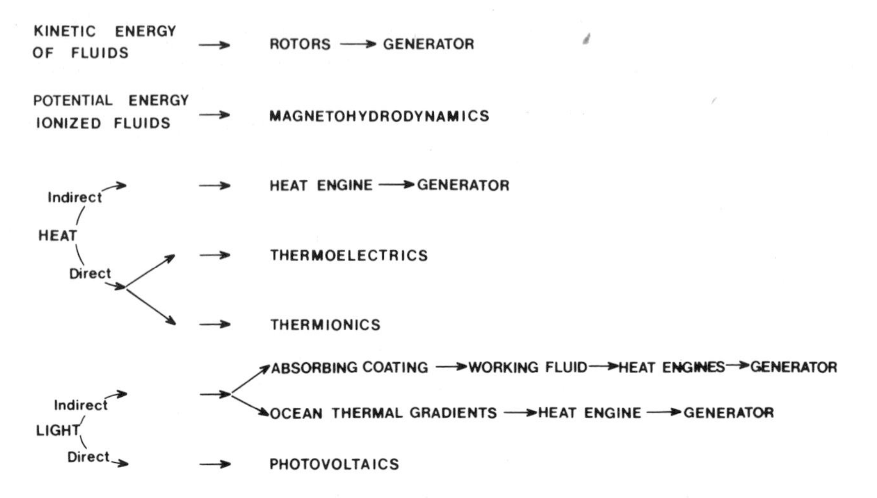

Fig. 5.2 Some methods of conversion of energy to electricity.

The fossil fuel energy source which has been predominantly used up till now, is solar in origin, from photosynthesis. Oil, coal and natural gas were formed in complex processes in which decaying vegetation was subjected to high pressures.

An alternate view of the formation of fossil fuels has been put forward by Dwyer.[1] In an early stage of the earth's history, the atmosphere was largely methane. This methane condensed to higher, and liquid, hydrocarbons by photochemical reactions arising from lightning and electrical storms, occurring in the first billion years of the earth's atmosphere. A layer of warm hydrocarbon oil covered the surface of the sea and may have been the locus in which early life forms developed. The origin of the energy here is the electrostatic charge separation which gave the electrical discharges which drove the polymerisation reaction.

Other methods of converting energy from various sources to electrical energy are indicated in Fig. 5.2.

There is not yet confidence that energy from controlled fusion can be obtained: those who are optimistic about the prospects, project that a pilot plant could be built in the 1990s.

Nuclear energy from fission has been possible since the 1950s, but only 1% of electricity in the U.S. arises from nuclear sources. There

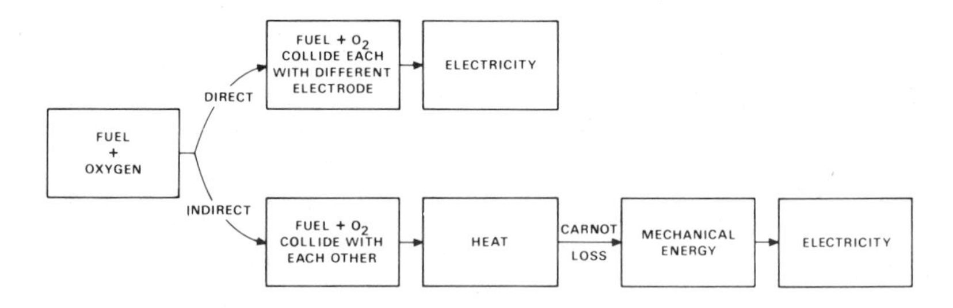

Fig. 5.3 Direct and indirect energy conversion.

have been, and are, hold-ups in building atomic plants in the U.S.A., partly because those aware of health hazards of such reactors have taken out injunctions, the granting of which has halted construction.

Breeder reactors are a *possible* source for the future. They are still only experimental, and no orders are being accepted for them. They would take about 10 years to get to the full breeding stage.[2]

Gravitational energy is manifest in hydro-electric power. Tidal energy is being used at Kislaya Guba in Russia and at Rance in France.[3]

Geothermal energy is used in small amounts *via* hot springs and its development from dry hot rocks is under study.[4]

OTHER SOURCES OF ENERGY

There could be other sources of energy not shown in Fig. 5.1. They are speculative. For example, the magnetic and electric field gradients which surround the earth contain much energy.

Other suggestions from Science Fiction may be mentioned. Could gravitational fields in the solar system be tapped? Could objects be orbited which would encompass the sun and earth? Would it be possible for the earth to draw energy from their regular passages near the earth?

Only a tiny fraction of the sun's energy strikes the earth. Will it eventually be feasible to engineer (from the material of asteroids?) a super-giant collector in space to focus a higher fraction of the sun's energy than can now be collected, by covering, say, 1% of the earth's surface. Is the beginning of such a concept to be seen in that of orbited satellites acting as solar collectors, beaming back to earth (Chapter 7)?

FORMS IN WHICH ENERGY IS MANUFACTURED

The form of energy (light, electricity, magnetism and mechanical energy) are interconvertible, but each time a conversion is made, some energy, nearly always more than one-half of the total concerned, is lost to the surroundings as heat (Fig. 5.3).

DIRECT AND INDIRECT ENERGY CONVERSION TO ELECTRICITY AND MECHANICAL ENERGY

In direct energy conversion, one goes from, for instance, solar energy in the form of light, to, say, electricity, in one step. This is what occurs in a photovoltaic cell. There are several methods for direct energy conversion, which tend to be simpler in the machinery necessary than indirect methods which involve several steps. Direct methods include photovoltaics (light to electricity), thermoelectrics (heat to electricity), magnetohydrodynamics (heat → ionized particles → electricity), thermionics (heat to electricity), etc. The basic principles of all these methods are well known. The development to practicality was greatly advanced in the 1960s in the U.S.A., but reduced in rate when NASA largely withdrew from the energy conversion field, having successfully developed photovoltaic cells and electrochemical fuel cells and batteries for space vehicles.

Indirect energy conversion is a method which involves two or more steps between the primary form of energy (usually heat) and the final useful form (such as mechanical energy), Fig. 5.3. Thus, in the working of an internal combustion motor, solar energy from a remote time, stored in vegetation which became oil, is reacted with air to give CO_2 + water + heat. Heat is the medium of energy which we are using, but the conversion is indirect, because heat by itself has only space heating use and must be converted to mechanical or electrical energy by well known processes in heat engines and generators.

The fact that man has so far used indirect energy conversion almost exclusively was an historical accident, due to the limits of knowledge in the late 19th Century when our present course was set. Indirect energy conversion is obviously disadvantageous. It is often less efficient, and involves more stages than direct methods. At the time energy conversion from coal got under way, the early processes were not seen in the context of the many possibilities for energy conversion which are now known. The engineering developments by the beginning of the century had gathered considerable momentum, making changes difficult. Interest in direct methods lessened, partly because (for example, as with fuel cells), their basics were not understood, so that the initial low performance expected to arise in the first few years of attempts to realise any device, was not corrected.

EFFICIENCY OF CONVERSION

Efficiency varies substantially with the method used. It may be as low as 1% for photosynthesis. For internal combustion, the figure is in the region of 25%, and increases to 38% for highly efficient and very large electric power stations. The most efficient devices for energy conversion are electrochemical fuel cells and here the energy conversion is more than 50%. If hydrogen (rather than natural gas) were the fuel, fuel cell efficiency of more than 70% at medium current densities would be possible.[5]

THE BURNING OF FOSSIL FUELS

Small attention will be paid to this subject in this book, because the method is classical, and there are many accounts of it in other works.[6] The fuel is exhausting and there are difficulties in obtaining it in a clean form. It is important to point out a basic principle: the fraction of the heat produced by the reaction (the result of the difference of potential energy between the bonds in the molecules before and after the reaction) which can be converted to mechanical energy is the Carnot efficiency factor (0.1 – 0.4).

There are many ways of achieving the conversion of heat to mechanical energy: the devices concerned are all called 'heat engines'. The Wankel engine is a variation on this in that the internal pulse is connected to a moving part which is not a piston but a rotor. The Australian Sarich engine bears a similarity to the Wankel engine.

A combustion engine which pre-dates the Otto internal combustion engine is the Sterling engine. The fuel burns *externally* and heats a compartment which contains a working fluid, which may be air, or a volatile gas. This expands and thrusts a piston, etc.

The lack of emphasis given to avoiding the Carnot cycle efficiency loss[7] (as in the use of Carnot-free electrochemical conversion) in the first half of this century, has wasted about three-quarters of the fossil fuels used and simply dissipated the heat into the surroundings. Had these fuels been converted electrochemically to energy from the beginning of the century, the average efficiency of conversion could have been above 50%, instead of less than 25%.

Due to the delay in giving sufficient R & D funding to develop inexhaustible, abundant, clean energy, the remaining fossil fuel supply will have to continue to be used and, as this is already in short supply[8], it is desirable to apply the most efficient way of doing this.

The fossil fuels should obviously not be used to exhaustion. There is a need for carbon-based chemicals for synthetic products, for example, textiles for clothing and synthetic food production. Using coal and oil as fuel is like burning the wooden walls of the farmhouse to keep the occupants warm. Once the situation is realised -- that there are inexhaustible clean sources available, whilst we spend our research dollars to find out how to use up the last of our polluting carbon-containing compounds - the present using up of the limited amounts of fossil fuels begins to seem intensely irrational.

SOLAR

The earth intercepts the energy of the sun at the rate of about 1.7 10^{17} watts. So long as world population growth can be stopped at less than 10 billion, it can be shown (Chapter 6) that solar energy alone could provide enough energy for an affluent world of abundant, clean energy.[10]

AREAS INVOLVED IN COLLECTION OF SOLAR ENERGY

The method by which one may gain a concept of the area of solar collectors involved for a given population is to give this population

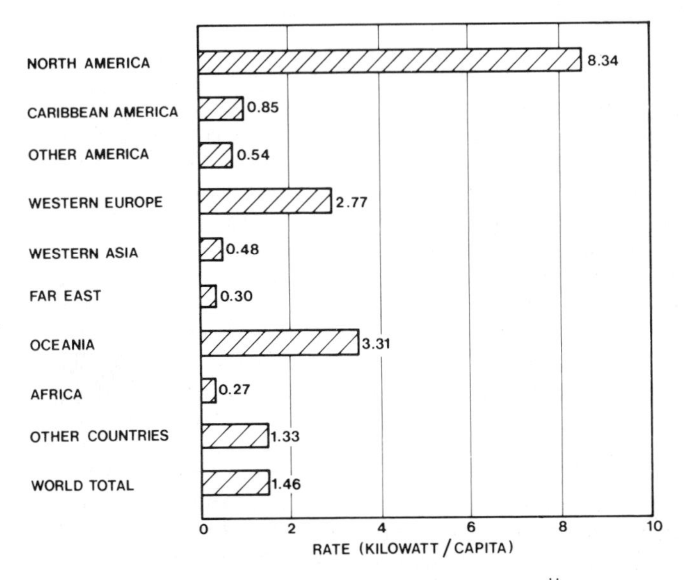

Fig. 5.4 World energy consumption by continental areas (1966).[11]

an energy rating. The energy ratings as given by Hammond[11] are shown in Fig. 5.4. They will increase and it is reasonable to count them as about 10 kW per capita. The efficiency of energy collection from known methods is 1-20%, and it may be assumed 10%. The solar collectors (Fig. 5.5[12]) should be counted as covering one-half of the space provided to allow for cleaning and repair ducts, and also for the tilting of panels so that they remain orthogonal to the sun over hours of exposure. One must multiply the solar constant by a factor which allows for cloudiness figures which are available for specific areas of sunlight per year.[13]

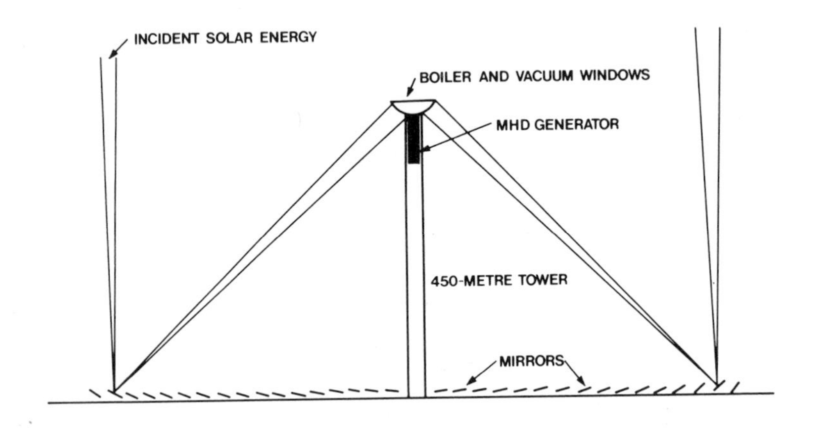

Fig. 5.5 Solar collector.[12]

The number of square miles necessary for a solar collector to supply Japan with energy, assuming a population of 120 million and a consumption of 10 kW per person, can be found. The overhead sun at the equator gives 1kW m^{-2}. With a 10% collector, at midday at the equator, 0.1 kW m^{-2} can be collected. However, walkways and space for the function of sun-tracking panels are needed so that, say, twice the space is required, that is 0.05 kW m^{-2}. Finally, let it be assumed that the sun shines about half the time (in a tropical region) and cloud reduces the incident light by another half. Take 0.01 kW m^{-2} as the basic figure for smoothed out continuous collection. Then, because 1.2 10^9 kW is necessary, the space needed is about 10^{11} sq metres, or 10^5 sq km. If laid out in the form of one single square, the side would be $10^{2.5}$ km, or there would have to be a square with side about 300 km or 200 miles.

The arrangement of the area would be subject to local considerations: it might not be best to have it in one site. Solar farms could be broken down into small areas. They must have a water supply (hydrogen for storage and transmission), that is, be near the sea or artesian sources. Solar collectors on the sea (Chapter 7) are to be encouraged. Space is then of lesser consequence; and the largest number of cloudless days is attainable on Pacific Ocean areas (Fig. 5.6[14]).

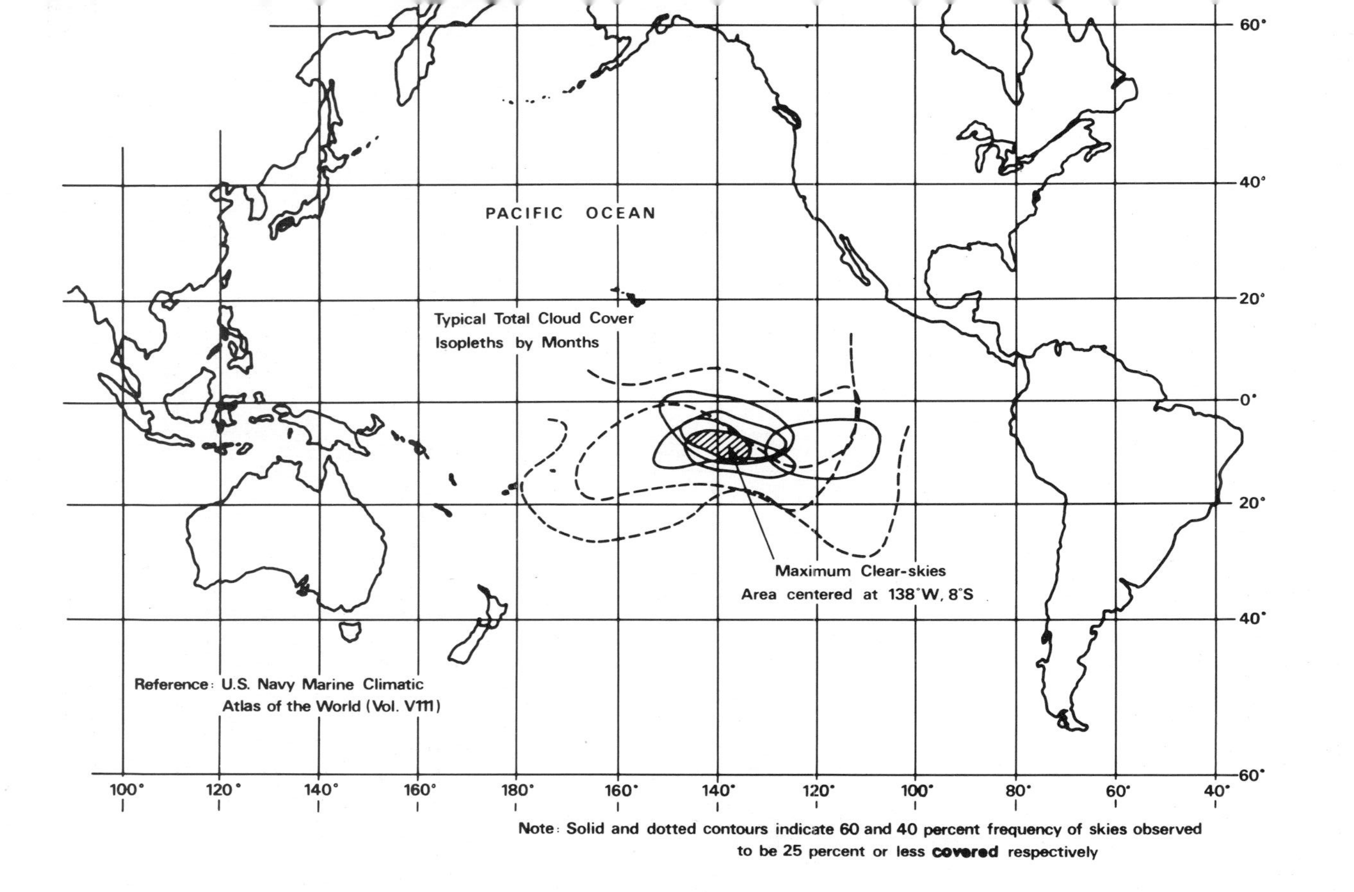

Fig. 5.6 Cloud-cover minimum trends in mid-Pacific Ocean.[14]

METHODS OF CONVERSION OF SOLAR ENERGY TO ELECTRICITY AND MECHANICAL ENERGY

These matters are discussed in Chapters 6 and 7.

CONTROLLED FISSION

(a) *Fission reactors.* Fission reactors are the reactors in use on a small scale now. These reactors do not all under consideration for a future path for energy conversion, because they use materials which would not last for more than a few decades if used in simple fission.

Two solutions to the problem of uranium supplies can be put forward. There is abundant uranium in the sea and it is more concentrated (*c.* 10^{-9} mol l^{-1}), than, say, gold (10^{-12} mol l^{-1}), which has sometimes been considered extractable. Marchetti[15] has suggested that fission reactors could be placed on islands of the Pacific Ocean, processing seawater to extract their fuel. The fission plants would give energy which could decompose water to hydrogen and this could be shipped to sites of use.

Extraction from rocks, where there is uranium at 20 ppm, is possible, but the energy needed to do this may approach the energy contained in the uranium extracted. There are much better possibilities for obtaining abundant, clean energy in the future than either of these suggestions.

(b) *Breeder reactors.* This mode of generating energy need not be discussed because it is well known.[16] A 250 megawatt plant is operating in England, and a 150 megawatt plant, with a distillation capacity, began operating in Russia in 1972. In the U.S.A., a contract for the construction of a breeder reactor of 350 megawatts has been given for completion by 1980 (two small experimental breeder reactors have already been built).

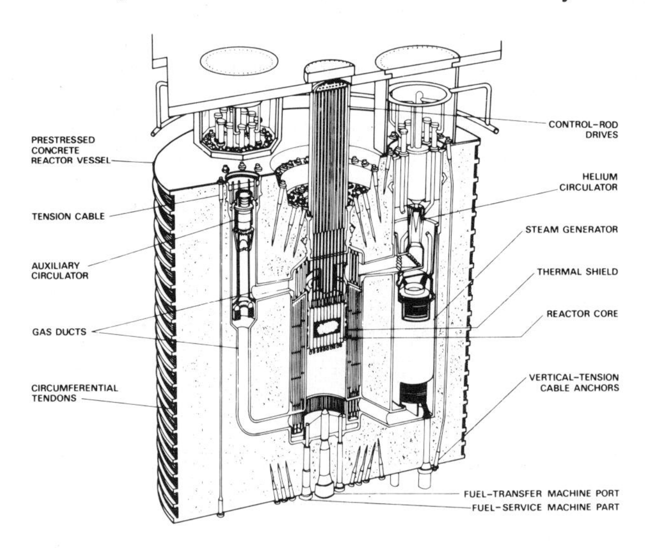

Fig. 5.7 Fast breeder reactor of the gas-cooled type. Such a reactor differs from a liquid metal-cooled reactor in that the helium coolant is pressurised to obtain adequate cooling. The reactor is enclosed by a massive pre-stressed concrete vessel.[17]

Breeder reactors will not be built in any significant quantities until post 2000. After installation, several years are needed for the reactor to reach the steady state of breeding. There may be difficulties in respect to breeder reactors on the pollutional side. An account of breeder reactors (Fig. 5.7[17]) is given in the book by Hammond *et al.*[2] The pollutional aspects of atomic reactors and a reasoned case for the termination of the nuclear energy programme is given by Ford *et al.*[18] According to Tamplin and Cochran[18a], the pollutive dangers from breeders have been underestimated by 100 times. The key paper, however, is by Petkau[18b], which showed that radiation damage may increase with an decrease [sic] of concentration of pollutant.

CONTROLLED FUSION

Reactions which might be used in fusion reactors are two:

$$\text{deuterium} + \text{deuterium} \rightarrow \text{helium}^{-3} + \text{neutron} + \text{energy}$$

$$\text{deuterium} + \text{tritium} \rightarrow \begin{array}{l}\text{tritium} + \text{proton} + \text{energy} \\ \text{helium} + \text{neutron} + \text{energy}\end{array}$$

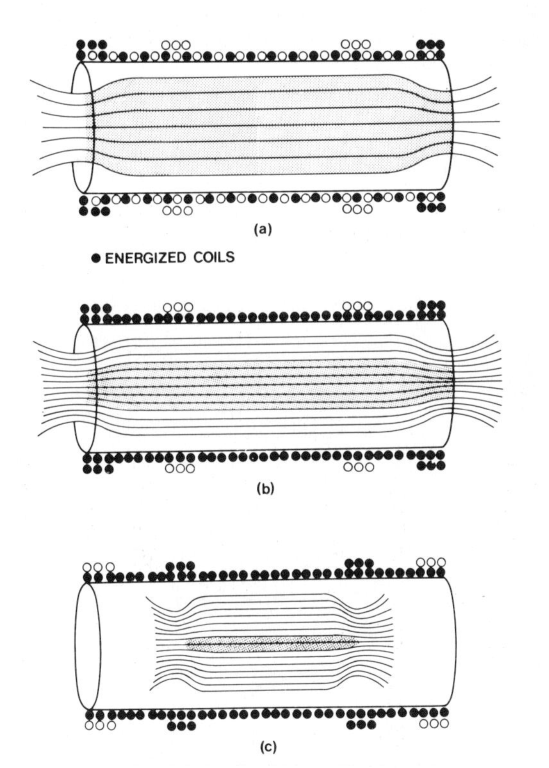

Fig. 5.8 Adiabatic compression of the plasma. (a) Plasma is injected into the chamber while the field is weak. (b) The magnetic field strength is then increased, compressing the plasma toward the centre and raising its temperature. (c) The magnetic mirrors may also be moved axially inward to provide additional compression and further increase in temperature.[19]

The deuterium-tritium reaction releases more energy, and has a lesser temperature at which it begins, so it is more favourable for the early fusion reactors. However, tritium sources are exhaustable indeed and such fusion reactors could not supply more than *c*. 100 years of energy. It is the more difficult deuterium-deuterium reaction which gives the utopian prospects, for deuterium is effectively inexhaustible.

To start up the fusion reactor practical energies of 10-20 keV must be achieved, otherwise electrostatic repulsion of the nuclei occurs. Thus the fuel must be heated to a *very* high temperature to achieve the energies required. For the deuterium-tritium reaction, the ignition temperature is about 10^{8}°K, ten times the temperature of the sun's interior.

Secondly, particles having this temperature must be confined for a time so that one of the fusion reactors occurs, and heat or electricity can be extracted.

Materials at such temperatures are plasmas, ionised atoms. The particles can hence be held in a vessel by a magnetic field which ties the ions within magnetic field lines. Contact with the walls of a vessel would, of course, be impractical. Fig. 5.8[19] shows the type of magnetic confinement contemplated.

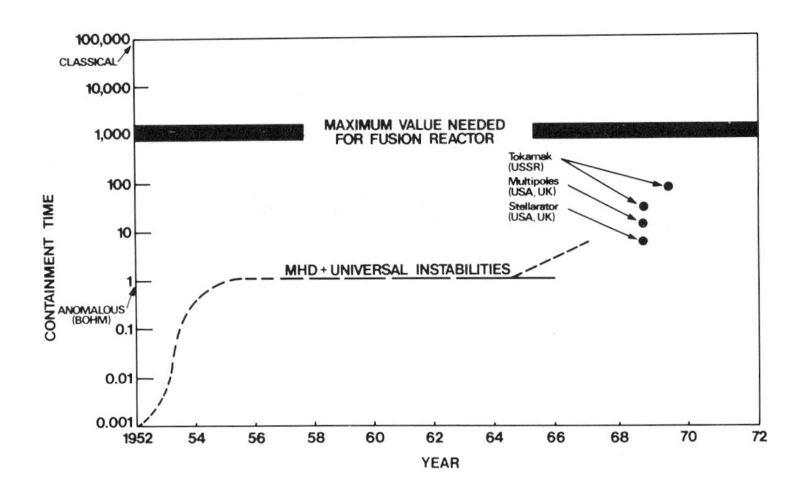

Fig. 5.9 Improvement in containment time in toroidal machines. The classical diffusion time varies as $(kT)^{1/2}B^{2}n^{-1}$; the anomalous diffusion time ascribed to Bohm varies as $b(kT)^{-1}$.[20]

A difficulty is the unstable motions which a plasma exhibits. In Fig. 5.9[20] the times of stability so far attained are shown, compared with that which must be attained for practical conversion. A well-known landmark of research is a Russian one where, in a device, 'Tokamak', a reverse of the town of Kamakot, where it was built, 0.2 10^{8}°K was attained. The containment time was some one-tenth that required for the production of power. To scale-up a machine such as that built at Kamakot, costs $10 million, and such a machine would still be only an experiment (Fig. 5.10[20]).

The Russian Tokamak machine has been reproduced in the U.S.. Much of the rest of planning for fusion is in the systems analysis stage. The energy of the deuterium-tritium fusion would be released as 14 meV neutrons which would contact a system of heat exchangers and generate steam. Tritium would be bred out.[20]

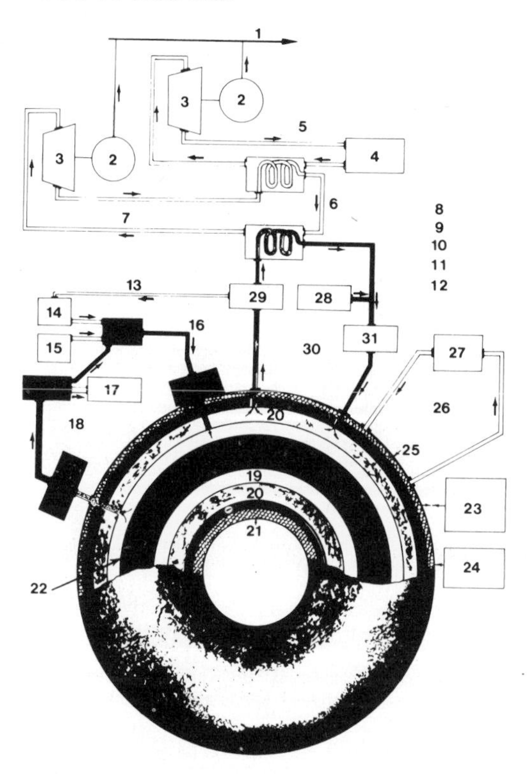

Fig. 5.10 Schematic design of a 2500 MW (e) fusion reactor.[20] (1) electrical energy; (2) generator; (3) turbine; (4) condenser pump; (5) steam stage; (6) heat from exchanger; (7) potassium stage; (8) measurements of the core reactor; (9) core external diameter; (10) tube diameter; (11) performance value; (12) N_{heat} = 5,000 MW, $N_{electric}$ = 2,500 MW; (13) tritium – 0.6 gr per minute; (14) tritium supply; (15) deuterium supply; (16) D-T mixture; (17) He gasometer; (18) combustion circle; (19) vacuum; (20) blanket; (21) superconductor; (22) plasma; (23) excitement and discharge of the magnet; (24) energy storage for starting; (25) magnetic coil; (26) helium cooling circuit for superconducting coils; (27) cooling apparatus for 4.5° K; (28) lithium supply; (29) tritium separator; (30) lithium circuit; (31) lithium pump.

Materials problems are very important in fusion. The materials must show tolerance to neutron bombardment. Lithium would blanket the reaction vessel. Next would be a shield of a solution of borates and finally the magnetic coils, superconducting, to produce the confining field of 100 kilogauss.

In another, different, concept (laser fusion), particles of hydrogen would be injected into a chamber. A laser fires upon them and induces in the very small volume of the hydrogen pellet a very high temperature which begins fusion and therefore gives out heat. No difficulties of plasma retention would be present (Fig. 5.11[21]).

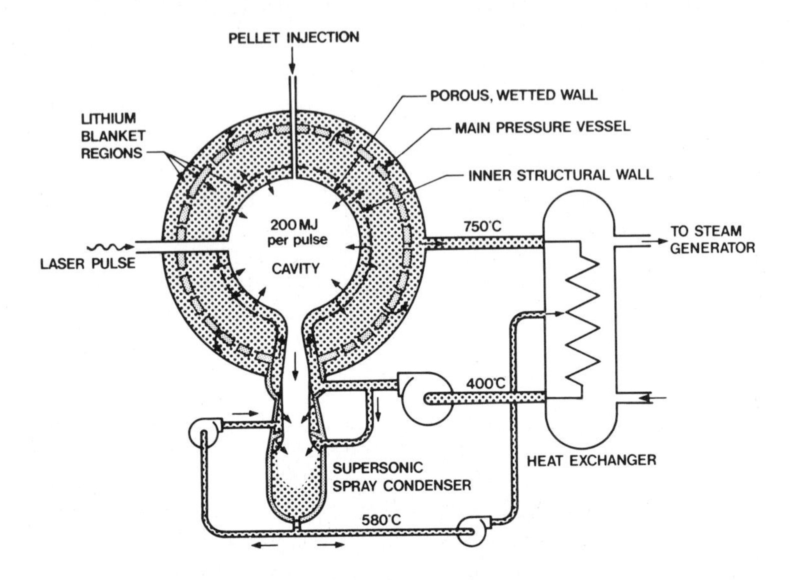

Fig. 5.11 Laser fusion reactor with a 'wetted-wall'. This proposed design employs a porous wall to define a cavity at the centre of a sphere of liquid lithium. Fuel* pellets are injected into the cavity through a channel at the top and heated by a laser beam aimed through another channel at the side. A layer of lithium about 1 mm thick on the inside of the porous wall, formed as pressurised lithium flows through the wall, would vaporise and protect the wall from melting. A larger steel pressure vessel absorbs most of the shock pressure generated by each explosion. The spray condenser absorbs the energy of the vaporised lithium. Hot lithium recirculated through a heat exchanger provides steam for a turbine. If a new pellet were ignited every second releasing 200 megajoules (MJ), the thermal power output would be about 200 MW.[21]

* Solid hydrogen.

General Situation with Fusion[22]

A fusion reactor represents an attractive *possibility* for the mid-21st Century. It would have to be competitive with *clean* fast breeders and solar collectors of all kinds, in particular ocean thermal gradient collectors. It will have large fuel reserves (and, if the D-D reaction were used, these reserves could be regarded as inexhaustible). The pollution is leaked tritium (relatively harmless) and the thermal pollution. The latter factor would tend to make fusion reactors be placed on the deep ocean.

The engineering problems of fusion reactors will be too great for attainment of trial devices before 2000. If the rate of progress of the U.S. atomic energy programme is taken as a criterion, *practical* devices will not be available until several *decades* later. Thereafter, one must allow for the building into systems. Some properties of plasma are not understood, and the materials science of the interaction of neutrons with the surrounding matter has been little researched. $300M is being spent on the fusion reactor in 1974 in the U.S.A., about one-tenth of that being spent on flights into outer space. The priority in degrees of funding belongs to the 1960s when the energy shortfall had not been perceived.

GRAVITATIONAL

Hydro-electric

Many hydro-electric sources have been tapped. A Hydrogen Economy would improve prospects for a number of others. Previous considerations needed sources to be within a few hundred miles of sites of use, whereas if hydrogen is produced at the source (fresh water, no desalination necessary) and piped to user areas, remote, large hydro-electric sources, such as those in Rhodesia, could become economic in supply to distant industrial areas.

Tides

When a tide enters a bay or estuary, a part of this may be closed after the tide has reached a maximum, and opened again when the tide is at a minimum. The height difference allows the operation of a periodically functioning hydro-electric generator. Tidal power may be worth developing when the tide is 20 feet or more. Rance, on the Channel Island coast of France, has a 240 megawatt generator (24 units, each of 10 megawatt capacity) which works on this principle.[3]

An alternative way of tapping tidal energy in estuaries and rivers would be to make incoming and outgoing tides turn rotors. This approach has been used in an experimental Russian power station, Kislaya Guba, at the Islaya Inlet of the Barents Sea (Fig. 5.12[23]). However, some excellent tidal sources are in Western Australia.[23a]

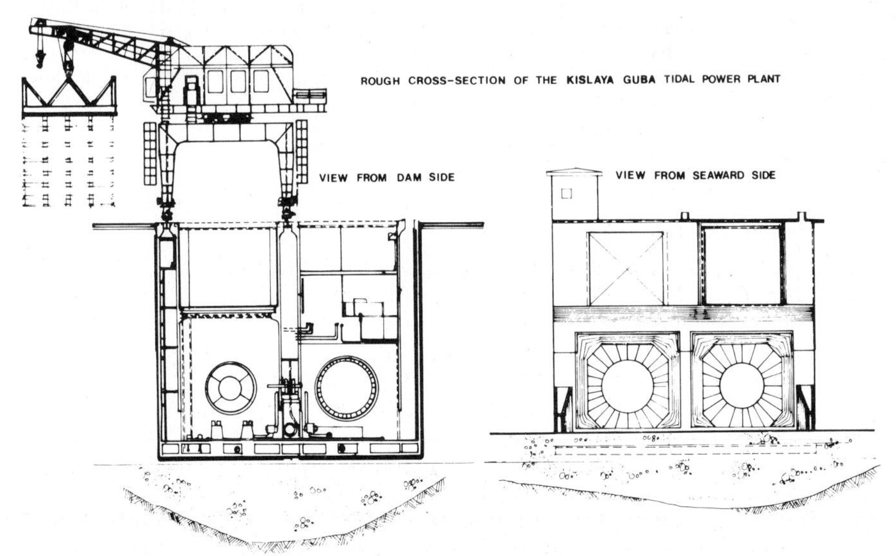

Fig. 5.12 Aspects of the tidal generator at Kislaya Guba.

There are numerous situations in the world where such arrangements could be exploited. Thus, consider the generating power which could be produced by placing rotors across the lower reaches of the Amazon, and other great rivers. If the water in the Amazon moves at 5 mph, and the propeller diameter is 200 feet, *c.* 0.1 MW per rotor could be developed. With a 3 mile width the river could produce 5 MW per set of rotors across the river. In 40 miles of river, with propeller barrages every 1,000 feet, the generator would give 1,000 MW – enough to support 100,000 people.

Use of streams in the sea

Currents circulate around the sea due to the motion of the earth. Thus, rotary currents several miles in diameter exist in the depths of some oceans.[24]

The total energy in these currents is immense. Whether they could prove useful sources depends on the energy density which they show, and the economics of the machinery necessary. The field is little investigated.

WIND POWER

Introduction

Wind was the first energy harnessed by man, windmills having appeared in England and France in the 12th Century. The use of small aerogenerators in remote areas, particularly in Australia, is well known, the machines producing 1-2 kW in a 20 mph wind and becoming energy producing in winds above 4 mph. Storage of electricity is usually made in lead-acid batteries.

Aerogenerators have seldom been considered as a massive source of power because: (i) Wind is generally regarded as entirely sporadic and unreliable; (ii) If strong winds exist, they are in remote areas (Alaska), and of little interest as a power supply; (iii) The equation which connects the dimensions of a rotor to the wind velocity is little known[25] and so trial calculations of what might be in steady relatively high winds are seldom made.

The relevant equation for instantaneous power is[26]:

$$\text{Power} = \frac{16}{27} \cdot \frac{1}{2} \rho v^3 \tag{5.1}$$

per unit area swept out by a rotor, where ρ is the density of the fluid and v the velocity of the wind. $\frac{1}{2}\rho v^3$ is the kinetic energy of the wind per unit volume and $\frac{16}{27}$ a hydrodynamic factor for the extraction of energy.

The empirical equation[27] is:

$$\text{Power} = \frac{16}{27} \cdot \frac{1}{2} c\rho v^3, \tag{5.2}$$

where c is a parameter, generally about $\frac{1}{2}$. The dependence of v^3 is noteworthy.

It should be noted that the equations 5.1 and 5.2 require the mean of the cubes of the instantaneous wind velocities over the year. If the mean of the velocities is cubed, results are 2 to 3 times too small.

Winds

Winds are in fact constant, that is the pattern of their velocities as a function of time repeats itself year after year. Thus, Fig. 5.13[28], from the work of Mullett, shows the mean wind velocity at the Adelaide Observatory, Australia, from 1878 to 1954. After 1900, the effective mean variation is 1.4%, i.e. the wind velocity over the 54 years was 98.6% constant. The winds in towns are diminished, they are about double the town velocity in open areas.

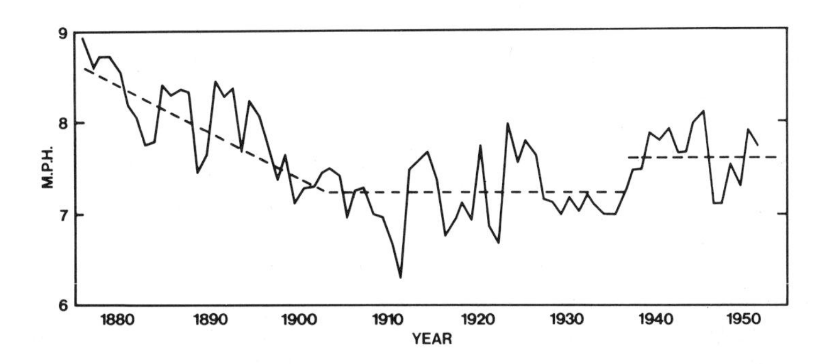

Fig. 5.13 Wind annual mean velocities, as derived from Adelaide Weather Observatory data. The change in 1938 was caused by a change in position of the anemometer.[28]

South of the 30th parallel of latitude, there is a system of anticyclones, a group of vortexes of subsiding cold air, drifting constantly around the planet. The group travels west to east at 25-30 mph.[28] It travels across the coast of South Australia, New Zealand, Chile and the Cape of Good Hope. It is the presence of this kind of high velocity wind system over the sea which makes the massive use of wind energy feasable. This Great Southern Wind which blows constantly round the world with a breadth of about 1,000 km[28], is identical with that known to mariners as the Roaring 40s, and is unique in constancy of velocity and breadth. However, there are many other regions of the world where the wind at ground level averages more than 15 mph around the year[29 30], and mountain areas where greater velocities are frequently continuously available for the transduction of energy.[31]

Apart from global factors*, there are local ones. Strong winds are found on hill crests. Nellie Carruthers[32] derived an equation for the dependence of velocity on height. It is[28]:

$$\frac{v_1}{v_2} = \left\{ \frac{h_1}{h_2} \right\}^{0.17} \qquad (5.3)$$

* The Great Southern Wind arises from the heat at the equator. Air masses rise there, then later cool and fall, to become the origin of the wind described.

TABLE 5.1[33]

WIND VELOCITIES MEASURED BY MULLETT IN SOUTH AUSTRALIA*

Site No.	Measured mean speed (mph)	Measuring height (ft)	Height factor (100 ft)	Adjusted speed for 100 ft (mph)	Site description
101	17	30	1.18	20	Sea level
102	12	30	1.23	15	Flat country
103	16	30	1.18	19	130 ft hill
104	16	30	1.20	19	Sea level
105	15	30	1.23	18	Flat country
106	17	30	1.10	19	600 ft ridge
107	22	30	1.10	24	800 ft ridge
108	13	30	1.23	16	Flat country
201	17	10	1.30	22	Sea level
202	21	10	1.15	24	400 ft hill
203	17	10	1.30	22	Sea level
204	16	10	1.40	22	Flat country
206	18	10	1.30	23	800 ft hill
207	20	10	1.15	23	1,200 ft hill
208	17	10	1.20	20	400 ft hill
209	17	10	1.30	22	Sea level
210	14	10	1.30	18	Sea level
211	14	10	1.30	18	Sea level
212	19	10	1.25	24	Sea level
213	16	10	1.40	22	200 ft hill
214	16	10	1.30	21	Sea level
215	15	10	1.30	21	Sea level
216	18	10	1.30	23	100 ft hill
217	17	10	1.20	20	1,200 ft range
218	20	10	1.15	23	1,200 ft ridge
219	12	10	1.40	17	Sea level
220	17	10	1.25	21	200 ft hill
221	18	10	1.20	22	300 ft hill

Note: 'Sea level' refers to coastal sites less than 100 ft above sea level.

* Australia is an energy rich country with high coal reserves, see Watson-Munro.[43]

Table 5.1[33] shows results obtained by Mullett for South Australia. A hill of over 200 feet in open country is a good site. Flat country sites must not be more than a few miles from shore. Vegetation does not occur in windy areas.[28]

A typical velocity and frequency graph is shown in Fig. 5.14.[33]

Aerogenerators

With the ideal equator for an aerogenerator, using $\rho = 1.4\ 10^{-3}$ .g. cc^{-1}, R being the radius of the rotor in metres, v in km per hour:

$$\text{Power} = 2.7\ 10^{-5}\ v^3\ R^2\ (\text{kW}). \tag{5.4}$$

The equation is illustrated in Table 5.2. In practice, one can expect a power of $\frac{1}{3}$ to $\frac{1}{2}$ of that given by this equation.[27]

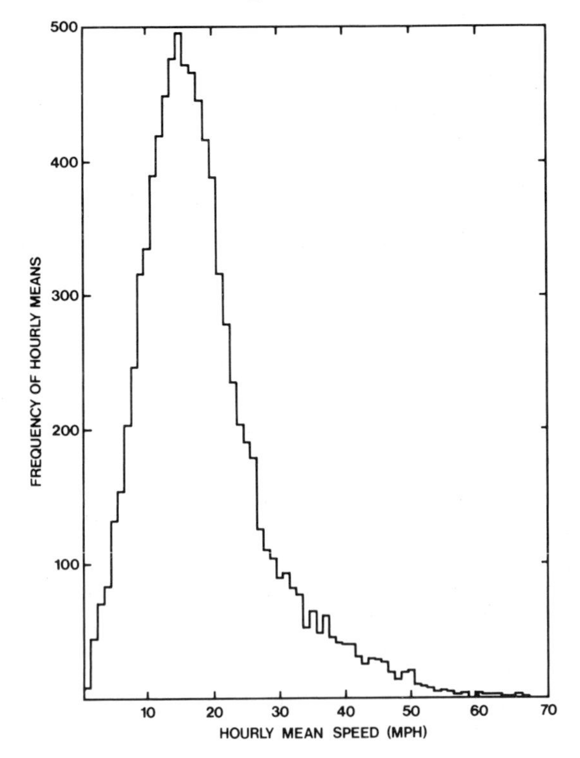

Fig. 5.14 Typical annual velocity-frequency of wind for South Australia.[33]

TABLE 5.2

THE IDEAL WIND ENERGY EQUATION
(Radius of rotor: 50 metres)

Average wind velocity in $km(hr)^{-1}$	Power in megawatts (approx., rounded off)
10	0.1
20	1
30	2
40	4
50	8

The 'all or nothing' character of wind power is clear from Table 5.2. For wind velocities below 20 kph, there is little to be collected. Above 30 kph, a rotor of 50 metres radius would in practice deliver about 1 MW. 50 metres is too large for a practical rotor, the weight of which goes on an *axis*. Other designs are possible. One, due to Mullett, is shown in Fig. 5.15.[28] The Mullett design avoids the existence of a tower. It consists of a ring spinning in an axis normal to its plane and resting on a number of rollers. Generators are coupled to the axes on which the rollers rest.

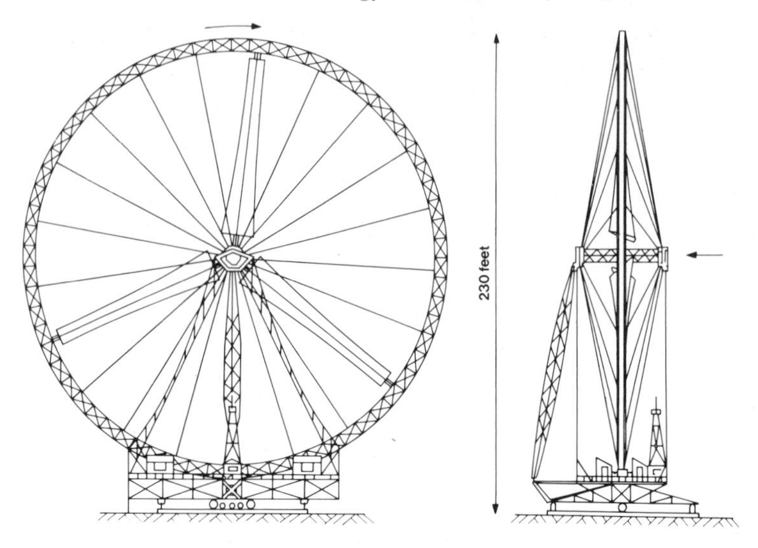

Fig. 5.15 A proposed aerostation.[28]

The Tracked Vehicle-Airfoil Concept
Powe, Townes, Blackhetter and Bishop[31] have devised an arrangement in which a tracked vehicle possesses a sail. The wind drives the vehicle on an endless track. The wheels drive generators and the electricity is communicated to the rails (Fig. 5.16[31]). The authors claim an enhanced efficiency over the windmill principle.

Wind power can be collected for the production of power on a massive scale in regions of the world with average winds of 20 kph or more.

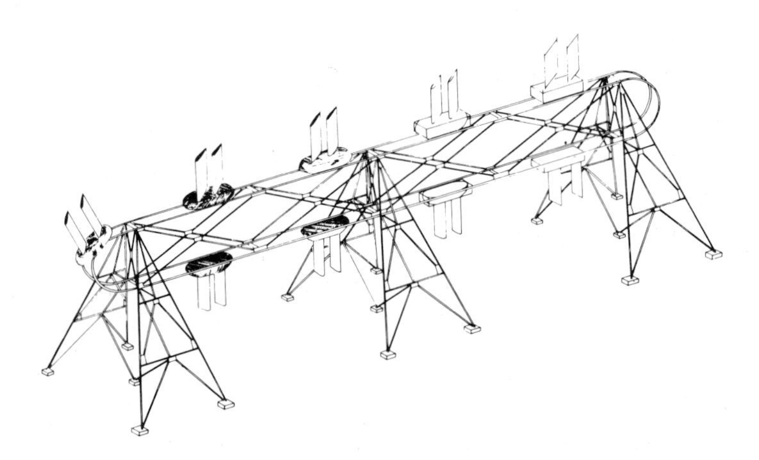

Fig. 5.16 Schematic of the vertical track concept for converting wind energy to electricity.[31]

Wind energy as the exclusive energy source for large population groups
Assuming the presence of a 20 kph average for winds throughout the year, one would need about 10^4 megawatts for a city of about 1 million and its associated industry and transportation, and hence some 10^4 rotors in the aerostation. With a diameter of 100 metres, about 100 rotors at 1km intervals could be spaced out along a coastal zone over 100km.

Seaborne aerostations could be anchored in the strong windbelts which exist in many areas of the world, for example, off the north-east part of the U.S.A.[32a]

The total potential energy in certain stable wind currents
At 30 kph, wind contains more than 1000 MW per cubic km. A wind such as the Great Southern Wind contains a very large amount of energy indeed. If the current were only 100 km wide, the world round, the energy in it up to 0.1 km could support the entire present population at a 10 kW per person level. In fact, the zone in which such winds exist is between latitudes 30° and 40°S., about 1000 km wide and up to an undetermined height. Thus this current has more than 10 times the entire needed world supply of energy if collected up to a height of 0.1 km.*

There are probably other circulating systems of the same kind, as well as some places where coastal winds average near to 30 kph. For example, much of the Western Coasts of Ireland and Scotland have average coastal winds of 28 kph. The Western Islands of Hawaii are in a similar situation. Greenland and Alaska are states parts of which have a high *average* wind velocity.[29]

Storage and transportation
Storage is essential. The winds discussed are averages and the velocities quoted are of mean annual winds in which there are periods of quiessence (see Fig. 5.14). Storage in gaseous H_2 under the sea would be feasible.[30] (See Fig. 10.6.)

The distance of the cities to be supplied to the location of the wind belt would be of diminished importance in a Hydrogen Economy.

GEOTHERMAL ENERGY

Low grade
The existence of 'low grade' geothermal energy is well known. It arises as shown in Fig. 5.17.[34] An aquifer in contact with deeper hot rock finds a fissure in the higher impermeable rock layer and ejects hot water and steam through the surface.

This source of energy has hitherto been developed largely in New Zealand at Wairecki[35], where it supplies 6% of the North Island's electricity. Such low grade sources have been thought to have only minor importance in the U.S. – a few per cent of the energy supply for a few decades.[36]

However, this picture could be revised upwards. For example, the

---

* Collection of only a small fraction will be feasible with present foreseen technology. However, the illustration indicates, at 1% collection efficiency, a much greater *energy density* than does solar energy. It seems reasonable to predict a far more economic attainment than that of fusion.

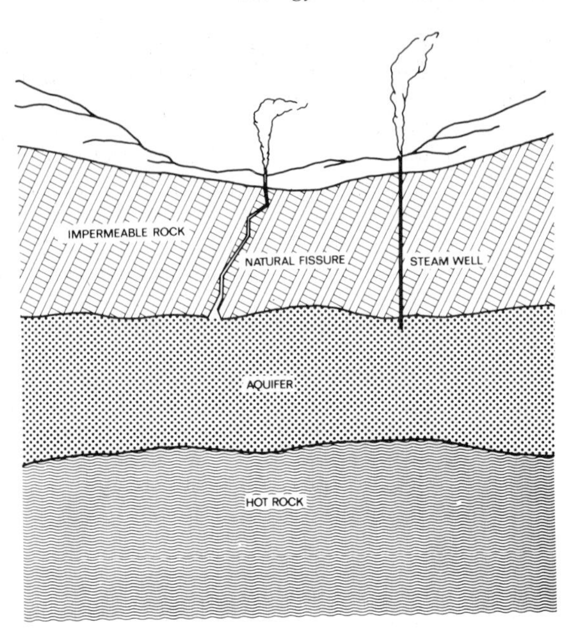

Fig. 5.17 Schematic diagram of a 'wet' geothermal reservoir.[34]

'low' estimate is made upon the basis of hot springs already known. It is now thought[36] that many sources of hot water may lie undetected near the earth's surface. A prospecting method for these may be developed. Correspondingly, only steam sufficiently hot to drive conventional electricity generating plants is considered in the low estimate. However, heat engines can be used with low boiling, stable organic liquid to work with, for instance, 70°C heat. Thermocouples could be used to collect the heat. Peck[36] estimates that, under these circumstances, the low grade energy sources would amount to 440,000 MW more than the 1974 electrical generating capacity of the U.S..

Correspondingly, some of the artesian water underneath large areas of Australia is at a temperature of about 100°C[37] and could be used to produce energy by the approaches mentioned.

High grade

Introduction. Nearly all of the earth consists of molten material, much of it siliceous, at about 900°C. The skin separating the surface from the molten interior is only a few tens of miles thick. To estimate the amount of energy involved, the planet may be regarded as a lump of liquid at 900°C, undergoing a temperature loss of, say, 10°. The earth has a volume of some 260 billion cubic miles. A specific heat of about 0.2 cal $g^{-1}{}^{\circ}C^{-1}$ can be assumed, and an average density of 3. If the consumers on the surface obtained all their energy from the earth's heat and number 10 billion people, and all live a 10 kW existence, then the temperature would take 0.1 billion years to drop 10°. Thus, geothermal energy is inexhaustible. However, the extraction of energy from deep hot rock has not yet been achieved.

Projected hot rock technology. A typical depth of the rock which separates us from the magma at 900° is 40 miles. A depth of about 5 km is sufficient to obtain contact with rock at 300°C. About half of this depth would be in the sedimentary rock and the second half in granite.

A difficulty is that a large cavity, at 3 miles depth, would have to be made (Fig. 5.18[34]). It is unclear at present just how big the cavity would have to be: suggestions have been made[37] that, for a generator of only 10 MW, the cavity would have to be several km in diameter. A nuclear explosion seemed to be a possible approach to forming the cavity. However, a Plowshare study[38] concluded that this method would become economical only if *many* nuclear devices were detonated sequentially in a three-dimensional array. There would be difficulties of containing

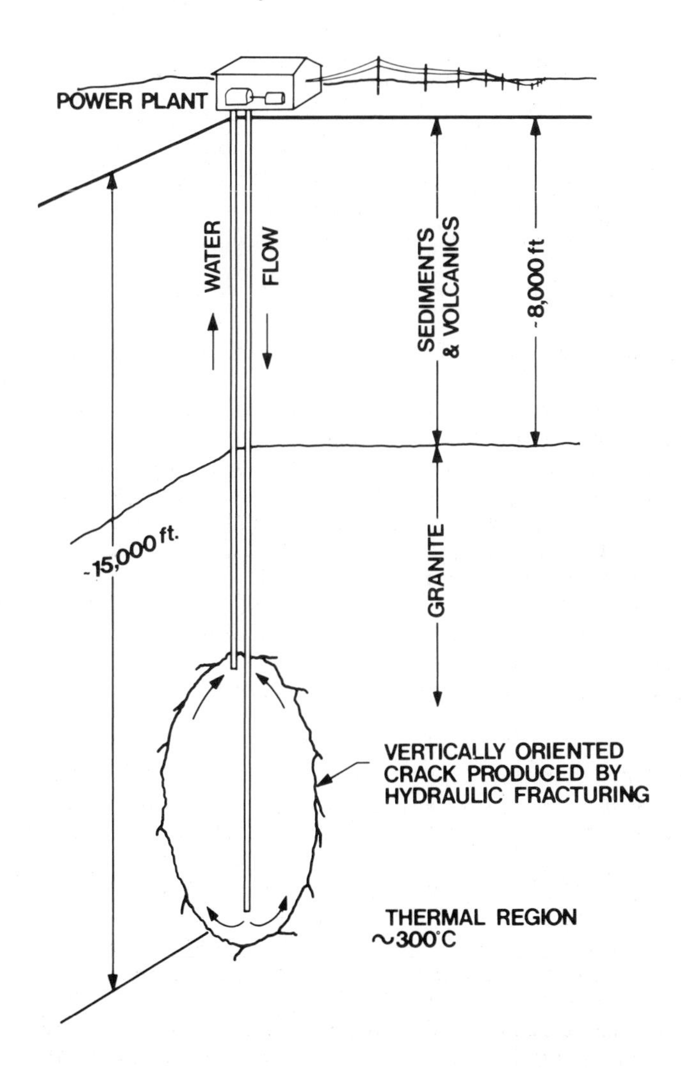

Fig. 5.18 A dry-rock geothermal energy system developed by hydraulic fracturing.[34]

the radioactive products and the detonation would be so intense that it could not be tolerated in most parts of the world without fear of causing subsidences.

Alternatively, fracturing could be used.[39] This is a technique known in the oil and gas industry. It is used to increase permeability in formations near the well. A high pressure pump is used at the surface to develop fluid pressure in the borehole sufficient to crack the rock and extend the crack. Pipes would flow cold water into the bottom of the crack and hot water would flow up another pipe (Fig. 5.18[34]).

A difficulty is the cooling of the rock face which will gradually occur.[37] The time constants would be in the region of 10-100 years. The hydraulic cracking might extend the stresses introduced by the cooling of the rock near the bottom, and new surface will be produced. If this could be arranged to happen at a rate which would compensate that of the cooling, a hot surface for energy production would be self-perpetuating.

Some estimates of geothermal electricity costs have been made.[34, 39] They are in the range 3-4 mils kWH^{-1}. A review has recently made out the prospects.[40]

Future development

Low grade surface sources are likely to be far from centres which use energy on a large scale and hydrogen transmission would be helpful. High grade sources are being analysed at the Los Alamos Laboratories of the A.E.C.[34]

However, the attainment of limitless hot rock geothermal energy, though seemingly an easier task than that of controlled fusion, does not offer a very attractive solution to our energy problems. The *size* of the hole needed for significant power is prodigious. For a city of 1 million approximately 100 holes would have to be made. If the cavities will not undergo fracture at an appropriate rate, each cavity would only last a few decades and the prospects of long-term hot rock geothermal sources would be reduced. The solid material which would be carried out with the steam at several thousand tons per day would be difficult to deal with. What of the expense of removing the solids *before* the steam strikes the turbine blades?[41]

SUMMARY

The most likely developments of the next decade or two are as shown in Fig. 5.19.

RESEARCH BEING DONE

In spite of the fact that the exhaustion of fossil fuels will occur in between one and two generations, and that the research and buildup of a new energy system will take about two generations, no action is yet being taken to research abundant, clean energy sources on a massive scale, except for nuclear sources. Very large sums are being fed into research on the conversion of coal to oil and methane. *Individual projects* in these areas are funded at a level greater than the entire U.S. solar energy programme. The developing of (necessarily temporary) energy sources from coal is the only energy field (in addition to the atomic) being researched on a massive scale. This situation arises partly

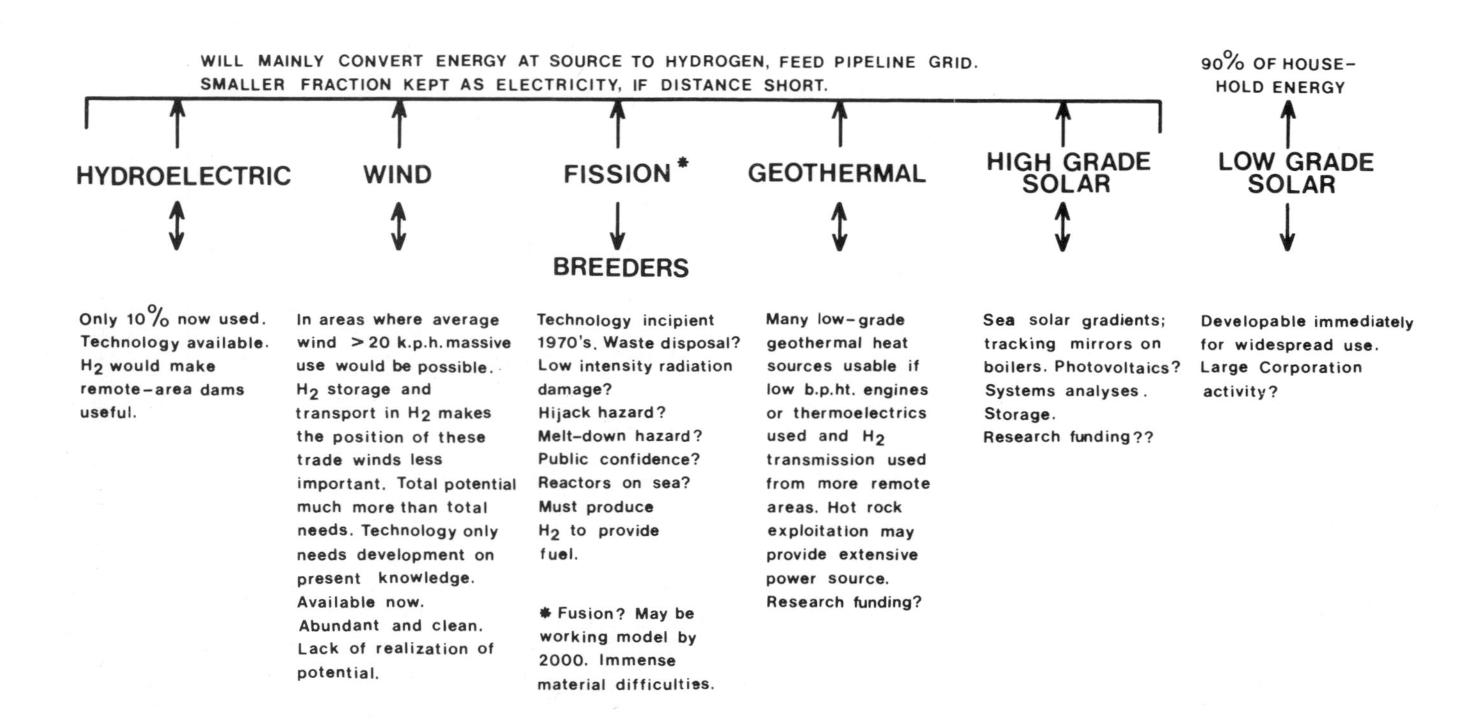

Fig. 5.19 Prospects in Abundant, Clean Energy Sources for 2000 AD.

as a result of the misapprehension that coal would last for many hundreds of years (see Chapter 4), together with a lack of recognition of the difficulties which may arise from Greenhouse effects.

What is the appropriate amount of money for research into Abundant, Clean Energy Sources? Income per capita is proportional to energy per capita (Chapter 3). Companies invest 1-10% of turnover in research. It seems appropriate, therefore, that countries should invest between 1 and 10% of the gross national product in research and development of inexhaustible, clean energy sources. This would be $(10-100) billion per year for the U.S.A. ($(0.3-3) billion for Australia).* These very large sums stretch, for the U.S., between the budget for space and the military budget. The attainment of Abundant, Clean Energy is a national goal in a class of priorities by itself: it must logically rate with a priority above that of defence.

THE NEED FOR A RECONSIDERATION OF PRIORITIES

Man requires abundant, clean energy from one of the inexhaustible sources developed *and built* within 30-60 years. The general view is that the gasification of coal is the first priority, then breeder reactors, and, finally, solar energy (by which is usually meant the direct collection of solar energy, for example, by photovoltaics).[42]

This order of priorities does not stand up if the following is taken as the basic question: How can energy which is cheap, inexhaustable and contributes nothing to air pollution, and negligibly to heat pollution, be obtained for a large proportion of the inhabitants of the planet?

Such a re-ordering of priorities in Energy Research is not only necessary because the present one is oriented towards supporting methods independently of long term pollutional considerations (that is, a 1960s programme), but also because new possibilities have become available. Essentially, these are the new and cheaper prospects in solar energy (see Chapter 7) – and the cheap transmission of energy in hydrogen over long distances. Lastly, because Energy Science has been narrowly fixated on fossil fuels and atomic energy, the number of scientists who have a comprehensive view of the panorama of choices is few, particularly in the U.S., where a supposed abundance of cheap oil as a fuel has been the basis of past planning, and prevented the formation of schools of energy research.

In Table 5.3, some methods under consideration in the 1970s are tabulated and then some subjective estimates are given, concerning characteristics which would be leading ones if the choice of the source for development is community oriented. The numbers represent no more than the author's hunches, but may serve to stimulate discussion of a more informed base than that of the present energy thinking in most countries.

The development of abundant, clean energy from the high velocity wind belts to provide electricity and hydrogen on a massive scale seems the first priority. It could be developed more quickly than the rest and could make a significant impact on the energy supply within 1-2 decades.

---

* Such sums would be the total budgets of the proposed Energy Research and Development Administrations. The amount for *research* would be perhaps 10% of the total budget; much would be spent in developing and building pilot plants.

TABLE 5.3

SUGGESTED WEIGHTS (-5 to +5) CONCERNING RELATIVE ADVANTAGES OF SOURCES OF ABUNDANT, CLEAN ENERGY

Source	Abundance?	Clean?	Guesstimated ease of development	Cost factor, unit fuel to consumer	Net (arbitrary) score
Fossil: coal gasification	−4	−3	+5	+5	+3
Nuclear: breeders	+3	−5	0	+1	−1
Fusion	+5	+5	−5	+0	+5
Geothermal	+5	+1	−2	+5	+9
Solar: low grade	+1	+5	+5	+2	+13
high grade, absorbers	+5	+5	+2	+2	+13
high grade, ocean thermal	+5	+5	+3	+5	+18
Solar: gravitational: wind (seaborne)	+5	+5	+5	+5	+25

The development of ocean thermal gradient collectors asks for rapid and massive research support.

Such considerations do not imply that the breeder programme, or even that of fusion, should be *abandoned*. But it is entirely inappropriate to fund them so exclusively when doubt is associated with their attainment – danger if they *are* attained – whilst several other more certain and far more environmentally sound possibilities exist, relatively unfunded.

What is stated here in respect to the comparison between the relative funding of nuclear and solar sources applies *a fortiori* to the funding of research on fossil sources, namely, coal, at present funded at a higher rate than the nuclear research (because of large expenditures by the oil companies as well as heavy government funding).

The reasons for the obverseness of this funding – the exhausting (and polluting) sources getting most research support and the clean, inexhaustible sources the least – is not only connected with the fact that the first sources are consistent with the present system, and are connected with organisations which have the money for lobbying for their funding. It is mainly connected with a lack of factual information on the alternatives, and the newness of the realisation of the advantages of the transmission in hydrogen, which much increases the distances over which one can economically send energy, that is, makes solar and wind sources more attractive despite their remoteness.

REFERENCES

[1] M. Dwyer, Ph.D. thesis, University of Pennsylvania, 1974; K. Lasaga, M. Holland & M. Dwyer, *Science*, **174**, 53-5 (1971).

[2] A.L. Hammond, W.D. Metz & T.H. Maugh II, *Energy and the Future*, American Association for the Advancement of Science, Washington, 1973.

[3] T.J. Gray & O.K. Gashus (edd.), *Tidal Power*, Plenum Press, New York, 1972.

[4] M.C. Smith, 'Geothermal Power', Publication LA-UR-74-233 of the Los Alamos Scientific Laboratory, 1974.

[5] J.O'M. Bockris & A.K.N. Reddy, *Modern Electrochemistry*, Rosetta Edition, Plenum Press, New York, 1973, p. 1350.

[6] Reference 2, p. 17, 25.

[7] J.O'M Bockris & D. Drazic, *Electrochemical Science*, Taylor & Francis, London, 1972.

[8] J.O'M. Bockris, *Search*, **4**, No. 5, 144 (1973); AMBIO, **3**, No. 1, 17 (1974).

[9] J.O'M. Bockris, N. Bonciocat & F. Gutmann, *Introduction to Electrochemical Science*, Wykeham Press, London, 1974.

[10] J.O'M. Bockris & Z. Nagy, *Electrochemistry for Ecologists*, Plenum Press, New York, 1973.

[11] R.P. Hammond, in *The Electrochemistry of Cleaner Environments*, ed. J.O'M. Bockris, Plenum Press, New York, 1972.

[12] A.F. Hildebrandt, G.M. Haas, W.R. Jenkins & J.P. Colaco, E. & S. Transactions of American Geophysical Union, **53**, 684 (1972).

[13] G.O.G. Lof, J.A. Duffie & C.O. Smith, *Solar Energy*, **10**, 27 (1966)

[14] W.J.D. Escher & J.A. Hansen, 'Ocean-based Solar-to-Hydrogen Energy Conversion Macro System', Escher Technology Assoc., St John's, Michigan, November 1973; also see Symposium on Non-Fossil Chemical Fuels, ACS, Boston, April 1972.

[15] G. de Beni & C. Marchetti, *Eurospectra*, **IX**, No. 2, 46 (1970).

[16] Reference 2, p. 31.

[16a] D.J. Rose, *Science*, **184**, 351 (1974).

[17] G.T. Seaborg & J.L. Bloom, 'Fast Breeder Reactors, *Scientific American*, November 1970.

[18] D.F. Ford, T.L. Hollocher, H.W. Kendall, J.J. MacKenzie, L. Scheinmann & A.S. Schurgin, Union of Concerned Scientists, M.I.T. Cambridge, Massachusetts, 1974.

[18a] A.R. Tamplin & T.B. Cochran, *Radiation Standards*, Washington, 1974.

[18b] A. Petkau, *Health Physics*, 22, 293 (1971).

[19] A. Bishop, *Project Sherwood, the U.S. Program in Controlled Fusion*, Anchor, 1960.

[20] M. Brennan, 'Energy and Fuels over the Next 50 Years', Flinders University of South Australia, March 1973; 'Physics and the Energy Industry', Flinders University of South Australia, January 1974.

[21] Los Alamos Scientific Laboratory Report LA-4858-MS, Vol. 1 (1972).

[22] Reference 2, p. 79.

[23] Figure obtained from *International Construction*, September 1973; see also Reference 3.

[23a] J. Lewis, I.E. of Australia, *Proceedings*, 1963.

[24] R. Radok, priv. comm., 1972.

[25] E.W. Golding, *The Generation of Electricity by Wind Power*, Spon, 1955.

[26] J.M. Noel, in 'Energy Conversion Systems', Workshop Proceedings, NSF/RA/W-73-006 (1973), p. 186.

[27] P.C. Putman, *Power from the Wind*, van Nostrand, New York, 1948.

[28] L.F. Mullett, 'Wind as a Commercial Source of Energy', presented at Engineering Conference, Canberra, 1956.

[29] A.H. Stodhart, in 'Energy Conversion Systems', Workshop Proceedings, NSF/RA/W-73-006 (1973), p. 62.

[30] W. Hausz, in 'Energy Conversion Systems', Workshop Proceedings, NSF/RA/W-73-006 (1973), p. 130.

[31] R.E. Powe, H.W. Townes, D.O. Blackhetter & E.H. Bishop, 'Technical Feasibility Study of a Wind Energy Conversion System Based on the Tracked Vehicle-Airfoil Concept', Annual Progress Report, NSF/RANN/SE/GI-39415/PR/73/4, 31 January 1974 (follow-up reports are those of 31 July and 30 September 1974).

[32] N. Carruthers, *Quart. J. Roy. Met. Soc.*, **69**, 1943.

[32a] W. Heronemus, paper, presented at the 8th Annual Conference, Marine Technology Society, 11-13 September 1972, Washington.

[33] L.F. Mullett, *Journal of the Institution of Engineers, Australia*, p. 69 (March 1957).

[34] M.C. Smith, 'Geothermal Energy', LA-5289-MS, Informal Report, Los Alamos, May 1973.

[35] J. Banwell & T. Meidav, 'Geothermal Energy for the Future', presented at the

138th Annual Meeting of the American Association for the Advancement of Science, Philadelphia, December 1971.

[36] D.L. Peck, Co-ordinator, 'Assessment of Geothermal Energy Resources', prepared by the Panel on Geothermal Energy Resources for the Committee on Energy Research and Development Goals, Federal Council of Science and Technology, 26 June 1972.

[37] S.H. Wilson, priv. comm., 10 September 1972.

[38] 'Feasibility Study of Plowshare Geothermal Power Plant, American Oil Shale Corporation, Battelle-Northwest, Westinghouse Electric Corporation and Lawrence Livermore Laboratory, 1971.

[39] R.G. Bowen & E.A. Groh, 'Geothermal – Earth's Primordial Energy', *Technology Review*, p. 42-48 (October/November 1971).

[40] A.L. Hammond, *Science*, 182, 43 (1973).

[41] A.L. Hammond, priv. comm., 8 May 1974.

[42] Hearing before the Sub-committee on Energy, U.S. House of Representatives, No. 10, 15 May 1973.

[43] C. Watson-Munro, Australian Resources Availability and Requirements, Science and Industrial Forum, 1974.

CHAPTER 6

Solar Energy: Basic Concepts

A SOLAR STILL to provide fresh water was operative in Chile in 1872; and a solar-driven steam engine was demonstrated in Paris in 1878.[1] In these early attempts – indeed until the 1970s – solar energy was seen as a source of energy for people in less developed countries. It was looked at as a source of low-grade energy largely for use in countries 30° S to 30° N of the equator; particularly as a source of heating.

Several events have recently transformed this pastoral concept of solar energy – if it is understood in a wider sense than that of the direct pick-up of solar radiation by flat collectors – until solar energy joins with atomic energy as the main prospective source of clean energy. These events have been:

(1) The development by Chapin and others in 1954 of silicon photovoltaic layers which could convert about 10%* of the incident light directly to electricity.[2] Since then, these cells have been engineered into arrays of tens of kW as a result of research funding from the U.S. and Russian space programmes of the 1960s. Thus, the Skylab A of the U.S. programme of 1972 used 25 kW of solar power from silicon photovoltaic cells. However, the Space Programme's silicon photovoltaic cells were still impossibly expensive for commercial application, and apart from cost, they would need enormous areas of land even where the insolation was greatest. The electric power developed in these areas would be required several 1000 km distant from the source and the cost of sending it through wires is regarded as prohibitive for distances of more than 500 miles.

(2) In 1962, a suggestion was made by Bockris[3] that solar energy could be collected by an array of photogalvanic cells, placed on platforms, floating on the sea in areas of high insolation. Hydrogen would be manufactured by electrolysis at the platform, and this hydrogen transferred to land.[4] In 1972, Gregory, Ng and Long[5] published calculations which showed that the transmission of electricity in the medium of hydrogen through pipes is cheaper than its transport through wires, when the distance is greater than a few hundred miles. The precise point at which the advantage begins depends on the transmission voltage (see Chapter 8). Alternatively, in 1973, Ehricke[6] suggested the use of power relay satellites, by means of which solar energy obtained on the ground in

* 1% efficient photovoltaic devices had been used earlier in light meters.

the areas of high insolation of North Africa and Australia is beamed in micro waves up to relay satellites and then down into areas of high usage of energy.

(3) Of the problems at the beginning of the 1970s facing those attempting the $10^2 - 10^3$ times cost reduction of silicon cells necessary before photovoltaic solar energy conversion becomes economically feasible, the formidable one was how to get from very pure silicon to silicon in single crystals, and in shapes from which photovoltaic couples could be made. An important contribution was made to the solution of this problem in 1971 by Mlavsky and La Belle[7] in a development known as the edge-defined-film-fed growth method for the production of highly pure silicon as single crystals in strips of great length. With this new method as a basis, Currin, Ling, Ralph, Smith and Stirn[8] were able to publish a detailed cost projection of a massive solar energy plant, using silicon photovoltaic cells manufactured by the Mlavsky and La Belle method. The predicted cost per unit of power collected fell in the same range as the actual cost of atomic energy plants (~\$1,000/kW), although a comparison on a per kW basis screens the fact that the atomic reactor works continually whereas the solar collector works intermittently.

Thus, in 19 years (a period commencing with the 1954 paper of Chapin, Fuller and Pearson[2]), but particularly between 1971 and 1972*, the prospects for solar energy as a source of a massive energy supply, were substantially improved. In terms of cost *estimates,* the cost of a unit of power had declined by more than 10^2. † The massive conversion of solar energy had become worthy of an international research effort of a magnitude corresponding to its potential benefit – inexhaustible, abundant and clean energy.

THE AMOUNT OF SOLAR RADIATION STRIKING THE EARTH

The solar spectrum is shown in Fig. 6.1.[9] The difference between the energy which reaches the earth outside the atmosphere, and that which reaches the earth's surface, is reduced in intensity by >25%, owing to the scattering by air molecules, and by selective absorption at certain wavelengths, by water and carbon dioxide molecules in the atmosphere.

If the sun is directly overhead, on a cloudless day, the rate of energy arrival on the earth is about 1 kW m^{-2}. Thus, if the efficiency of the collector is 10%, there would be needed (for the above condition), 10 sq metres to collect energy at a rate of 1 kW. Several factors reduce this amount for the usual practical situations which are not those of the ideals stated above:

(1) The latitude. This may be compensated by orientating the collectors appropriately.

---

* As late as 1972, some specialists in solar energy[10] were expressing markedly negative views concerning the part which solar energy would play in the future energy picture.

† The necessary inventions were forthcoming, at the time when pollution was beginning to be joined with the prospect of exhaustion of fossil fuels to make them economically and socially desirable.

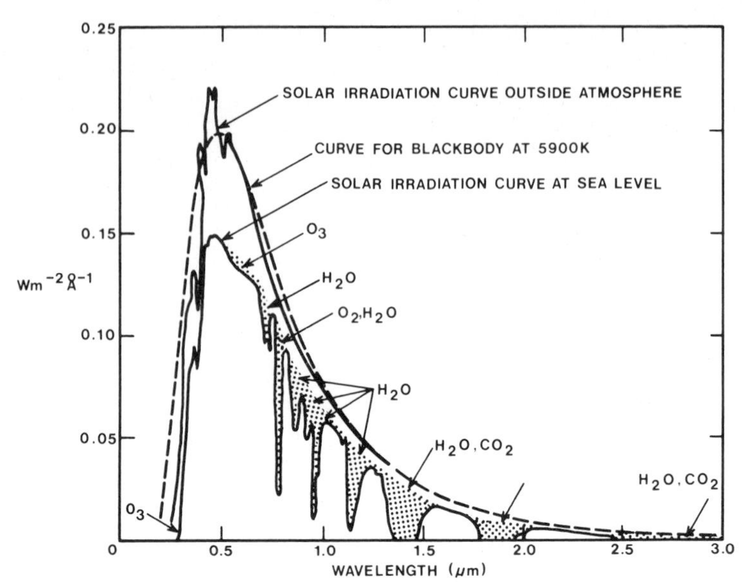

Fig. 6.1 Spectral distribution curves related to the sun; shaded areas indicate absorption, at sea level, due to the atmospheric constituents shown.[9]

(2) The rotation of the earth. This may be compensated by making the collectors track the sun.

(3) Obscuring of the sun by clouds: diffuse insolation. Fig. 6.2[1] exemplifies the variation of solar radiation in Madison, Wisconsin. The average 14 hours of sunlight for the latitude 43 was 3.5 times less for 23 December 1962, than for 15 June of that year. The maximum between 10-12 pm was about 1.25 kW per sq metre. Averaged over a 24 hour period, the results of Fig. 6.2[1] suggest figures of 0.24 kW m^{-2} for the June and 0.08 kW m^{-2} for the December data.

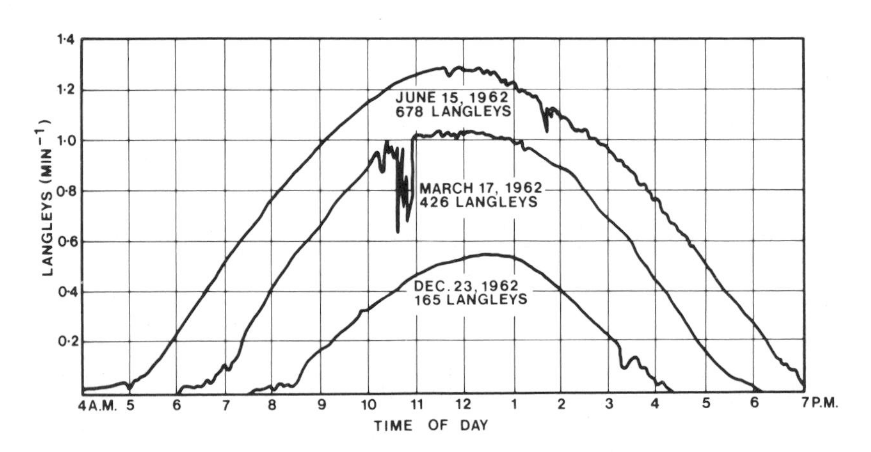

Fig. 6.2 Typical solar radiation on clear days at latitude 43°N.[1] (1 Langley is equal to 1 calorie $cm^{-2}min^{-1}$.)

Thus, a calculation of the energy which could be collected in a given location must refer to long-term data for insolation per year at the area. It must use average annual incidence figures. It is important to know how much of the light is diffuse (as on a bright, but cloudy day) and how much time the solar disc is unobscured. Thus, devices in which solar energy is collected through a lens focussing system are dependent upon the disc's visibility. Information on the total light flux is not sufficient. Hence, the advantage of cloudless skies may be greater than is indicated by information given by the total solar energy lines as a function of latitude.

A world map of solar insolation (but not of visibility of the disc) is given in Fig. 6.3.[11]

TABLE 6.1[6]

SOLAR-ELECTRIC ENERGY FROM EARTH'S HIGH INSOLATION AREAS

	Nominal Areas		Nominal Annual Thermal Energy Flux			
Desert	Sq km	Sq mi	(Gwh_{th}/km^2)	(Thousands of Gwh_{th})	Percentage of area assumed usable	Electric energy extracted at 25% efficiency (Gwhe/year)
North Africa	7,770,000	3,000,000	2,300	17,870,000	15*	670,000,000
Arabian Peninsula	1,300,000	500,000	2,500	3,250,000	30†	244,000,000
Western & Central Australia	1,550,000	600,000	2,000	3,100,000	25	194,000,000
Kalahari	518,000	200,000	2,000	1,036,000	50	129,000,000
Thar (NW India)	259,000	100,000	2,000	518,000	50	65,000,000
Mojave, S. Calif.	35,000	13,500	2,200	77,000	20	3,900,000
Vizcaino, Baja, Calif. (Mexico)	15,500	6,000	2,200	34,000	25	2,100,000
Total/Average	11,447,500	4,419,500	2,190 ave.	25,895,000	31 ave.	1,308,000,000

* Parts of Arabian and Lybian deserts.
About 60% of Rub' al Khali desert.

The solar-electric energy available from the most energy-rich parts of the world is shown in Table 6.1.[6] North Africa (particularly the Sahara), Saudi-Arabia, Australia and parts of South Africa and India are the great energy-rich land areas of the world. Australia combines the properties of affluence with a high degree of engineering education among the population, and the second largest harvest of solar energy falling on one nation.

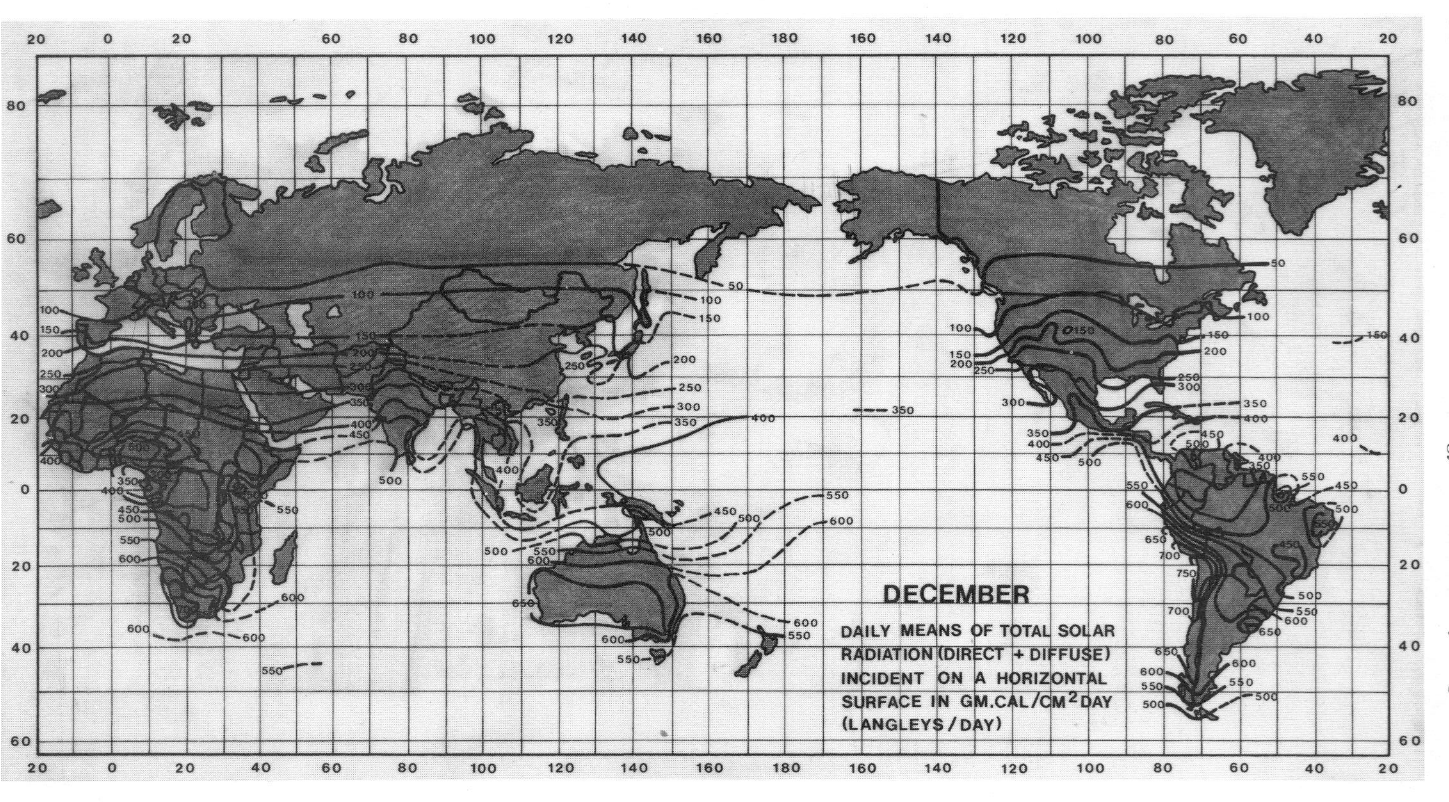

Fig. 6.3 Distribution of solar radiation throughout the world for December.[11]

It is important to distinguish the product of degree of insolation times free area available for collection from the degree of insolation. For example, some regions in the world obtain a greater yearly average intensity than does Australia. However, 59% of the Australian continent is a desert, mostly flat. It is this property, combined with the intensity of insolation, which makes Australia so attractive for the development of land based solar energy collection on a large scale.

POPULATION AND LIVING STANDARD, SUPPORTABLE BY SOLAR ENERGY

The sun radiates 1.7 10^{17} watts onto the earth continuously.[12] 10% of this on 1% of the earth's surface is 1.7 10^{11} kW. A world population of 10 billions, with a total power need per person of 10 kW (including all industry, transportation, etc.), needs 10^{11} kW, i.e., all could be supplied by 10% efficient collection on *c.* 1% of the earth's surface. Were this energy use to be exceeded by introducing substantial new heat (e.g. from fusion) then the earth's temperature would rise; i.e. there is no *feasibility* of producing a much greater amount of usable energy on earth than could be converted from solar sources.

THE USES OF SMALL-SCALE SOLAR ENERGY[13 15]

If roof-top collectors could contribute to the residential energy supply, particularly in the sunny areas of the world, the amount of energy from central sources which would have to be transferred over long distances would be reduced. When excess household energy can be collected from roofs the possible storage modes are as hydrogen or in batteries. Local transportation could be fueled by energy so obtained.

The main uses of small scale solar energy are as follows[1]:

(1) *Cooking*

Solar energy for remote areas was developed in Mexico in the 1950s.[1] It could not compete with oil, at that time cheaply available.

(2) *Use for household heating*[16]

Black-sheets, with copper pipes containing water soldered onto the back, are manufactured in Israel, Japan and Australia (Fig. 6.4[17]).

Solar space-heating of houses is incipient[18] (Fig. 6.5 and 6.6[17]). Air can be forced past the back-side of the roof-top heat collectors, which could house photovoltaic cells. Duct work takes the air to a storage system (a molten salt, rock pile, or a large water reservoir), and cold air from the house is circulated over the storage system.

(3) *Household cooling*

In one concept for solar space-cooling, there would be two connected vessels, one containing liquid ammonia, and the other a concentrated solution of salt in ammonia. The latter solution has a lower vapour pressure than has pure liquid ammonia, and the latter vaporises to the compartment containing a solution of the salt in liquid ammonia. The vaporisation causes cooling. On the reverse cycle, solar heat is focussed on the solution and liquid ammonia is made to evaporate back to the first compartment, where high pressure and lower temperature causes condensation. The cycle is repeated.

Another concept, as yet to be applied, would use radiation to the night sky. There are IR windows in the atmosphere. If the object to be cooled (a house roof for example) contains a surface made of a material, which radiates in the wavelength range of the windows, it will cool air in contact with it. Silicon on iron has been suggested.[19]

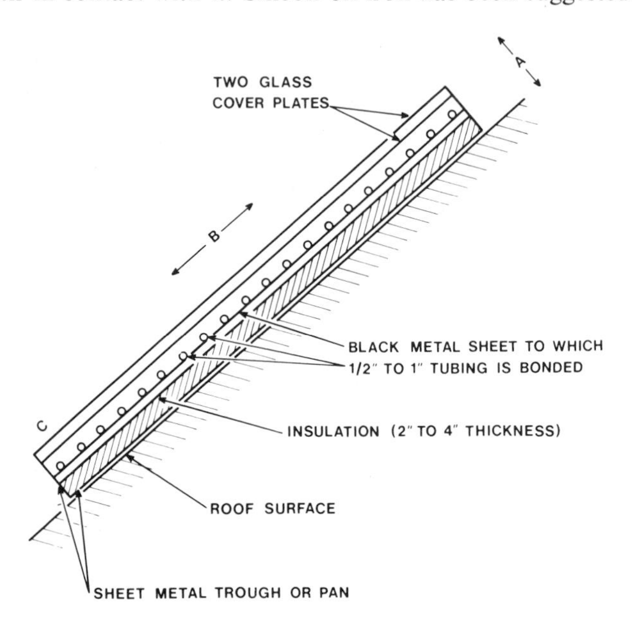

NOTES: ENDS OF TUBES MANIFOLDED TOGETHER
ONE TO THREE GLASS COVERS DEPENDING ON CONDITIONS
DIMENSIONS: THICKNESS (A DIRECTION) 3 INCHES TO 6 INCHES
LENGTH (B DIRECTION) 4 FEET TO 20 FEET
WIDTH (C DIRECTION) 10 FEET TO 50 FEET
SLOPE DEPENDENT ON LOCATION AND ON WINTER-SUMMER LOAD COMPARISON

Fig. 6.4 Solar collector for residential heating and cooling: diagramatic sketch of one alternative (elevation-section).[17]

(4) *Desalination*

Black-bottomed trays of water are covered with a transparent roof. The saline water heats up upon receiving solar radiation, and water condenses on the air-cooled roof (which may be laminated), runs down inside it and is collected in troughs and storage vessels. A problem is the life-time of the materials[14 15]; another is the large space needed. The production is in the region of one gallon m^{-2} per day in sunny climes, and the cost $2-3 per 1,000 gallons.*[1 20]

* Atomic distillation aims at 20 cents per 1,000 gallons.

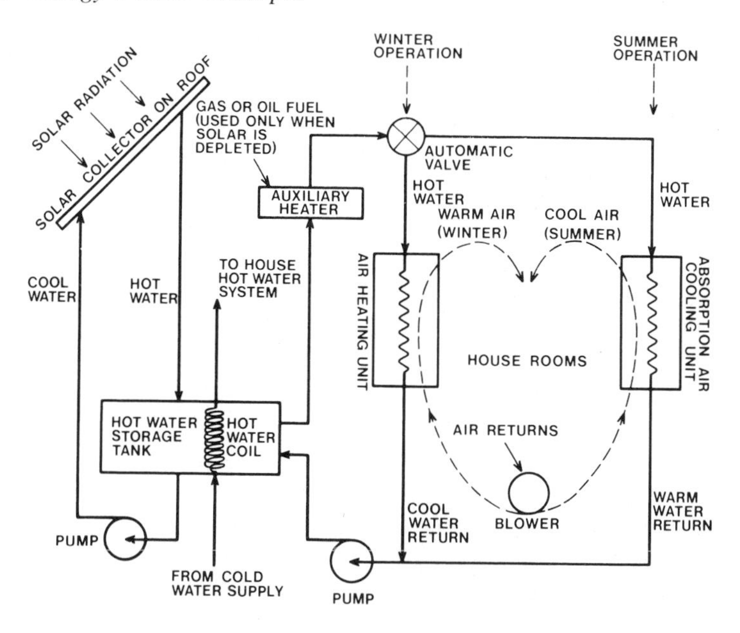

Fig. 6.5 Residential heating and cooling with solar energy: schematic diagram of one alternative.[17]

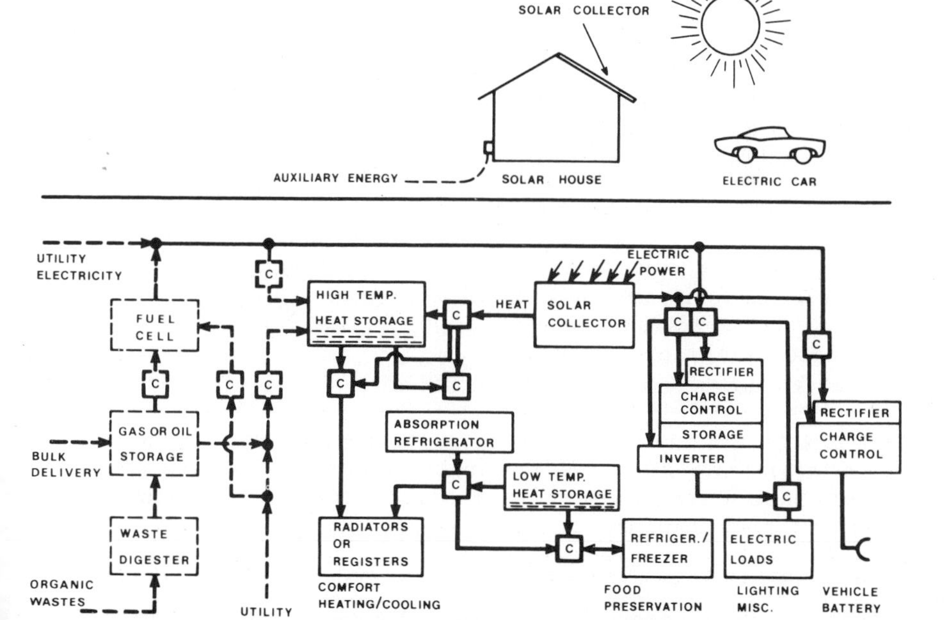

Fig. 6.6 Schematic of solar energy system for residential building.[17]

The solar method may be the most economic if only medium amounts of water are required, up to about 10,000 gallons per day, and in areas of high insolation.

Another concept, hypothetical as yet, proposes the extraction of moisture from the atmosphere. Even over the Western Australian desert, there is a relative humidity of 10%, i.e. about 0.2% of the air mass is, in fact, water vapour. If such air were passed through a tower containing a material such as calcium chloride (during the night), the water in the air would be retained in the solid. During the day, the solid could be heated by the use of tracking mirrors and could be made to give up the water, which would be condensed and collected. The method could have value in hot, dry areas with local populations of a few hundred.

If one accounts for drinking, cooking, laundry, bathing, sanitary and primary production (but not industry and commerce), about 300 litres per person per day is needed. For a community of 1,000 persons, a chimney 300 feet high and 6 feet in diameter would have to have air containing 0.1% of water passed through it about twice per minute (assuming a 12 hour day) to provide the necessary water (assuming 100% efficiency of water absorption, recovery and condensation). The order of the cost per person per year should be in the tens of dollars.

PROSPECTS

The prospect for low-grade heating and cooling applications of solar energy depends on the price of fossil fuels. This will continue to climb at amounts greater than the inflational amount as exhaustion is approached. When the cost of household solar energy becomes cheaper than that supplied by electricity and gas, conversion will be made rapidly. In latitudes between 30°N and 30°S that position has recently been reached. However, residential solar heating may become economic for many parts of the world within the next decade.

10% of new houses in the U.S.[21] could be solar heated and cooled by 1985, 50% by 2000 and 85% by 2020. Household energy is some 25% of the total energy budget and more than half of that is used in heating and cooling, thus substantial energy would be saved by solar means. However, apartment buildings need too much energy for it to be collected from their roofs: if solar-derived, the energy for them would come from a collector elsewhere; and, for the Northern Hemisphere, probably in the form of hydrogen.

Stackhouse[22] regards a solar climate control industry as an incipient industry of the time. He likens it to the time of Henry Ford with respect to the automobile. No break-throughs in research are needed for the massive growth of this industry (although *development* work must precede commercialisation, except for the roof-top water heaters which are practical now).

RESEARCH AND DEVELOPMENT IN SOLAR ENERGY

Why was the economic collection of solar energy not developed at an earlier time? One answer is that the needed seminal advance (the construction of a photovoltaic cell of acceptable efficiency) was not made till

1954. There was, of course, no call for solar energy among the one-third of the world's population which had the necessary technological research and development capability to engineer an economic conversion method, because the illusion existed that there was cheap oil and natural gas for far into the next century.

Support for solar energy-oriented work in the U.S. for 1974 was at *c.* $20 million, an increase from $4 million in 1973.* Compared with budgets for research into nuclear energy and the exploration of space (several hundreds of millions of dollars per year), support of solar energy research in the U.S. is still at levels of a lower order of magnitude than atomic energy. Solar energy has been no exception to the neglect of energy research in the U.S.. For example, there are three energy conversion laboratories in U.S. universities, but twelve for materials sciences. There is a Space Agency funded with billions and an Energy Agency funded (apart from coal and atomic energy) in millions.

One matter in assessing the future of solar energy research is that there are few solar energy researchers. 'Solar Energy' is not a University discipline; even Energy itself is not a subject which has been institutionalised into departments, as have, for example, Materials Science or Communications. The tiny number of people who do research in solar energy at present come from many disciplines, particularly from solid state physics, but also from electrical engineering, physical chemistry, electrochemistry, biology, and even architecture, etc. Correspondingly, in considering to whom funding would be given, it is not of primary importance that the scientists and engineers should have experience in the field. In 1958, there were no 'space scientists'. Space science is interdisciplinary, as is the group of sciences and technologies associated with solar energy. If massive research funding becomes available in an area, proposals will arise and groups of workers in the new discipline will form. *The rate of progress is determined by the availability of government research funding,* and a proper funding organisation with teams of research officers who can plan and co-ordinate research for the long term.

MODES OF COLLECTION OF SOLAR POWER

Due to the earlier lack of interest in the field, caused by the lack of appreciation of the nearness to exhaustion of fossil fuel supplies, most of the options in solar energy conversion are still open, i.e. insufficient research has been done upon them to indicate which approaches may be most worth pursuing.

Six possible pathways can be examined:

(1) *Photovoltaic*

This is the most widely known. The model is that of Fig. 6.7.[23] Photons enter the p-type semiconductor and activate electrons from the valence band into the conduction band. These diffuse to the junction and form

---

* Even the Australian Government (that of the country with the most to gain financially from the massive development and export of solar energy) spent only about $0.4 million on solar energy research in 1973, and this was principally in respect to research upon residential solar energy collection. At the same time, the Australian Government continued to fund an Atomic Energy Commission with a budget in the tens of millions.

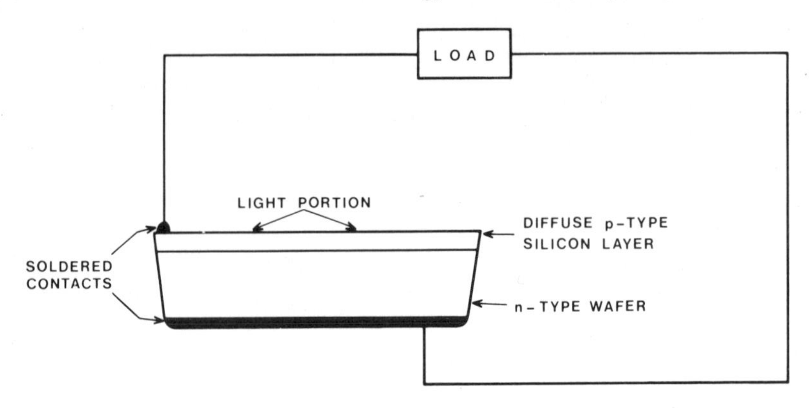

Fig. 6.7 Mechanism of function of photovoltaic generator.[23]

a potential difference across it. This potential difference can then be applied across a load.

(2) *Photogalvanic*
This is the least discussed method, although recognised for many years. An elementary theory of photogalvanic effects in electrodes was first given by Hilson and Rideal[24], assuming that light is absorbed by radical intermediates absorbed on the electrode surface (cf. Bockris[25]).

In the direct photogalvanic effect, light impinges upon one electrode and causes it to emit electrons. This drives the cell from which power can be drawn; correspondingly, for semiconductor electrodes, holes can be activated to be available as acceptors of electrons at an anode.

There are other electrochemical light effects, for example, light activates molecules in solution to higher energy states, and the molecules decay to lower energy states, transferring an electron to a conductor in solution during the process.[26]

These photogalvanic effects are less developed than photovoltaic ones. *They will not need pure single crystals* in the collectors, as photovoltaic devices do, and hence may lead to converters cheaper than those using the latter method.

(3) *Photothermic*
The photothermic method relies upon 'selective coatings', see Fig. 6.8.[9] The sun emits radiation at a temperature of about 6,000° K. The earth re-emits it at near 300° K. If a coating can be found which absorbs well in the lower wavelengths, but emits poorly at higher wavelengths, the coating should become abnormally warm if it is exposed to the sun. It could act as the source of heat to a working fluid.

(4) *Photosynthetic*
Plants could be grown, sugar canes for example, and burnt directly to work a heat engine. Alternatively, the material could be heated anaerobically and decomposed to give CH_4, H_2 and higher hydrocarbons. Algae could be force-grown, collected and decomposed by heat to form hydrocarbons.

Photosynthetic concepts are less attractive when the average efficiency

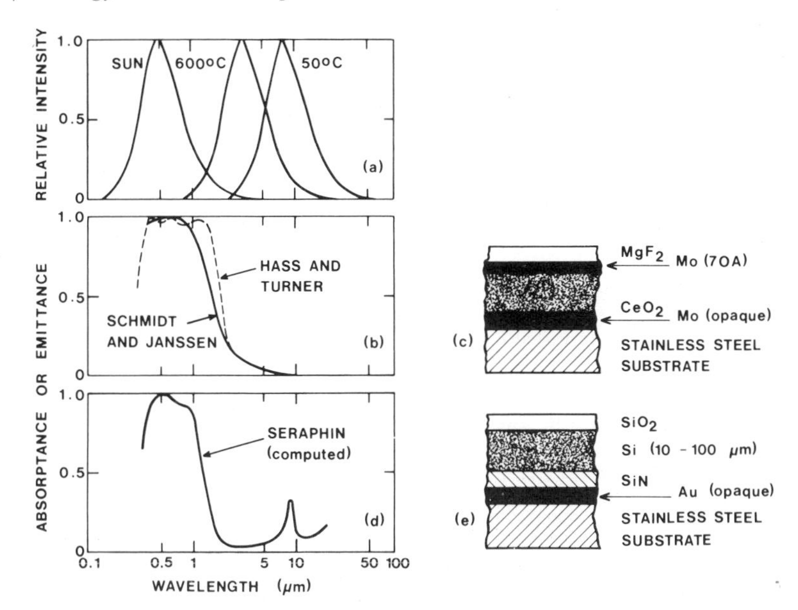

Fig. 6.8 (a) The spectrum of blackbodies at 50°C and 600°C compared with that of the sun; (b) (d) adsorbing/emitting properties of some selective surfaces; (c) (e) composition of Schmidt and Janssen's surface (c) and Seraphin's (e).[9]

of solar energy conversion which they attain is examined. It is about 1%, so that the overall efficiency of the conversion of sunlight to electricity would be about 0.3%, compared with 10% for the photovoltaic. The collecting areas would have to be, therefore, some 30 times greater than those envisaged for photovoltaic collector farms. The direction of research here is towards the breeding of bacteria and algae which contain enzymes which might catalyse a more efficient conversion of light to matter. This is a reasonable goal as can be seen from data on some maximum

TABLE 6.2[9]

MAXIMUM GROWTH RATES AND ESTIMATES OF EFFICIENCY OF SOLAR ENERGY CONVERSION FROM PHOTOSYNTHESIS

Crop	Location	Length of Growth Period (days)	Growth Rate $g.m^{-2}.day^{-1}$	Average Radiation $MJ.m^{-2}.day^{-1}$	Percentage Utilisation of total radiation
Bulrush millet[27]	Australia	14	54	21.4	4.2
Pindus radiata[28]	Australia	4	41	25.4	2.7
Maize	Calif., USA	14	32	28.8	2.0
Barley	England	n.a.	23	20.3	1.9
Sugar cane	Hawaii, USA	90	37	16.8	3.7

photosynthetic growth rates where the efficiency approaches 5% (Table 6.2[9]).

(5) *Ocean Thermal Gradients*
Fossil fuels represent chemically stored solar energy, originating from millions of years of insolation in the remote past. The withdrawal of oil from a field, and its combustion in engines to create energy is, therefore, equivalent to a using up in a few decades solar energy which arrived over millions of years. In spite of the disadvantages of using fossil fuels as an energy source, which arise because of the associated and unavoidable excess CO_2 injection into the atmosphere, the fossil fuels represented a unique form of concentrated stored energy from photosynthesis.

Similar advantages – but without the pollutive difficulties of fossil fuels – would arise if the heat contained in the sea, which is solar in origin, could be converted to electricity. Thus solar energy would be *mined* from the insolation of the recent past, and it would be the mining of a *stored* amount and tend to overcome a negative feature of conventional solar conversion, the diluteness of the incident radiation.

The temperature gradient in the tropical ocean exists between the surface of the sea in tropical regions ($T > 300°K$) and the depth at 1 km, where the temperature is less than 285° K. Thus, if the temperature of the surface water is used to heat some low boiling organic liquid, and work a heat engine, the exit vapour can be condensed by the use of cold water pumped up from the sea depth. One would then have a thermal cycle withdrawing heat from the surface water, converting a small fraction to mechanical energy, and rejecting the rest to heat up some of the water from the depth (Fig. 6.9[29]).

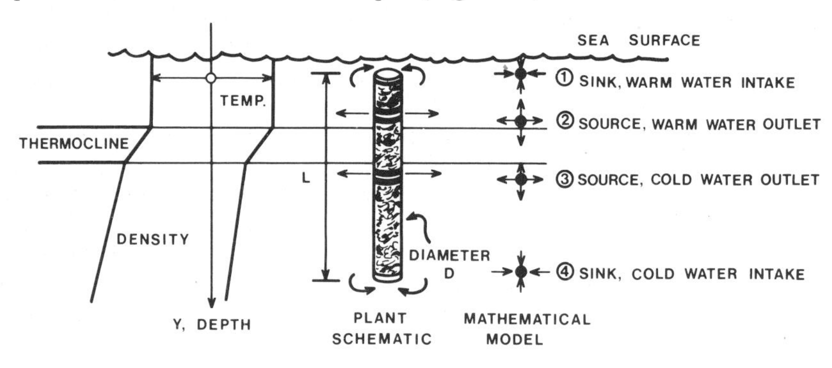

Fig. 6.9 Topology of a solar sea power plant.[24]

The efficiency expected from this (hypothetical) process is *c.* 2%. However, the collection of the energy occurs over a volume, and not from an area – as it does with photosynthesis for instance – so that the low efficiency does not imply inconveniently large areas of collectors. In the ocean thermal gradient approach, the heat engines would be placed on floating platforms (Fig. 6.10), and these would drive conventional electricity generators. The boiler would be just below the surface and the condenser below the boiler. Pipes would bring cold water from the deeper levels. The electricity could be used to generate hydrogen from

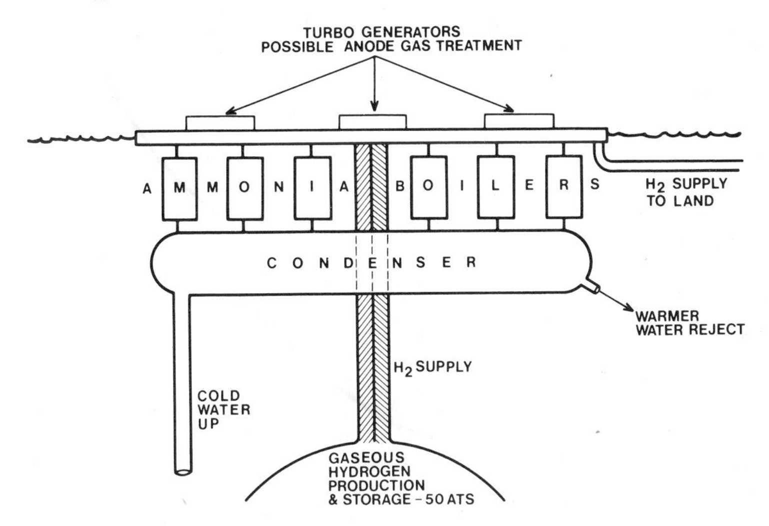

Fig. 6.10 Solar sea power plant.

sea water (but see Chapter 8) and this could be piped to shore and used thermally or electrochemically as fuel.

(6) *Wind*
Electrical energy, and hydrogen fuel collected from the kinetic energy of wind belts (see Chapter 5), is a form of solar energy conversion, because the density differences which cause the winds arise from thermal gradients.

SOME BASICS OF PHOTOVOLTAIC COUPLES

Principles
When radiation of frequency exceeding a critical value strikes a solid, it may cause positive and negative carriers in excess of equilibrium to be generated. One of these carriers must collect on a surface of the system, so that an excess interfacial charge develops, and therefore a potential difference is built up between the phase containing the excess carriers, generated by the solar photons, and that which is not so insolated, and hence contains no corresponding excess carrier concentration. The carrier generated by light must live for τ secs, where τ is the time to diffuse to the boundary.

Generation of carriers[30]
The number of photons in a beam which has traversed a distance x of material is:

$$N_{\mathrm{ph},x} = N_{\mathrm{ph},x=0}e^{-\alpha x}, \tag{6.1}$$

α being a function of wavelength. Fig. 6.11[30] shows how α depends upon frequency or wavelength. Thus, at frequencies less than the critical value, α tends to zero and there is no absorption. ν_{crit} is given by $h\nu_{\mathrm{crit}} = E_{\mathrm{gap}}$, where E_{gap} = the energy gap of the semiconductor. The importance of E_{gap} is that it determines whether solar radiation (which exists largely between 0.5 and 2 eV) will cause charge generation in a given semiconductor. Any semiconductor with an energy gap greater than about 1 eV is of lesser, and if greater than 2eV, very little, use as a solar energy converter. The thickness needed to absorb 90% of the incident radiation is:

$$\alpha x > 2.3. \qquad (6.2)$$

Thus:

$$x_{\min} > 2.3/\alpha \qquad (6.3)$$

For, e.g. GaAs[30]:

$$x_{\min} = 2 \,.\, 10^{-4} \text{ cm}, \qquad (6.4)$$

Larger thicknesses are dead weight.

Fig. 6.11[30], due to Loferski, shows that there are two kinds of behaviour for the $\alpha - x$ relation. That typified by GaAs has $\alpha > 10^4$ for most of the range above 1.3 eV. Another behaviour (Si) shows that a slow increase of α with v is also possible. Materials of the first type are called 'direct' and of the second type 'indirect'. The first type is more desirable, if available, because the needed thickness is less and the amount of the material is smaller. Hence the term 'thin film photovoltaics'.

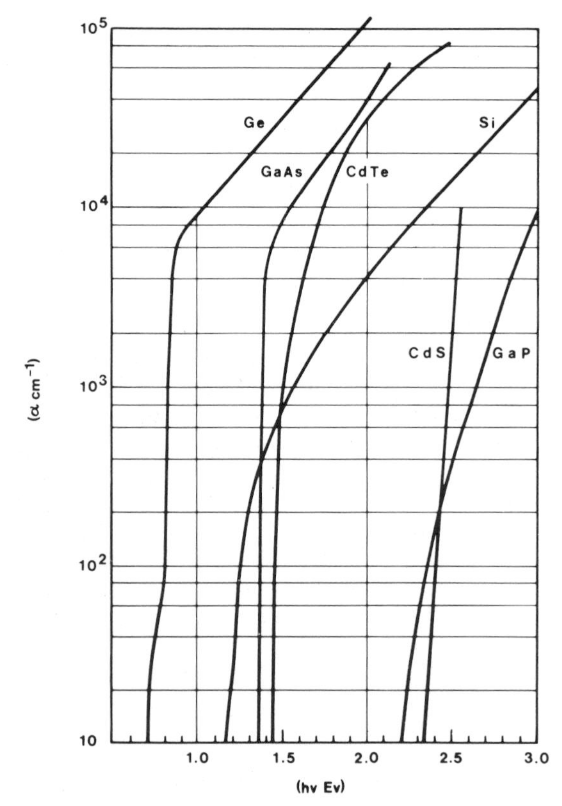

Fig. 6.11 Optical absorption constant α as a function of photo energy $h\nu$(ev).[30]

Types of photovoltaic junctions

There are two types of p-n junctions, homo and hetero. Silicon is the classic example of the first type, copper sulphide-cadmium sulphide a well-known example of the second type. Organic materials may also be considered.[31]

One possibility is to make a metal of one side and a semiconductor on the other: little knowledge of such couples exists at present.

Diffusion to the boundary[30]

Recombination of carriers is a principal cause of loss of efficiency. The direct gap materials, such as GaAs, have a short lifetime for carriers (*c.* 10^{-8} sec). The lifetimes of indirect couples are longer, e.g. 10^{-3} sec for silicon. This has important consequences. Thus, the indirect gap materials (thicker, longer lifetime) are more affected by the presence of recombination centres (impurities), for, when the carriers live for a longer time, they are more likely to collide with impurity atoms and thus undergo hole-electron pair recombination. Thus, for thin film photovoltaics, less purification is necessary, the requirement of single crystallinity is less, and thus the cost of a unit quantity of the semiconductor concerned (and the cost of unit energy produced) is reduced.

The central property of lifetime is not only affected by recombination in the semiconductor, but also by recombination on the surface. To raise the efficiency of a collector from 10%, often quoted for silicon to, say, 20%, one would need in a material, 100μ thick, and ρ = approximately 0.01 ohm cm, a surface recombination velocity of less than 100 cm sec^{-1}.

What materials are sufficiently available for the massive conversion of solar energy?

The rate of use of energy (both as electricity and in other forms) in affluent societies is about 10 kW per capita. Let it be assumed that this is to be supplied by photovoltaic collection of solar energy at 10%. If the collectors were made 10μ thick, the mass of the substance required would be in the order of tens of millions of tons. This would be reduced by ten to one hundred times if the material concerned was not Si, but a thin film photovoltaic, e.g. cadmium sulphide-copper sulphide. This large quantity of material needed reduces the possible choices: it is met, among present photovoltaics, only by silicon. The production of materials such as Ga, Sb and Te, is tiny by comparison with that which would be needed were they to be used on a massive scale: and it seems very unlikely that it could be increased by the necessary several orders of magnitude. Thus, cadmium had a production in 1968 of 15,000 tons.[30]

One substance plentifully available is aluminum. AlSb is a substance which is a possible candidate (production of antimony 68,000 tons in 1968).[30] Its energy gap is suitable for solar energy collection. Other details, such as surface recombination rate, are not known. AlSi may also be a photovoltaic of future importance.

Suitability as a Photovoltaic, and E_{Gap}

What is the relationship between the efficiency of conversion and the energy gap? The current-voltage relation across a small p-n junction is similar to that across an electrodic juntion.[32] Thus (see Fig. 6.12), following a formulation made by Loferski[30]:

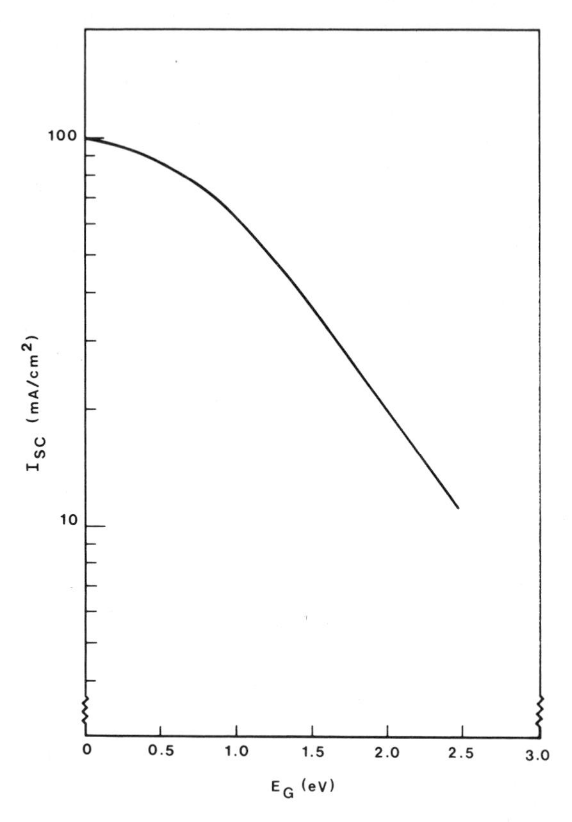

Fig. 6.12 The maximum possible short circuit current I_{SC} for solar cells made with semiconductors of forbidden energy gaps E_G. Air Mass Zero (AMO) solar illustration.[30]

$$I_{sc} = e_o \int_{E_G}^{\infty} Q(h\nu) N_{ph}\,(h\nu) d(h\nu), \qquad (6.5)$$

where I_{sc} is the current density across the couple, N_{ph} the number of incident photons of frequency ν; and E_G the energy gap.

However[30 32],

$$V_{oc} = \frac{kT}{e_o} \ell n \left\{ \frac{I_{sc}}{I_o} \right\} \qquad (6.6)$$

where V_{oc} is the open circuit voltage, I_o the current at equilit
and

$$I_o \ \alpha \ \exp(-E_G/kT). \qquad (6.7)$$

Equations (6.6) and (6.7) show that V_{oc} increases with increase of E_G. Correspondingly, I_{sc} decreases with increase of E_G. The product of I_{sc} and V_{oc} (namely, the power per unit area produced by the photovoltaic device), plotted against E_G, passes through a maximum. Fig. 6.13[30] shows the actual efficiency of a number of couples as a function of E_G. The efficiency maximum occurs for E_G in the region of 1.2 to 1.5 eV, and the maximum theoretical efficiency is 25%. It decreases with increase of temperature, and this reduces the advantage of using optical concentrators of solar energy, which, when used to increase the intensity of solar radiation reaching the photovoltaic several times, heats this up.

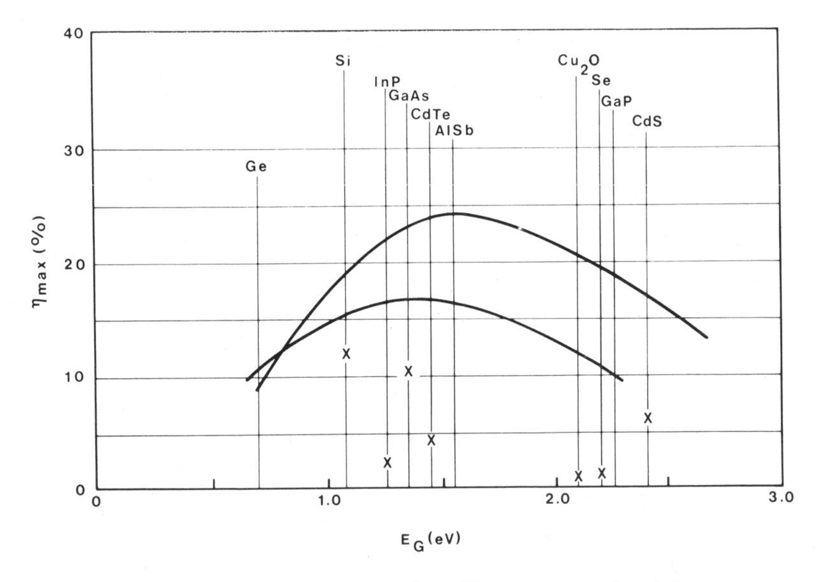

Fig. 6.13 Maximum solar energy conversion efficiency *vs* semiconductor energy gap. Solar spectrum outside the atmosphere, ideal junction behaviour.[30] X = attained; unbroken curved lines = theoretical maxima.

Silicon is the best prospect. Its E_G value means it is sensitive to most of the solar spectrum.* Silica is available in very large quantities. A negative aspect is that it is an indirect semiconductor and cannot be made into thin films. Further, its carriers have a relatively long lifetime, so that it has to undergo super-purification, and must be highly single crystalline. Few alternatives exist. AlSb and AlSi are prospects.[33]

The photovoltaic known as 'cadmium sulphide' requires some explanation. It is the hetero-couple Cu_2S-CdS and the energy gap which determines the activity is that of cuprous sulphide, 1.1 EV, near to that of silicon. (This contrasts with that of cadmium sulphide itself, 2.3 EV, higher than that of silicon.) The maximum efficiency of cadmium sulphide-cupric sulphide junctions should theoretically be 20%, as for silicon. However, the maximum efficiency observed for the sulphide couple is 8%.

* 0.5 to 2.0 eV.

Directions of fundamental research in silicon[9 22 23]

The main direction is in controlling non-radiative recombination. Impurity and defect levels in the forbidden gap may be responsible. Auger effects may be present. Even very highly purified silicon contains a sufficient concentration of impurities to lower the lifetime. These do not affect *transistor* performance, but may affect lifetime relevant to photovoltaic use, so that the manufacture of photovoltaics is a more demanding technique that that for transistors. Could materials be obtained which would tolerate higher doping? If so, this would achieve a higher output voltage, yet maintain lifetimes enough for efficiency in collection.

There are many questions which have to be tackled before progress can be made with silicon[33], for example by what mechanism do doping elements contribute to recombination? Technology to obtain large area p-n junctions must be improved, with respect to impurity pick-up, whilst remaining highly single crystalline. Vacancies, and other defects, must be reduced; annealing techniques must be examined. Could surface recombination be counteracted with an applied potential difference to reduce diffusion of undesirable centers?

The attainment of 10% efficiency for the conversion of light to electricity in silicon is 20 years old and the theoretical maximum is 20%. Striving for an increase in efficiency may not be the most economically effective research goal. Lower production costs of highly single crystalline and pure materials are needed to transform the prospect for commercialisation. What of cheaper techniques for polycrystalline junction formation? Can large *strips* of semi-conductor material be grown in single crystal form, but remain uncontaminated? How is it best to keep the temperature of the cell low in high insolation? What of cheap mirror systems and tracking mechanisms for increasing the light falling upon the cell? What about the join-up of solar collectors with storage systems?

Organic semiconductors, such as pthalocyanines, have several positive and negative characteristics when compared, with Si. They often have greater values of α (see eqn. 1) so that the thickness required is less. They do not need to be made in single crystal form. The disadvantage is their relative lack of efficiency (about 1% has been attained).[33 34] On the positive side, there is a loose coupling, whereas in the inorganic systems, there is a high electron-lattice interaction. Thus, in organic substances there is less trapping of the photo-produced electrons in the lattice, and this should lead to higher efficiencies: this is one reason why, as with the work of Lyons[34], there is interest in organic semiconductor research for solar energy converters. However, organic materials are poorer in the transport of excitation energy, involve higher effective masses, lower carrier mobility and sharper absorption bands. These factors tend to reduce efficiency for the intake of radiation.

PHOTOELECTROCHEMICAL TRANSFORMATION OF SOLAR ENERGY

The photogalvanic possibilities are the least known of the methods for solar energy conversion and equations which relate the power production to the characteristics of an electrochemical cell have not yet been formulated (but see ref. 35).

Several physical phenomena underlie photoelectrochemical methods for the production of electric energy from solar power. Thus, the work of Casey[36] showed that, if electrodes were irradiated with gamma radiation in the presence of ferrous sulphate, current is produced. The current lasts for a relatively short time. Casey's cell is shown in Fig. 6.14[36] The mechanism is probably one involving radiochemical reactions in solution, with secondary interfacial reactions, e.g. the production of hydrogen peroxide from the irradiation of water, followed by the oxidation of ferric ions to ferrous and their subsequent oxidation at the electrodes. At the other electrode, a cathodic reaction occurs, the evolution of hydrogen or the reduction of oxygen, depending on the pH and the oxygen concentration in the solution.

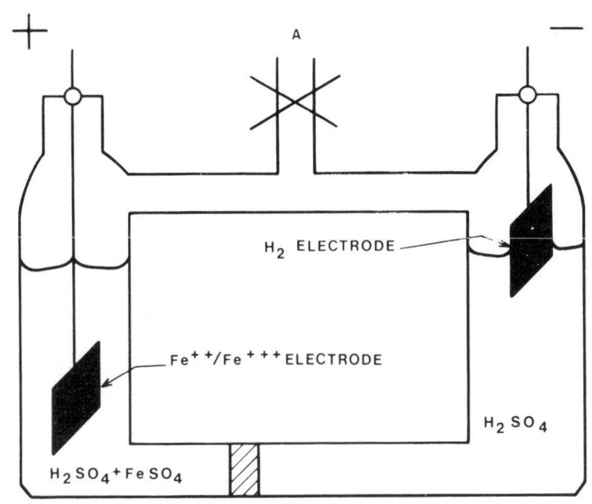

Fig. 6.14 Casey's cell for electrochemical conversion of light to electricity.[36] The left-hand electrode was irradiated.[36]

Conditions for the passage of a current through a cell, on one electrode of which falls radiation, have been deduced by Matthews and Khan.[35] They suggest that there are 3 uses of the photons instant upon the electrode. Firstly, they activate some electrons in the metal to emit through the energy barrier at the electrode to be accepted by H_3O^+ ions in solution: this part of the current, however, is small. A larger current is produced by photons which enter the metal and emit electrons which are too energetic to be captured by ions in the double layer and form instead solvated electrons in solution. A characteristic law which pertains to such a situation is Brodski's, which states that the current produced from such photons raised to the power 2/5 is linear with the potential of the electrode.

Thus, it may be possible to irradiate two electrodes, and drive an electrochemical cell which would then produce hydrogen and oxygen, with electricity as a by-product. Fujischima and Honda[37] used a doped titanium dioxide electrode and found that they could get 1% efficiency in the absorption of light and its conversion to hydrogen and electricity. Results which they obtained are shown in Fig. 6.15.[37] Some ideas pertinent to the model are discussed by Calvin.[37a]

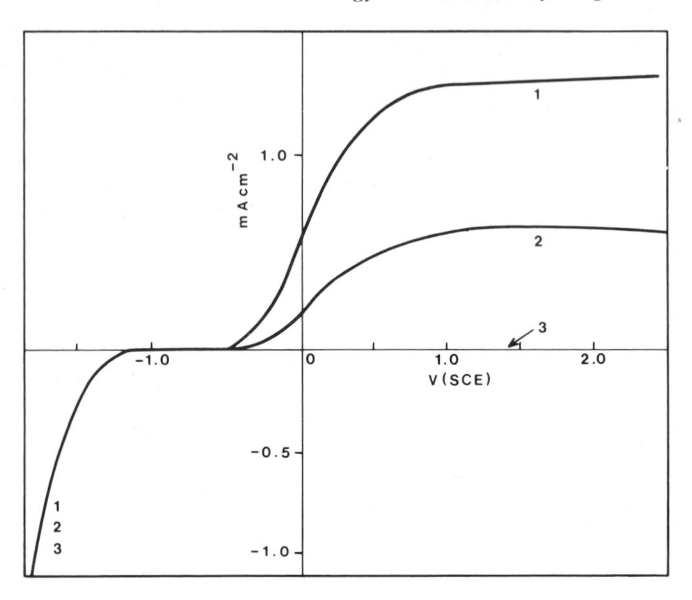

Fig. 6.15 Current-voltage curves for TiO_2 n-type semiconductor. A single crystal wafer of n-type TiO_2 (rutile) was used after treatment at 700°C at $10^{-4} \sim 10^{-5}$ torr for roughly 4 hours to increase the conductivity of the crystal.
Curve 1: Under irradiation (relative intensity of light, 100%).
Curve 2: Under irradiation (relative intensity of light, 50%).
Curve 3: Without irradiation.

SOME BASICS OF SELECTIVE COATINGS

Principles

The temperature of a surface exposed to the sun's radiation rises until the rate of heat loss equals the rate of heat adsorbed. In the relation between the intensity of energy emitted in radiation and the wavelength of the emitted energy, a maximum occurs, but the wavelength at which it occurs is strongly temperature dependent. The sun has a spectrum which looks as in Fig. 6.8(a), with a maximum in the region of $0.5\mu m$. However, heat loss by radiation from the collector takes place at 100° to 400°C, and the maximum of the relevant spectral distribution curve for these lower temperatures is in the region of 5μ.

Consider a material which is a good absorber in the region around 0.5μ (the maximum of the solar radiation). Suppose that its wavelength-dependent emission coefficients fall towards zero in the higher wavelength region. Then, there will be minimal black-body radiation, i.e. a low rate of loss of heat by radiation, but a good absorption of the solar radiation. Such a body – or coating on some other and cheaper substance – will come to a steady state at a high temperature.

Now, any one type of material will not absorb all the sun's radiation with equal efficiency. It is desirable, in manufacturing such selective coatings, to use several materials in sandwich form. Thus, a diffraction grating, with its ability to smooth out absorption in various wave length regions, is formed.

The same idea of absorbing in the low wavelength region characterising the high temperature solar spectrum, but *not* emitting in the higher wavelength region characteristic of the temperature of the collector, can be obtained in another way. Micro-surfaced structures are made in such a way that they are in effect wave guides.[38] The shorter wavelengthed solar radiation is absorbed and the collector gets hot. In respect to its radiation, this would be at higher wavelengths (spectrum of cooler body) and here the wave-guide properties of the structure cut off the emission of such radiation, so that loss of heat by emission is much reduced.

The attainment of the principle of selective coatings

In one early application of selective coatings, the coating was made by electroplating silver onto nickel on a metal plate, and then electroplating a layer of copper, 5×10^{-5} cm thick. When heated in air, the copper turns to black cupric oxide and adheres to the metal, even though it is undergoing large temperature changes each day. (Were black paints containing organic binders to be used, they would be decomposed by the high temperature reached whilst in contact with air.)

This early model has been improved, in particular by Schmidt and Jensen, as shown in Fig. 6.16.[39] It is possible to obtain higher ratios of absorption to reflectance with complex coatings.

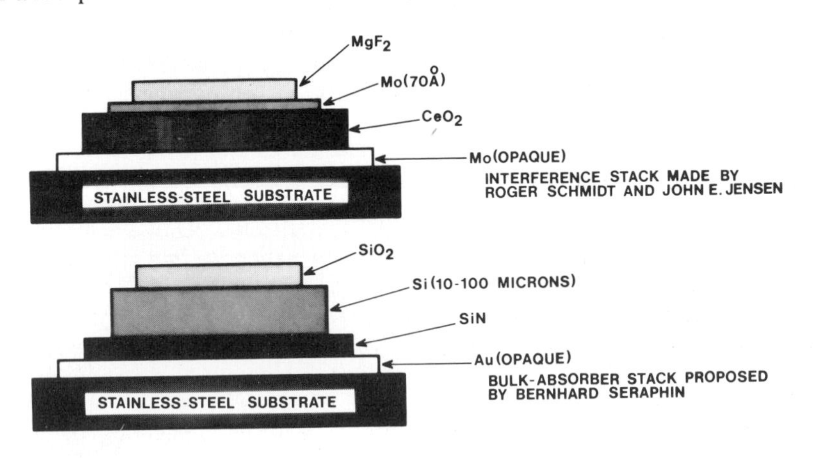

Fig. 6.16 Selective coatings.[39]

Materials problems of high temperature selective coatings

The realisation of the selective coating method involves arrangements of the type shown in Fig. 6.16.[39] Thus, a vacuum in the collector tubes is made necessary so as to avoid the decomposition of the coatings upon heating them, e.g. oxidising molybdenum, silicon, etc. It would seem an expensive matter to maintain a vacuum in miles of tubing. Chemical deterioration would seem probable.

There is also daily expansion and contraction, and the mechanical stability of the coatings may be strained. It may be that the microstructure surface approach of Watson-Munro and Horowitz[38] would be, therefore, longer lived.

The fluid to be used in the heat engine associated with a selective coatings approach could be air, water, sodium, etc. Low boiling organic liquids could also be used, if there were a suitable trade-off between the lower cost of coatings which reached a lower temperature (100°C?) and the decreased Carnot efficiency caused by the lower temperature.

Directions of research in selective coatings
The position of photovoltaics is more advanced than selective coatings. Investigation of the spectra of substances, in the range 350-550° has to be made more extensively, particularly of inorganic oxides. The absorption maximum sought is in the region 0.2-2μ. Materials research on the thermal, chemical and mechanical stability of suitably absorbing coatings in low pressure air atmospheres is needed.

Complex sandwich-type coatings seem unlikely on a big scale. A single coating, evaporated, electrodeposited, or electrophoretically grown, would be more likely to be stable than the complex sandwich coatings, because of the mechanical instability introduced into these by the different coefficients of expansion (cf. the daily cycle of heating and cooling). A stainless steel layer onto which an iron layer had been electrodeposited, and the black property introduced by anodic oxidation, could be of use.

Optical collectors and heat transfer systems also require systems and engineering studies, but not with great funding until the feasibility of maintaining photothermic coatings as chemically stable during the cyclic heating and cooling for prolonged times in poor vacua have been established.

USE OF OCEAN THERMAL GRADIENTS

This method was suggested in 1881 by D'Arsonval.[40] Claude[41] attempted to build such a plant, aimed at 40 kW, in Cuba in 1930 (Fig. 6.17[41]). His apparatus was wrecked by a storm.

Temperature differences in the ocean
Tropical oceans contain a 20°C differential between the temperature of the surface and the depth. A typical temperature distance profile is shown in Fig. 6.18.[42]

The machinery necessary
One proposal for a solar sea plant, from Lavi and Zener[42], is outlined in Fig. 6.19. A boiler takes in warm water from the surface and uses it to boil ammonia. The ammonia works a turbine which generates electric power. The NH_3 is passed through a condenser which uses cold water at 5°C. pumped from the depths.

The power density calculated by Lavi and Zener[42] is about 12 kW per sq metre of the surface of the sea occupied by the collector. (The kW per unit volume is not of importance for the apparatus is seaborne and most of the volume would be undersea.) The cost per kW should not exceed $165 (compare $300-$600 for fossil fuels and $500-$1,000 for nuclear) and the cost of electricity from the solar sea plant would be 3 mils $(kWH)^{-1}$ at source. Hydrogen could be used to transport the electricity produced over long distances to areas of high use.

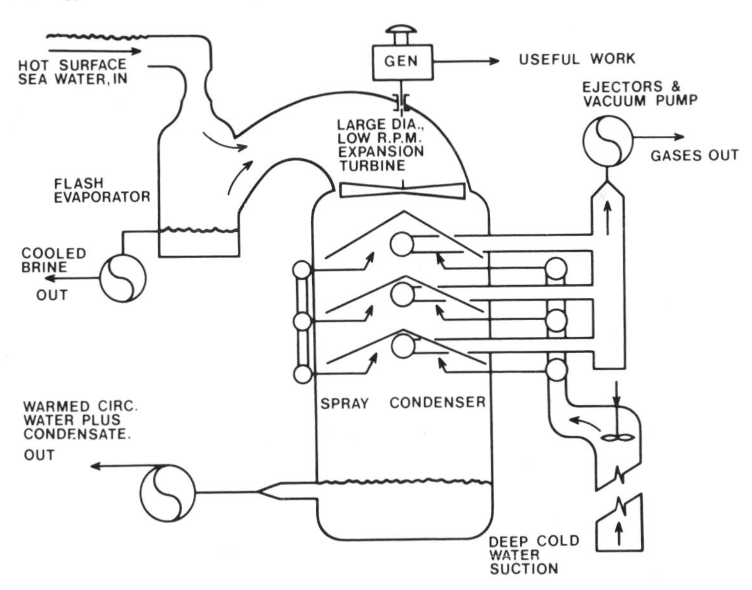

Fig. 6.17 The Claude ocean thermal differences process.[41]

A founding paper in this field is that by Anderson and Anderson.[43]

The efficiency of the solar sea power method

Solar sea power is inefficient because of the low temperature difference over which it must work. Thus, the efficiency is less than 3% compared with the 38% of conventional steam plants. Solar sea power boilers must produce more than 10 times as much heat as the boiler in a conventional plant, and this leads one to conclude that the solar sea power boiler tube core must be more than 10 times greater than that of the conventional plant. However, two facts compensate this potential difficulty.[42] Vapour pressure of liquids rises exponentially with temperature: thinner tubes must be used for lower pressures. The solar sea power plant will work at low vapour pressures so that the tube will be thinner than normal boiler tubing. Correspondingly, the decrease in strength of metals with rise of temperature makes it necessary to use special alloys in many boilers, whereas in the solar sea plant, cheaper alloys may be used at the lower temperature and pressures concerned.

Practical difficulties

Difficulties which have been discussed for solar sea plants include corrosion. This can be reduced by the use of aluminium and new aluminum alloys, such as alclad. Microbial fouling could be reduced by causing a small chlorine concentration to exist in the water.[42] The problem of mooring the plant to be stable under storm conditions needs study.

Comparative costs of solar sea power

Fig. 6.20[42] suggests that solar sea power would be competitive with liquefied natural gas, e.g. from U.S.S.R. Solar sea power is one of the most attractive methods of transforming solar energy to electricity.

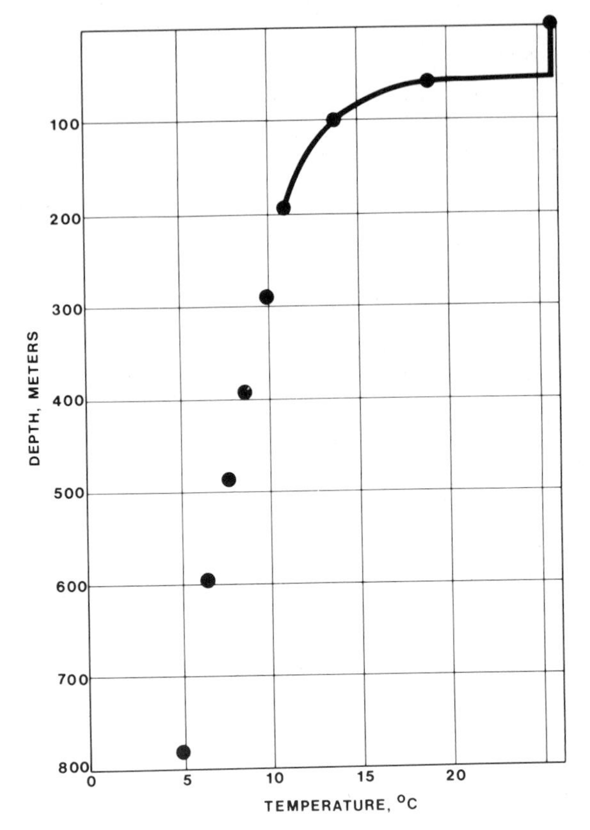

Fig. 6.18 Typical ocean temperature profile at various depths. The surface layer, which is about 200 metres deep, takes its heat from the sun and stays at about 25°C. The cold water at the lower depths comes from the Arctic region and can be as low as 5°C.[42]

A small prototype power plant near the island of Hawaii, or at St Croix in the Caribbean, has been recommended (1974).

BASICS OF PHOTOSYNTHETIC COUPLES

The principles of photosynthetic energy conversion are extensively discussed in textbooks of Biology. It is a process, the details of which are not fully understood.

A newer concept[43a] is that of the production of hydrogen directly from a photosynthetic process (see Fig. 6.21[9]).

Benemann's work

The photosynthetic method has been worked upon by J. R. Benemann at the University of California.[44] Qualitatively, hydrogen can be produced upon insolation of water in the presence of a particular type of algae. It may be possible to manufacture the enzyme outside the algae and use it without the surrounding plant.

The information obtained from Benemann[44] suggests that the enzyme

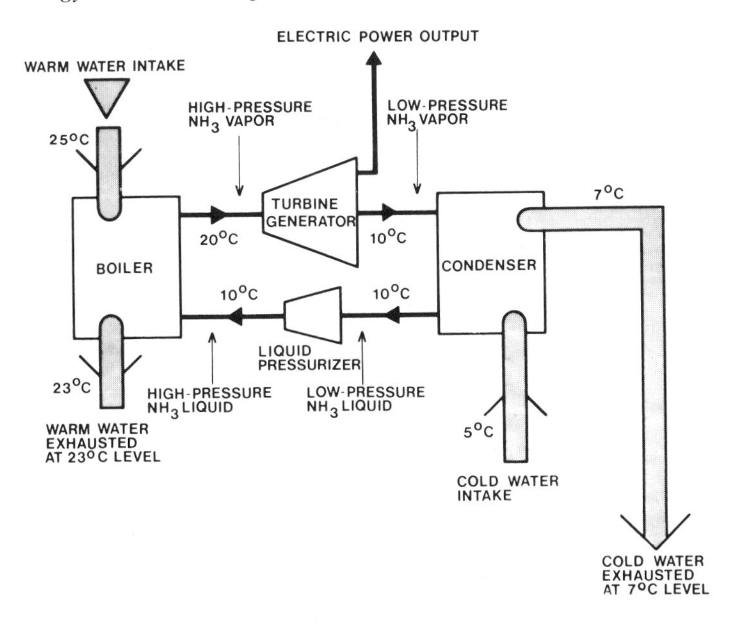

Fig. 6.19 The basic components of a solar sea power plant.[42]

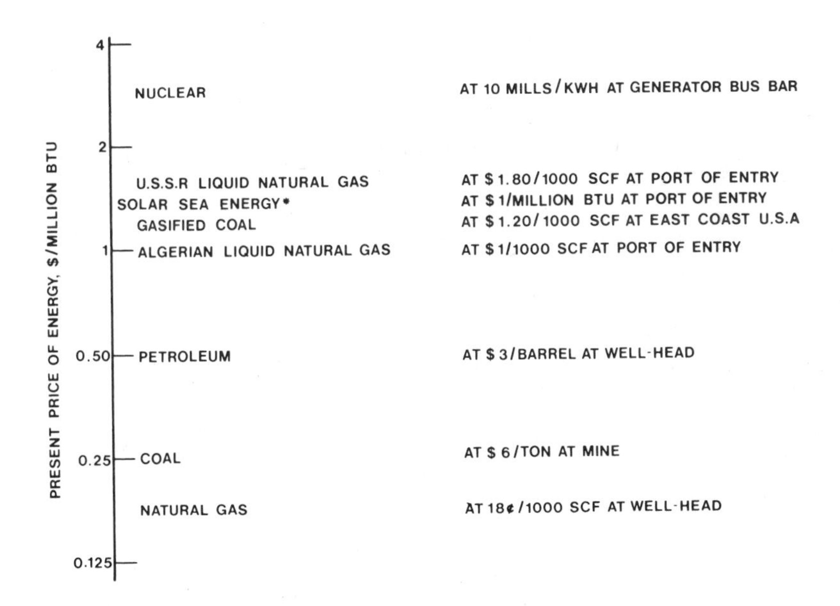

Fig. 6.20 Comparative prices for energy produced by fossil fuels and other sources.[42] The prices correspond to those of pre-October 1973. The fossil fuel price has escalated 150-300 per cent (late 1974). The estimate for the solar sea plant should be changed only by inflational considerations of 10 to 20%.

system, hydrogenase-nitrogenase, is appropriate. The hydrogenase is the vital section as it is that part which is directly involved in hydrogen evolution. The algae used by Benemann, anabaena cylindrica, contains cells called heterocysts which protect the enzyme from O_2. This algae will produce H_2 from water without noticeable decrease over about one day.

What is lacking at present is a clear definition of the mechanism and intermediates involved. A measurement of the rate of hydrogen evolution has recently been made by Neal and Bockris.[45] It is adequate for practical solar energy farming on the fulfillment of two as yet unresearched conditions: that the present short life of the algae is lengthened by proper feeding with CO_2 and light; and that scale up from low to high light intensities follows a linear law.

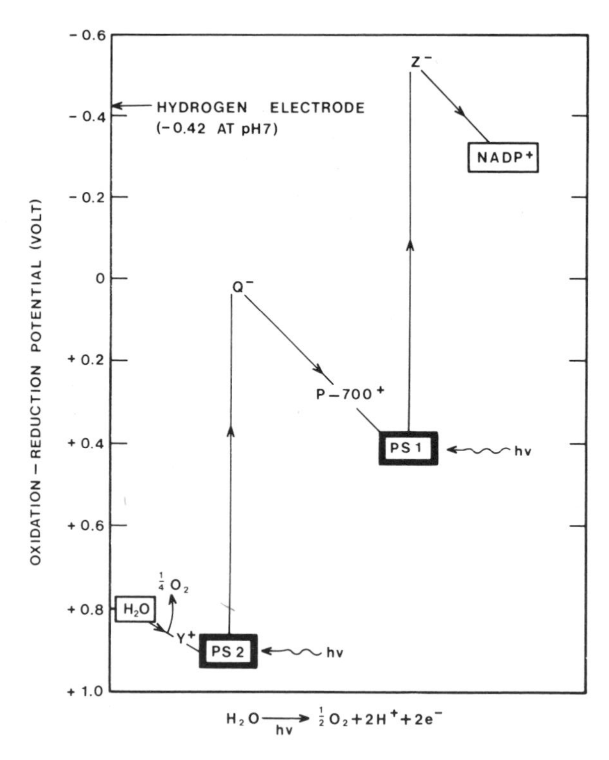

Fig. 6.21 Schematic diagram of the biological photolysis of water. The light-reactions of photosynthesis. Photosystem II generates a strong oxidant (Y^+) and a weak reductant (Q^-). Photosystem I generates a weak oxidant (P-700$^+$) and a strong reductant (Z^-). Q^- interacts with P-700$^+$ via a chain of electron carriers (not shown). Z^- is used to reduce a pyridine nucleotide ($NADP^+$). The photosynthetic production of H_2 is simpler than the photosynthetic production of $(CH_2O)_n$ and O_2.[9]

Possibilities of the photosynthesis of hydrogen

The gain which would come from a successful photosynthesis would be similar to that from any other light-based method – there would be no cost of fuel. As the costs of the fuel (electric or thermal) are the main costs of producing hydrogen by electrochemical or thermal methods, the cost of hydrogen produced photosynthetically would be low. The necessary enzymes could be cheaply produced because they are built up biochemically in bacteria, which in turn are grown using sunlight and CO_2 in aqueous solution. Principal costs would be the apparatus to separate hydrogen from oxygen and to store the hydrogen.

BASICS OF WIND GENERATORS

These have been discussed in Chapter 5.

SUMMARY OF SCHEMES FOR SOLAR ENERGY CONVERSION

A summary is shown in Fig. 6.22.[17] Thus, there are many quite different approaches to the collection and conversion of solar energy, in addition to the classical photovoltaic method usually discussed.

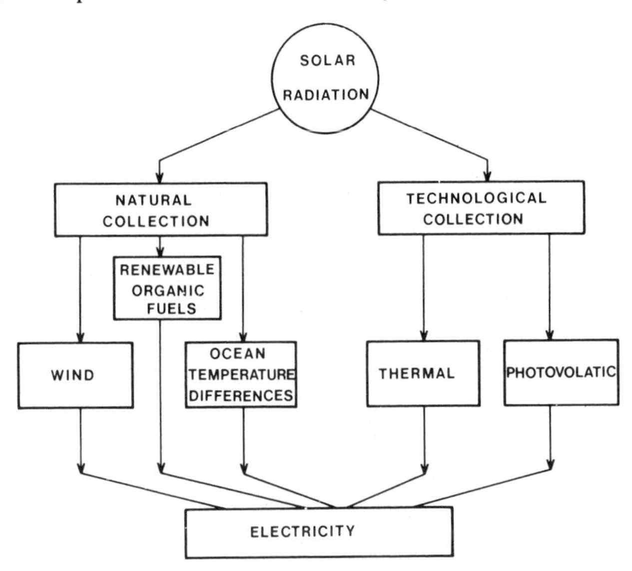

Fig. 6.22 A summary of schemes for solar energy conversion.[17]

REFERENCES

1 F. Daniels, *American Scientist*, **55,** No. 1, 15 (March 1967); Photovoltaic Terrestrial Conversion of Solar Energy for Terrestial Applications Vols. 1 and 2; October 23-25, 1973, Cherry Hill, N.J.; N.S.F. RANN Grant AG 485.

2 D.M. Chapin, C.S. Fuller & G.L. Pearson, *J. Appl. Phys.*, **25,** 676 (1954).

3 J.O'M. Bockris, Internal Report to Westinghouse Corporation, 1962.

4 J.O'M. Bockris, *Encironment,* 13, 51 (1971).

5 D.P. Gregory, D.Y.C. Ng & G.M. Long, in *The Electrochemistry of Cleaner Environments,* ed. J.O'M. Bockris, Plenum Press, New York, 1972.

6 K.A. Ehricke, 'The Power Relay Satellite Concept of the Framework of the Overall Energy Picture', North American Aerospace Rockwell International, E73-12-1, December 1973.

7 H.E. LaBelle & A.L. Mlavsky, Proceedings of the 9th IEEE Photovoltaic Specialists Conference, Silver Spring, Maryland, 2-4 May 1972.

[8] C.G. Currin, K.S. Ling, E.L. Ralph, W.A. Smith & R.J. Stirn, Proceedings of the 9th IEEE Photovoltaic Specialists Conference, Silver Spring, Maryland, 2-4 May 1972.

[9] Report of the Australian Academy of Science on Solar Energy Research in Australia, Report No. 17, p. 22, September 1973 (Chairman: C.N. Watson-Munro).

[10] E. Cohn, priv. comm..

[11] G.O.G. Löf, J.A. Duffie & C.O. Smith, *Solar Energy*, **10**, 27 (1966).

[12] J.O'M. Bockris & Z. Nagy, *Electrochemistry for Ecologists*, Plenum Press, New York, 1974.

[13] 'Solar Water Heaters; Principles of Design, Construction and Installation', Division of Mechanical Engineering (Circular No. 2), C.S.I.R.O., Melbourne, 1974.

[14] E.T. Davey, presented at International Solar Energy Society Conference, Melbourne, 1970 (Paper No. 4/71).

[15] D. Proctor, presented at International Solar Energy Society Conference, Melbourne, 1970 (Paper No. 5/55).

[16] K. Boer, 'Direct Solar Energy Conversion for Terrestial Use', Institute of Energy Conversion, University of Delaware, 1974.

[17] 'Solar Energy as a National Energy Resource', prepared by the NSF/NASA Solar Energy Panel, December 1972.

[18] K. Boer, priv. comm., 1974.

[19] B. Baker, priv. comm., 1974.

[20] W.R.W. Read, presented at International Solar Energy Society Conference, Melbourne, 1970 (Paper No. 5/52).

[21] P. Glaser, priv. comm., 1974.

[22] J. Stackhouse, four articles on Solar Energy in the *Financial Review*, Sydney, January 1974.

[23] J.O'M. Bockris & S. Srinivasan, *Fuel Cells: Their Electrochemistry*, McGraw-Hill, New York, 1969, p. 15.

[24] R. Hilson & E.K. Rideal, Proc. Roy. Soc., 199A, 295 (1949).

[25] J.O'M. Bockris, in *Modern Aspects of Electrochemistry*, Vol. 1, ed. J.O'M. Bockris, Butterworth, 1954 (Chapter 4).

[26] B. Quickenden, priv. comm..

[27] G.A. Stewart, *J. Aust. Inst. Agric. Sci.*, **36**, 85 (1970).

[28] O.T. Denmead, *Agric. Met.*, **6**, 357 (1969).

[29] 'Solar Sea Power', Semi-Annual Progress Report covering period 1st November, 1973, to 31st January, 1974, prepared by Carnegie-Mellon University, NSF Grant No. 39114.

[30] J.J. Loferski, 'The Principles of Photovoltaic Solar Energy Conversion'; 'Some Considerations Affecting the Choice of Semiconductors for Use in Solar Cells Intended for Large Scale Solar Energy Conversion', Brown University, Rhode Island, March 1972.

[31] F. Gutmann & L. Lyons, *Organic Semiconductors*, Wiley, 1967.

[32] J.O'M. Bockris and A.K.N. Reddy, *Modern Electrochemistry*, Rosetta Edition, Plenum Press, New York, 1973.

[33] 'Solar Cells: Outlook for Improved Efficiency', National Academy of Sciences, Washington, 1972.

[34] L. Lyons, priv. comm..

[35] D.B. Matthews & S.U.M. Khan, *Aust. J. of Chem.*, to be published.

[36] E.J. Casey, Bournemouth Battery Conference, 1960.

[37] A. Fujischima & G. Honda, *Nature*, **238**, 38 (1972).

[37a] M. Calvin, *Science*, **18**, 375 (1974).

[38] C. Horowitz & C.N. Watson-Munro, International Solar Energy Concept Symposium, Brisbane, June 1973.

[39] A. Meinel, *Phys. Today*, p. 44 (February 1972).

[40] J. D'Arsonval, *Revue Scientific*, 17 September (1881).

[41] G. Claude, *Mech. Engr.*, **52**, 1039 (1930).

[42] A. Lavi & C. Zenet, *IEEE Spectrum*, **10**, 22 (October 1973); Report RANN, Grant No. 39114, 1 July 1973.

[43] J. Hilbert Anderson & James H. Anderson, *Mech. Engr.*, 41 (April 1966).

[43a] H. Gaffron & J. Rubin, *J. Gen. Physiol.*, **26**, 219 (1942).

[44] J.R. Benemann, 'Hydrogen Production from Water and Sunlight by Photosynthetic Process', University of California, San Diego, La Jolla, California, December 1973.

[45] G. Neal & J. O'M. Bockris, unpublished; Gordon Neal, Hon. Diss., Flinders University, 1975.

CHAPTER 7

Solar Energy—Approach to Technology

COST REASONS FOR THE DELAY IN THE DEVELOPMENT OF PHOTOVOLTAIC CONVERSION

THE AVAILABILITY of cheap oil and natural gas and the earlier confidence that atomic energy would solve problems of the exhaustion of fossil fuels, is one of the factors by which one can rationalise the small funding which has been given to cheapening the manufacture of photovoltaic cells with efficiencies of 5 to 15%. There was a great discrepancy between cost estimates for photovoltaic cells, and costs of practical devices (oil, steam plants; $300-$350 (1974); atomic plants: $700-$1,000 per kW (1974). Thus, consider the analysis of Spakowski and Shure[1] (March, 1972). These workers set up as a standard for comparison with their estimates for solar plants, two conventional plants, and the cost analyses of these is given in Tables 7.1 and 7.2.[1] (For a more general discussion of costs, see refs. 2 and 3).

TABLE 7.1[1]

CONVENTIONAL ELECTRIC POWER PLANT: PLANT COSTS – 1971

	American Electric Power (coal)	Davis-Beese (nuclear)
Power capacity, MWe	2600	870
Cost, dollars	488×10^6	270×10^6
Specific cost, dollars/kW	188	310

Spakowski and Shure[1] considered a solar plant having a module containing a square mile of solar cells (Fig. 7.1[1]). The array consists of panels 100 feet long by 10 feet wide, separated by a walkway, 2 feet wide. The photovoltaic considered is silicon. The electricity for the panels is brought together at a station containing the power conditioning equipment and controls. The dollars (1971) include the cost of the arrays, power conditioning, buildings and interest. Land costs are neglected, as are taxes and profits. No allowance is made for a storage facility, it being assumed that the electricity is fed from the solar collector directly into the grid.

TABLE 7.2[1]

CONVENTIONAL ELECTRIC POWER PLANT: POWER GENERATION COSTS – 1972

	Power generation cost mils/kW-hr	
	Coal	Nuclear
Plant carrying charge	2.70	3.40
Fuel carrying charge	–	.46
Fuel cost	2.25	1.34
Operation and maintenance	.30	.42
Sulfur dioxide removal	.38	–
Thermal pollution control	.12	.17
Total cost	5.75	5.79

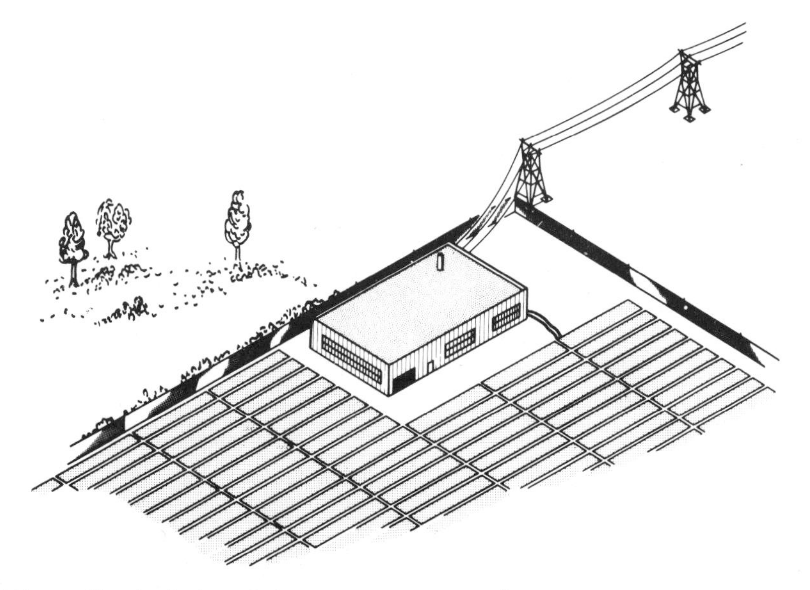

Fig. 7.1 Sketch of solar cell electrical generation farm.[1]

On the technology available to them (1971), these authors[1] estimated that a 2 × 4 cm silicon cell would cost $6, if obtained in large quantities. The mountings and connections for one of these cells is estimated at $5. They put the limit of the cost for 8 sq cm, with an efficiency of

14%, as $1.60. The assembly cost per cell would be $1.75; and the cost of the inverter, $250 per kW. They suggest a black-top surface with drainage of the solar arrays to steel bands cemented to the black top ($0.76 per sq ft, see Fig. 7.2[1]). The lifetime is assumed to be 10 years for the solar arrays and 30 years for the rest. Debt financing is over the useful life and taken in 1971 to be at 7%.

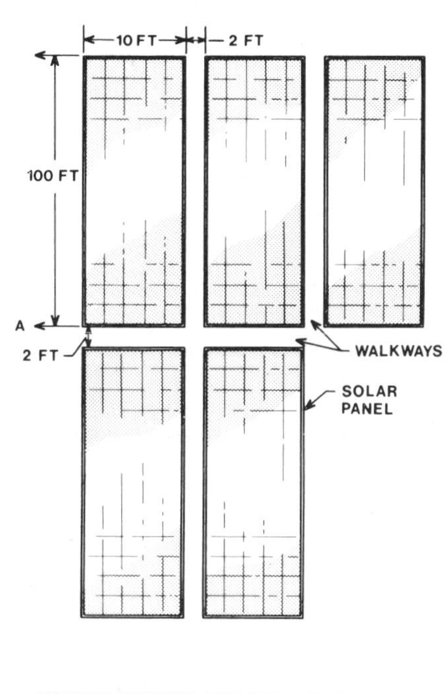

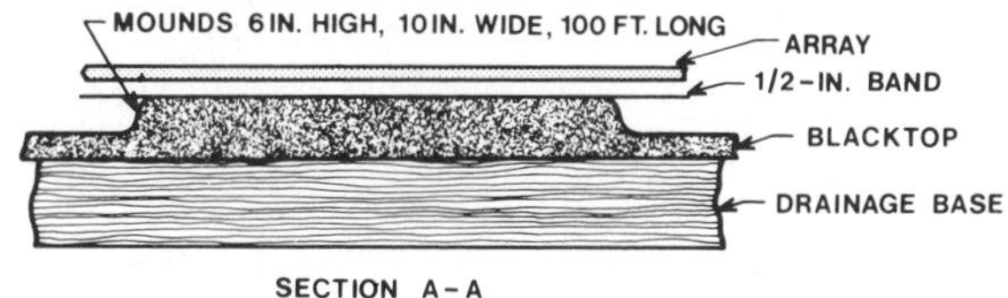

Fig. 7.2 Solar array site construction. Total array area, 2.79×10^7 sq ft per sq mile; array packing factor 81%; blacktop area, 3.42×10^7 sq ft.[1]

Spakowski and Shure used the solar data available for Cleveland, Ohio, and Phoenix, Arizona, and their conclusions are shown in Table 7.3.[1]

Thus, the electricity costs from such a plant were predicted to be 300 times more expensive than that from fossil fuels in 1971. Estimates such as this have contributed to a lack of funding of research and development aimed at massive electricity production from solar energy.

TABLE 7.3[1]

ESTIMATE OF SOLAR CELL SYSTEM POWER COST WITH NASA CONDITIONS AND 1971 TECHNOLOGY

	No degradation		50% degradation in 10 years	
	Cleveland	Phoenix	Cleveland	Phoenix
	Power cost, dollars/kW-hr.			
Array:				
Cells*	1.40	0.90	1.85	1.20
Assembly*	1.55	1.00	2.05	1.30
Subtotal	2.95	1.90	3.90	2.50
Solid-state power conditioning**	.0023	.0015	.0030	.0020
Facility:				
Site construction**	.0039	.0001	.0050	.0033
Buildings**	.0001	.0001	.0001	.0001
Maintenance, operation	.0004	.0002	.0005	.0003
Subtotal	.0044	.0028	.0056	.0037
Total	2.96	1.90	2.91	2.51

* Depreciated over 10 years; interest rate, 7%.
** Depreciated over 30 years; interest rate, 7%.
† Current revenues.
†† Land costs, taxes and profit not included.

A BRIGHTENING OF THE DARKER VIEW OF COSTS FOR PHOTOVOLTAICS

Spakowski and Shure's 1972 analysis[1], by its close attention to the minutiae in costing, obscures the fact that the *predominant cost item* is the one item – the photovoltaic silicon single crystals and their assembly. Thus, their cost for what Spakowski and Shure consider as the optimistic estimate quoted here, including assembly, is $3.35 per cell of 8 sq cm with 14% efficiency.

Thus, one would need about 10 sq m or about $\frac{10^5}{8}(1.4)$ cells to produce 1 kW, a cost of about $\$3.10^4$ per kW.

As the cost per kW of a fossil fuel or nuclear plant is of the order of $\$10^2$ per kW, the cost of the single crystal alone accounts for the tremendously high estimate made for solar energy by Spakowski and Shure in 1972. Correspondingly, their allowance for only 10 years of life for silicon cells on earth seems pessimistic: although degradation of Si cells occurs in space, due to micro-meteorites and to greatly increased density of insolation, the degradation rate on earth should be far less.

The efforts of Spakowski and Shure[1] excluded a number of factors which would have made the outlook for solar cells seem somewhat better, even before the new technology had been reported. Thus, the principal successful application of Si solar cells to date, provision of power for space vehicles, sets an exceedingly stringent requirement on reliability, and this requires extensive and expensive testing of each cell.

Limits on the area available in space vehicles require that the cells be cut in particular shapes, and this provokes extra costs in individual labour, and waste of much of the highly purified Si.

Ralph[4], writing in 1970, pointed out that simply changing the shape of the cells would make it possible to produce 11% efficient cells at a cost of $15,000 per kW, some 6 times decrease from cost levels which had for long been suggested, and which do not allow for automation and scale-up. 1970 U.S. production of photovoltaic Si was 100 tons per year, but, if the cells were used to provide the entire electrical power needs of the U.S., several million tons would be required. A drop in price of several times should be attainable by this factor.

Correspondingly, Ralph[4] pointed out that a simple conical concentrator could increase the power cell by 2.5. With a $15,000 per kW cell, this would reduce costs to some $6,000 per kW, and perhaps less with further concentration. An indirect comparison with reality can be made with the situation at Odeillo in France, where a solar furnace of 1,000 kW inel costs $2,000 per kW[3]. Reduction for scale would probably bring this to *c.* $1,000 $(\mathrm{kW})^{-1}$.

Thus, the long-standing impression that the photovoltaic conversion of solar energy to electricity would be 'impossibly expensive', detailed in recent times by Spakowski and Shure[1], has arisen partly from the maintenance of assumptions, which would be irrelevant to large-scale manufacture[5], and lack of sufficient investigation of cost reductions which occur with the massive production of cells for solar farms, rather than for space vehicles.

Nevertheless, even the more optimistic modifications of the usual position, initiated by Ralph[4, 6] in 1970, suggest that costs need still further lowering, i.e. there is a need for new technology. Thus the $1,000 per kW which could be inferred from Ralph's figures and the set-up at Odeillo, was about five times the cost of the fossil fuel burning generation of the time. Even with trebled fuel costs for these, the cost of electricity using silicon photovoltaics would be more expensive than that from oil or coal, but the rising costs of these have made the photovoltaic situation seem far more worthy of research.

NEW TYPE OF TECHNOLOGY NEEDED FOR PHOTOVOLTAICS

Consider the process of obtaining pure silicon. This is outlined in Table 7.4.[7] This shows that the greatest cost reductions needed are in preparing single crystals from the pure Si, but *particularly in fabricating the solar cells from the single crystals.*[8]

The effect of volume of production of the single crystals on the price can be estimated.[7] From 1961 to 1971, the single crystal size for solar cells increased from 30-80 mm in diameter and cost fell from 80 cents per gram to 25 cents per gram. If this information is utilised as the base of a learning curve, and the effect considered of a three orders of magnitude increase in the volume of silicon produced, the resulting reduction of cost of the fabricated single crystals would be too small for photovoltaics to compete with the coal-steam plants by twice times (1974 coal costs).

TABLE 7.4[6]

ECONOMICS OF 1972 SOLAR CELL SILICON (CLASSICAL TECHNOLOGY)

1972 silicon cost

Form of silicon	Purity of form	Prime use	\$/kg-Silicon	\$/kW capacity*
Sand, gravel	~90%	Construction	<0.005	<0.01
Metallurgical silicon	>95%	Steel	0.60	1.40
Chlorosilane	>99.999%	Silicones	6.00	14.00
Polycrystalline silicon	>99.999%	Semi-conductor crystals	60.00	140.00
Czochralski crystals	>99.999%	Small signal devices	250.00	600.00
Solar cell blanks	>99.999%	Solar cells	1300.00	30000.00

* At 2.3 kg silicon/kW capacity (0.01 cm thick cell with 10% efficiency at AM 1.0).

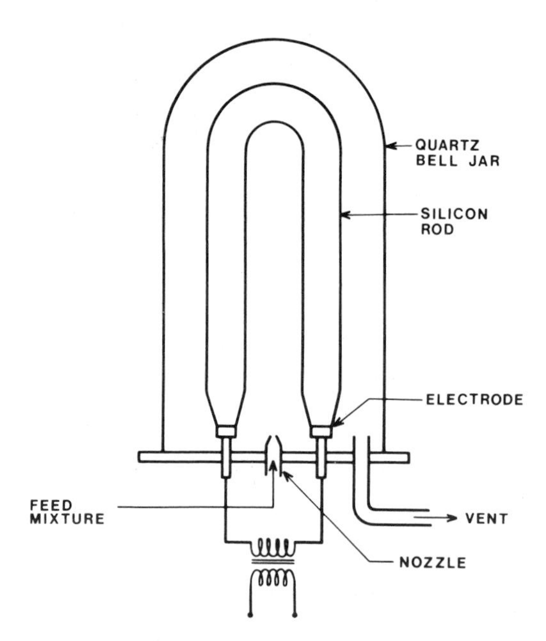

Fig. 7.3 Silicon deposition process.[7]

What of the cost of the pure silicon before it is made into crystal slabs suitable for photovoltaics? The process is described in Fig. 7.3.[7] The cost of this process is largely that of electrical energy and is not likely to decrease.

After the manufacture of pure silicon, and crystal growth, the crystals have been sliced into wafers, 0.2-0.4 mm thick, and in doing this one loses about half the silicon. Thereafter in successive handling, sawing and polishing – and these have been hitherto done by hand and involve high labout costs, – 70% of the silicon is lost in sawing. With this kind of base – no new technology – Currin *et al*[7] (see Wolf[9]) confirm the conclusion of Ralph[4] that $1,000 per kW would be the minimum for massive production of photovoltaic Si. In spite of the 50 times improvement on the estimates of Spakowski and Shure[1], new methods should be sought to give cost competitiveness with fossil fuel and nuclear alternatives, although, in respect to the former, fossil fuel cost increases would probably make even the old technology economic within less than the time needed to carry out further research for cheapening and building plants.[10]

PHOTOVOLTAIC SILICON USING THE EDGE-DEFINED, FILM-FED RIBBON GROWING TECHNIQUE[7,9]

Reference to Table 7.4[7] shows that the greatest cost is processing the pure Si into single crystals of appropriate shape (see also ref. 8).

Because the earlier (impossibly) heavy costs for photovoltaic crystals were due largely to the fabrication (rather than purification) work (see Table 7.4), the main point for the new process is the growth of suitable crystals in a way which is subject to automation. The principle is to place the sheets of very pure graphite at a capillary distance apart. The sheets are placed in liquid silicon and this rises. At the top of the plates radiation cooling from the ribbon, which extends beyond the top of the die, freezes the silicon. A single crystal seed is placed at the top and raised at a suitable rate, bringing the crystal slab of

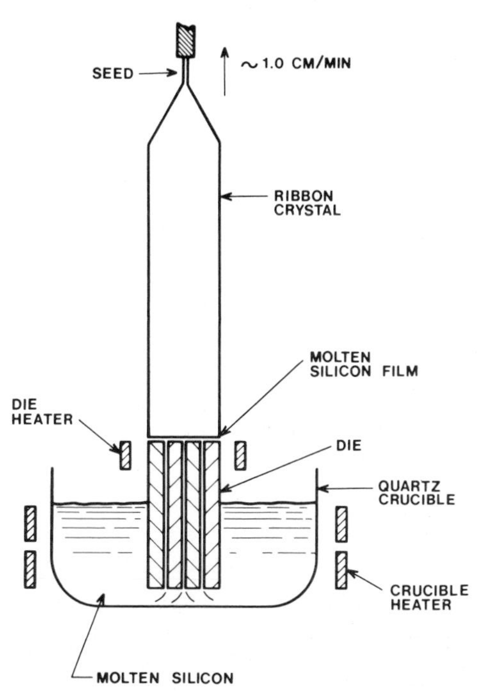

Fig. 7.4 Silicon EFG ribbon growth process.[7]

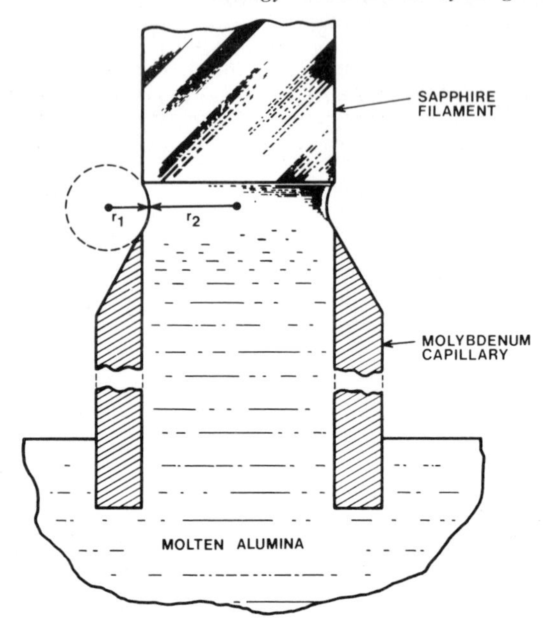

Fig. 7.5 The growth process.[11]

silicon behind it, see Figs. 7.4[7] and 7.5[11], and the Appendix. The aim is continuous growth for many feet, the 1975 attainment being 6 inches.

The ribbons obtained by the edge-defined, film-fed growth technique are at present about 1 inch wide, but it is thought that 10 inch width would be possible. Techniques which allow continuous feed are being researched. There is clearly a possibility of continuous growth (the ribbons grow at about 1 inch per minute at present); and also for mass production with many ribbon growing apparatuses working together.

The development of the silicon ribbon approach is one key for the more economic silicon solar cell (Fig. 7.4[7]). Estimates developed by Currin *et al* for development of this process are shown in Table 7.5[7], where the change in costs is from 30 cents per sq cm down to 0.25 cents per sq cm.

The rest of the estimate for the costs produced by Currin *et al* is based upon the diffusion process for forming a p-n junction. After diffusion, the ribbon would be cut into lengths much greater than those of present cells. Electron beam sources would be used for the metal evaporation and making of contacts. Aluminum would be the metal used.

The total reduction in cost with the ribbon technique, compared with earlier techniques, is some 300 times.

Collins[12] has described this process and stated that the U.S. National Science Foundation expects a practical capability for roof-top collection by 1985-1990. The cost per kW in 1974 dollars would be in the region of $250 (however this does not allow for making up the array of cells, the electrical conditioning, and storage system), and the fact a solar collector functions for a fraction of the 24-hour cycle, whilst an atomic source functions the whole time.

TABLE 7.5[7]

ESTIMATED EDGE-DEFINED FILM-FED GROWTH RIBBON COSTS

Based on successful process development, volume of greater than 1000 km^2/year; 1972 dollars

Ribbons grown simultaneously	1	5	20
Ribbon width, cm	15	8	4
Ribbon growth rate, m/hr	1.5	1.5	1.5
Growth yield, percent	0.80	0.85	0.90
Area growth rate, m^2/hr	0.18	0.51	1.08
Ribbon furnace	$25,000	$40,000	$100,000
Depreciation cost, \$/m^2	5.56	3.14	3.70
Personnel cost, \$/m^2	7.39	5.59	5.55
Silicon cost, \$/m^2	8.72	8.32	7.78
Services, supplies, \$/m^2	1.11	0.63	0.74
Direct costs, \$/m^2	22.78	17.68	17.77
Indirect costs, \$/m^2	10.38	6.87	7.42
Total ribbon costs, \$/m^2	33.16	24.57	25.19
Ribbon cost, ¢/cm^2	0.33	0.25	0.25

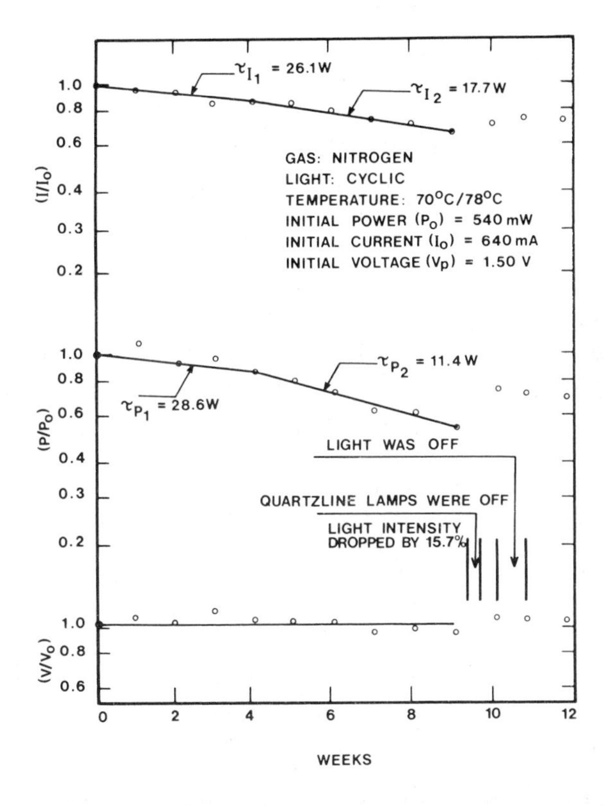

Fig. 7.6 Typical degradation of CdS/Cu_2S solar cells at 78 to 95°C with cyclic illumination.[16]

LONG LIFE-TIME CADMIUM SULPHIDE PHOTOVOLTAICS

These have been reviewed by Boer.[13] He reports that CdS/Cu_2S layers might be made in which the lifetime is more than 10 years, so long as the films are encapsulated. The conversion factor is 5%. Encapsulation is between two glass sheets, with hermetic sealing. They may be filled with N_2. Two copper foils protrude to provide electrical contacts. The cells would be mounted on roofs of houses, over parking lots, roadways and particularly in desert areas.

Life-time of Cadmium Sulphide Components

10 to 20 years life is the aim. The criticism of CdS/Cu_2S cells is that they have a shorter life than is acceptable for practical purposes. Cadmium sulphide degrades under insolation in moist air. However, after encapsulation against moisture, and if an inert gas such as nitrogen is circulated over the cells[14], life testing and extrapolation on a log (rate of decay) basis, predicts a rate of loss of efficiency which would give a life of up to 20 years. Bismuth doping has indicated an increased high temperature stability[15]. Part of the degradation arises from ion migration and diffusi(which can be reduced by doping near the p-n junction.

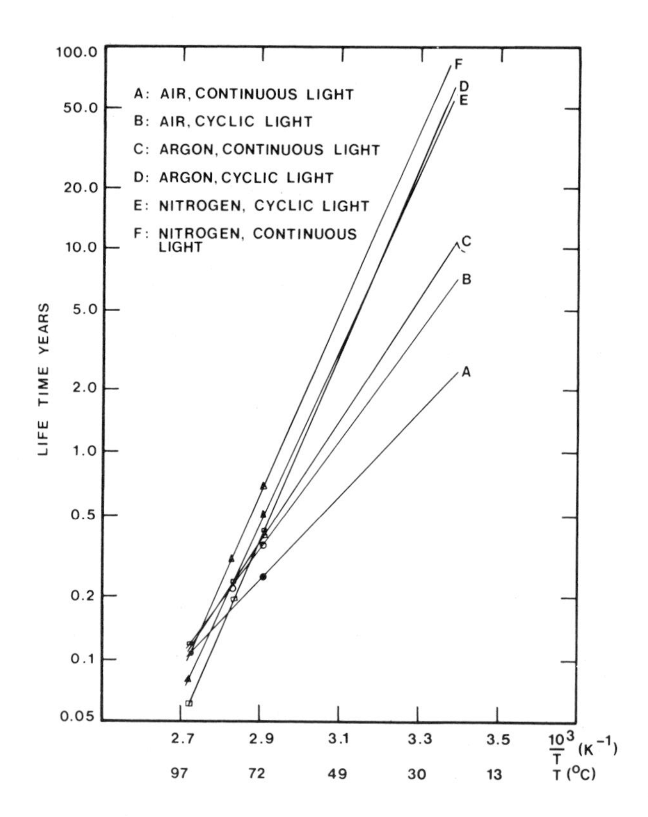

Fig. 7.7 Effect of ambient gas on the lifetime of CdS/Cu_2S solar cells. Temperature range 70 to 95°C.[16]

Examples of degradation when the cells are encapsulated in nitrogen is given in Figs 7.6 and 7.7.[16] In the absence of O_2 and water, one can anticipate usable lives of greater than 20 years at 50°C and greater than 100 years at 25°.

Cost Estimate for Cadmium Sulphide Cells
Alternatives have been described for the mass fabrication of solar cells, e.g. photo-etching of grids or using glass shingles of about 1 cm thickness. The deposition rate of cadmium sulphide layer occurs at about 1 μm per sec., and Table 7.7[13] gives a variation in costing due to Aaron and Isakoff.[17]

TABLE 7.6[16]

SUMMARY OF CADMIUM SULPHIDE CELL LIFETIME
(Cf. Fig. 7.7)

(Calculated from [extrapolated] rates of decay measured at higher temperatures)

Conditions	at 25°C	at 50°C
1. Air, continuous light	2.1 years	.67 years
2. Air, cyclic light	5.4 years	1.2 years
3. Argon, continuous light	8.0 years	1.5 years
4. Argon, cyclic light	38.0 years	3.2 years
5. Nitrogen continuous light	32.0 years	3.2 years
6. Nitrogen cyclic light	60.0 years	5.1 years

TABLE 7.7[13]

PROJECTED COSTING FOR THE PRODUCTION OF CdS/Cu_2S SOLAR CELLS*

	Concept 1 (film)	Concept 2 (film)	Concept 3 (glass)
Essential materials $/sq. ft.	1.03	0.90	1.05
Labour and supervision	0.55	0.25	0.47
Maintenance, utilities, taxes) insurance, depreciation) $/sq. ft.)	0.13	0.13	0.36
	1.71	1.26	1.88
Mill Cost			
Total investment ($/sq. ft./yr)		(1.52)	(2.90)
Selling, administration, research	0.23	0.18	0.28
Operating earnings (10% of working capital)	0.17	0.13	0.19
Operating earnings (10% on perm. investment)	0.17	0.19	0.42
$/sq. ft.	2.28	1.76	2.77

* The evaluation is based on an assumed volume of production of 100 million sq. ft/yr.

Boer considers that it will be possible to reduce the costs below those of Aaron and Isakoff[17] within five years (from 1973). They would come to $1.00 per sq ft (*c.* $10 per sq m, i.e. about $200 per kW).* The cost of a fault indicator and short protection system might be $40 per kW, equivalent to adding 27 cents per sq ft. Amplification by focussing has not been taken into account.

The overall costs, including installation, are shown in Table 7.8[13].

Using the installation costs, fixed charges and other financing charges, it is possible to evaluate (Boer[13]) the cents per kWH. It is 1.2-4.2 cents per kWH (1973), i.e. in average, about twice the range being paid in the U.S.A. by consumers for electric power in 1974.

TABLE 7.8[13]

TOTAL COST OF ELECTRIC POWER INSTALLED USING CdS-Cu_2S PHOTOVOLTAICS

System	A Sc	B Sc	C Sc	D Sc	AS	BS	CS	DS
Total costs $/kW	236	286	366	386	323	373	453	473

	$/ kW for Total System	
	Shingle	Continuous Sheet
A. Use of dc only	(SA)	(CsA)
B. Use of 115 Volt ac	(SB)	(CsB)
C. Use of 115 Volt ac with some storage	(SC)	(CsC)
D. Use of phase controlled synchronous dc to 115 V ac inversion	(SD)	(CsD)

Use of Combined Systems

Boer[13] has suggested solar panels, which are also heat collectors. There is a network of CdS/Cu_2S thin film solar cells between two glass plates. This sandwich is cemented to a plexiglass plate and backed-up with a styrofoam profile. Panels about 3 ft × 4 ft will produce about 110 volts at 0.6 amps in full sunlight. Were the cells sealed in inert gas, a 20 year lifetime is predicted by Boer.[13]

PHOTOVOLTAICS: HIGHER EFFICIENCIES

The presentation made in this chapter on photovoltaics has been concerned hitherto with costs and lifetime. Efficiency increase has been less stressed, because theory indicates (see Chapter 7) that an increase in efficiency of more than about twice times those commonly obtained with photovoltaics is not to be expected, whereas cost reductions for Si cells of several orders of magnitude seem possible. However, a doubling of the efficiency from the 5-8% of CdS and the 10% of Si would clearly be of interest.

---

* In photovoltaic and photothermic solar energy collecting devices, the cost in $ per kW cannot be compared directly with that of fossil fuel or nuclear devices. It is the final cost of a unit of energy, delivered to the consumer, which counts. Nuclear and fossil fuel devices can be worked 24 hours per day and solar only during the sunlight hours; and the variation due to weather must be taken into account. Conversely, solar devices have no fuel costs.

Woodall and Hovel[18] have reported efficiencies of 16% for heterojunction solar cells consisting of $pGa_{1-x}Al_xAs$-pGaAs-nGaAs. 20% would be obtainable in space. The authors attribute the higher efficiencies of these couples to a lesser surface recombination, which occurs as a result of the influence of a heavily doped $Ga_{1-x}Al_xAs$ layer. Open circuits give up to 1.0 volts. When the solar input is 98.3 mW cm^{-2}, the current density is 18-21 mA cm^{-2}. Couples are 6-7μ thick.

Large scale use of Ga would not be feasible because of a lack of the material.

Similarly, Loferski, Crisman, Chen and Armitage[19] have tried to obtain improvements in efficiency by covering the cell with fine collector junctions, separated by a distance small compared with the minority diffusion length. n-Si wafers were used and onto these a grating pattern of Al was introduced. Results were not improved compared with the conventional Si cells but the response was 'blue-shifted'.

AIMS OF RESEARCH AND DEVELOPMENT IN PRACTICAL PHOTOVOLTAICS

The cost projections reported here are based upon the assumption of very large scale use. *No components at prices near the estimates reported are commercially available at this time.* The considerable improvement in the *projected* situation between 1970 and 1975 arises because:

(a) An analysis of the silicon solar converter[4] has shown that, by invoking cost reduction effects for massive use, cost savings of about one order of magnitude could occur.
(b) New crystal growth technology has been developed which could lead to at least one further order of magnitude reduction in costs.[7,10]

However, there has been hitherto no experience in building solar arrays apart from the small ones used in space vehicles. Prediction is poor until pilot plant stage has been reached. Experimental knowledge of lifetime is very limited.

Researches now being funded at the fundamental level are exemplified by the photosynthetic production of hydrogen from nitrogenase in water.[20] Were this method successful, the best projections for photovoltaic and photothermal developments would both be uneconomic. At the developmental level, can single crystals grown in the edge-defined-film-fed-growth method be made with an impurity level (e.g. impurities picked up from the dye)? At the materials science level, an improvement needs to be made in lifetime studies, for example, what would be the lifetime of the Si photocells generally proposed? The funding of research in these branches of solar work depends on parallel advances made in other areas, such as thermo-electrics.[22]

Basic research in solar energy is still thin and new. Simple research on the spectroscopic and photovoltaic properties of new compounds for photovoltaics is still a big field, cf. the work of Loferski *et al*[23] on Cu_2S and $CuInS_2$. If the work is sufficiently intensive and on a world scale, it may be possible by the late 70s to pick out major directions for intensive development work and pilot plant building.

In respect to level of research funding, the present level in the U.S.A. (e.g. about $20 million Government funding in the U.S. in 1974) is small compared with investment in atomic energy ($500 million in the U.S. in 1974). Even in Australia, where a larger amount of solar energy falls compared with other countries, atomic energy development received *c.* 40 times more funds than did solar energy in 1974; and the principal (but small) government-sponsored development was in low-grade solar power. The policy of the Australian Government (1974), when the exhaustion of the indigenous oil was 7-15 years away, was *not* to research the production of electricity and transportation fuel from solar energy, but to consider the gassification of coal.

THE POSSIBLE COLLECTION OF SOLAR ENERGY FROM PLATFORMS IN ORBIT[24-26]

Introduction

The greatest drawback to solar energy collection on earth is its diluteness. A concept due to Glaser[24], entitled 'Satellite Solar Power Station', attempts to overcome this. The satellite is in synchronous orbit at a height such that it remains illuminated 24 hours per day. The satellite would be largely covered with photovoltaic collectors. The satellite's central part would condition the electricity produced to microwave and in this form the energy would be beamed back to the earth. The collection there would be on a floating platform or upon a collecting centre near a town. It could be stored by conversion to hydrogen and the hydrogen used as the medium of the energy concerned.

The orbit[25]

Glaser[24] suggests a stable node of a synchronous equatorial orbit. Such a collection would pass into the earth's shadow only around the time of the equinoxes, when it would be eclipsed for 72 minutes per day.

Advantages in concentration of solar energy[26]

With the orbit mentioned, the advantage is 6-15 times compared with the amount received per unit area of an earth collector. This advantage helps to compensate the principal disadvantage – the cost of putting the platform into orbit.

The weight[24]

It would be important not to use silicon for the present concept, because Si is a 'thick photovoltaic' and the weight (vital to the cost per kW of an orbited collector) would be some ten times greater than that of a thin film photovoltaic. Thus, Si cells can be reduced to 50-100 microns thick, but gallium-arsenide cells need be only a few microns in thickness. Concentration by mirrors could help the situation by twice times. Glaser submits that about two pounds per kW would be a possibility for the solar cell *array* (i.e. including circuitry and transducing equipment). Thus, a 10,000 MW converter would weigh $20 \cdot 10^6$ lbs on about 10 thousand tons. It can at once be appreciated that the orbiting of such a tremendous body would have to be done in small steps and that it is not to be regarded as feasible until the cost of putting something into orbit is reduced by an order of magnitude from the present $5,000 per lb.

Microwave power generation[26]

The frequency selected by Glaser would be in the region of 3.3 GHz. Brown showed in 1963[27] that microwaves could transmit much power. The reception has to be for the main lobe, 90% of the power, and the side lobes can be controlled.

The ionosphere only slightly absorbs microwaves (0.1%).[28] Troposphere absorption would be 1%. Rainfall attenuates microwave radiation about 3%. Overall efficiency, including some rain, would be >75%.

The efficiency is shown in Table 7.9[26].

TABLE 7.9[26]

MICROWAVE POWER TRANSMISSION EFFICIENCIES

Characteristic	Efficiency		
	Presently demonstrated*	Expected with present technology*	Expected with additional development*
Microwave power generation efficiency	76.7**	85.0	90.0
Transmission efficiency from output of generator to collector aperture	94.0	94.0	95.0
Collection and rectification efficiency (rectenna)	64.0	75.0	90.0
Transmission, collection and rectification efficiency	60.2	70.5	85.0
System	26.5# =	60.0	77.0

* Frequency of 2450 MHz (12.2-cm wavelength).

** This efficiency was demonstrated at 3000 MHz and a power level of 300-kW CW.

This value could be immediately increased to 45% if an efficient generator were available at the same power level at which the efficiency of 60.2% was obtained.

The microwave generator would be a cross field device.[29] A pure metal, self-starting, secondary, emitting cold cathode gives broad band gain. Use of samarium-cobalt alloys would reduce weight, but perhaps introduce problems of availability. The individual microwave generator would weigh fractions of a kg per kW of power output.

The diameter of space antenna would be about 1 km in diameter and the receiving antenna should be 7 km in diameter for 90% transmission. Very important would be a master phase control to keep the beam on target.[30]

Keeping station

For an orbited satellite collecting 5,000 MW at 1 kW per sq m, the area would be about 4.10^7 sq m, and could consist of two panels each 4.3 by 5.2 km. It would have to keep station exactly and need fuel to do this. The propellant needed would weigh about 30,000 lbs per year.[26]

Economics

An economic analysis for the orbited satellite depends much more upon future projections than with other solar collectors. Glaser gives the following suggestions for a 1990 projection (1973 dollars).[26]

Solar cells and solar collector array	310 dollars per kW generated by the prototype
Microwave generators and transmitting antennae	130
Rectifiers and receiving antennae	100
Transportation to orbit and assembly	1380-800

The projected value of about $1,600 per kW *assumes* a reduction of about one order of magnitude in the cost of putting substances into orbit. The projection is more favourable than at first sight because of the absence of diurnal and weather factors.

The main doubts depend on the reduction of the cost of putting materials into orbit; and the availability of sufficient thin film photovoltaic materials. It seems that at present the first hurdle might be less of a problem than the second.

Summary

The satellite concept for collecting solar energy is very attractive because of its future-look and large-minded solution to the energy problem. It depends for its feasability on a lowered cost of orbitting and the availability of thin film photovoltaic materials in sufficient quantities. At present, one can affirm only that the approach *does* come into the alternatives for a technology which will not be built on a large scale for one to two decades.

ECONOMICS OF THE PHOTOTHERMAL APPROACH

Economic projections are more uncertain in the photothermic than the photovoltaic method. This is due to lack of experience of lifetime of the components. Two alternative assumptions can be used[21]: (a) a 5 year lifetime system, and (b) a 30 year lifetime system. An early cost estimate for each is given in Table 7.10.[21] Thus, costs vary from about $100 to about $1,000, but do not allow for a storage system.

THE MIRROR CONCENTRATOR METHOD FOR THE CONVERSION OF SOLAR ENERGY TO HEAT

The solid state physics (and cost) involved in the development of photovoltaic cells can be avoided by using tracking mirrors to concentrate solar energy on to a small target and work a heat engine. An arrangement of this kind was proposed by Lenitske[31] in 1949. The qualitative view of this approach is shown in Fig. 7.8[32] (see also Fig. 5.5[32]).

TABLE 7.10[21]

MEINEL'S ESTIMATES OF THE COST OF PHOTOTHERMIC DEVICES

Short lifetime system	
Lifetime to replacement	5 years
Loss mass	4-8 kg/m^2
Low component cost	1.00 \$/kg
(High maintenance cost during 5 years)	
System efficiency	0.040 W/m^2
Initial collector cost	4-8 \$/m^2
5 year collector cost	28-56 \$/m^2
Initial collector cost	100-200 \$/kW
Long lifetime system	
Lifetime to replacement	35 years
High mass	15-30 kg/m^2
High component cost	1.50 \$/kg
(Low maintenance cost during 35 years)	
System efficiency	0.040 W/m^2
Initial collector cost	22.5-45 \$/m^2
Initial collector cost	560-1120 \$/kW

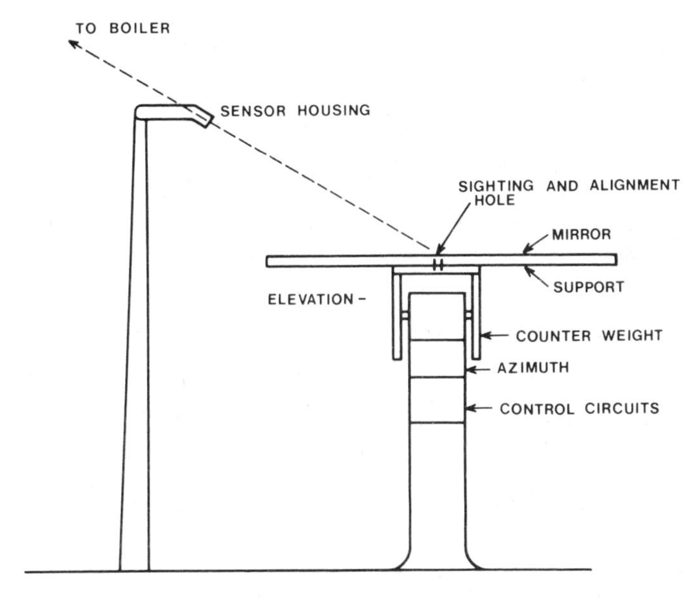

Fig. 7.8 Diagram of a single-mirror pedestal system.[32]

Henderson and Dresser[33] have studied a modified Fresnel reflector which consists of a number of planar segments and have found that the performance approached that of parabolic segments if the number of serrations were large enough.[32] The local atmospheric conditions have to be taken into account.[34]

Hildebrandt *et al* [32] have based the schemes which they propose on small heliostats (3-5 metres square) mounted on a pedestal and positioned by an automatic control system (Fig. 7.9[35]). The concentrator would be flat and would image the sun on to a fixed surface. To avoid image spreading, the front surface of the mirrors must be curved by 0.05 cm.[32]

The mirror reflector material should be abundant and inexpensive as it must be made upon a large scale. Aluminum is appropriate. The reflectivity is 85%.

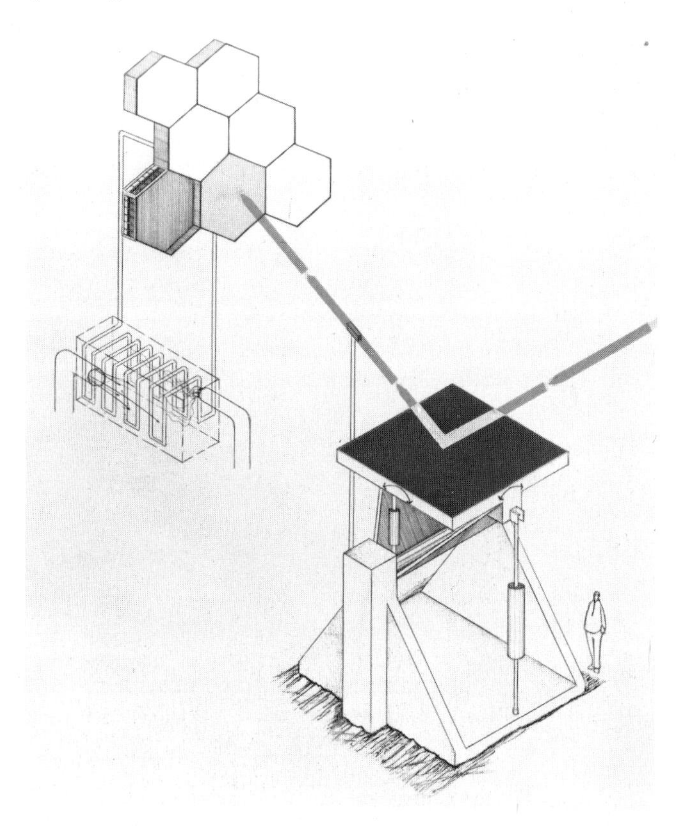

Fig. 7.9 15 ft square heliostat with optical sensor directing energy into a segment of the boiler.[35]

Large amounts of the material would be necessary. Hildebrandt[32] calculates that one would use about 10^8 tons of aluminum to produce 10^{20} Joules (U.S. Al production in 1969 was $3 \cdot 10^6$ tons). The energy needed to make this aluminum would be about one-tenth of the energy produced per day by the set up and is thus tenable.

The protection of the aluminum surface from the atmosphere[36] could be achieved by plastic film covering, and there would have to be a corresponding washer-wiper arrangement. Standard silver glass plate could also be used for the mirror and this has been used by Trombe[37] in solar furnaces in France.

The mirrors will concentrate their energy on a boiler, the height of which is given by Hildebrandt *et al*[32] as:

$$h \simeq \sqrt{\frac{4Q}{\pi E_o \rho \phi a^2}} \tag{7.1}$$

where h = the tower height in metres; Q = power concentrated at the boiler; E_o = incident solar radiation to the earth's surface in watts/m^2; ρ = efficiency of mirror utilisation; ϕ = effective area utilisation coefficient; a = scaling parameter (function of boiler temperature); $ah/2$ = radius of the concentrator area.

The height of the tower used must be chosen to optimise the solar energy provided: a tower with a constant fraction of the concentrator energy provided: a tower would optimise at about 450 metres in height, assuming a 2.6 sq km concentrator (Fig. 7.10[35]).

The boiler can be an MHD boiler or a more conventional one. Hildebrandt[32] evaluates conditions for both. Hydrogen would be the storage medium.[38 39]

Fig. 7.10 Square mile mirror array with boiler atop 1,500 ft tower.[35]

TABLE 7.11[32]

COST SUMMARY OF MIRROR CONCENTRATOR METHOD

Item	Cost (1972)
Square-mile concentrator (2.6 km^2) (coverage factor of 1/2 and mirror cost $20/$m^2$)	$26 $\times 10^6$
Tower (450 metres)	$15 $\times 10^6$
5% Interest for 30 years (amortized)	$41 $\times 10^6$
Total for concentrator	$82 $\times 10^6$
Fuel (gas at $0.22 per 1,000 cubic feet and 40% efficiency)	$0.0019/kWH
Concentrator (effective fuel cost)*	$0.0063/kWH
Electrolytic production of LH_2 due to concentrator alone at 20% efficiency overall	$0.16/lb

* Total electrical energy per sq. mile per 30 years (at 28% efficiency) equal to 1.3×10^{10} kWH.

Costs of the mirror concentrator method
Hildebrandt *et al*[32] gives a table which is their cost summary and this is reproduced in Table 7.11. Complete automation could reduce manufacturing cost of the mirrors by ten times. Conversely, the method must have the solar disc in view so that the hours it can operate are less than a method which can function on diffuse light.

SOLAR SEA POWER TECHNOLOGY

The definition of what solar sea power is comes from the work of Anderson and Anderson.[40] The Gulf Stream is the body of water usually envisaged, because of its movement, and the avoidance thereby of cooling effects between surface and bulk (which would take place were heat removed in between hot and cool parts of a static system).

Analysis of the temperature gradients in bodies of water of this kind show that it may go as low as 40°F for several months during the year.

In Fig. 7.11[41] is shown an area-power output plot as a function of ΔT. At what temperature is the optimisation of the operation of the cycle? If one runs with a ΔT between the working fluid and the water in both the boiler and the condenser as small as possible, the thermal efficiency of the cycle is small. Increase of ΔT makes one have a large heat exchanger (Fig. 7.11). As a compromise, there should be a fairly large ΔT between the working fluid and the ocean fluid. If the overall heat transfer is greater than 5,000 to 10,000 Btu/ft^{2}hr°F – then a ΔT between 5-8°F could be used. An overall heat transfer coefficient of only 300 to 500 Btu/ft^{2}hr°F can be expected with technology practical today.

Some typical turbine operating parameters (cf. Zener *et al*[42]) as suggested by Heronemus and McGowan are shown in Fig. 7.12.[41]

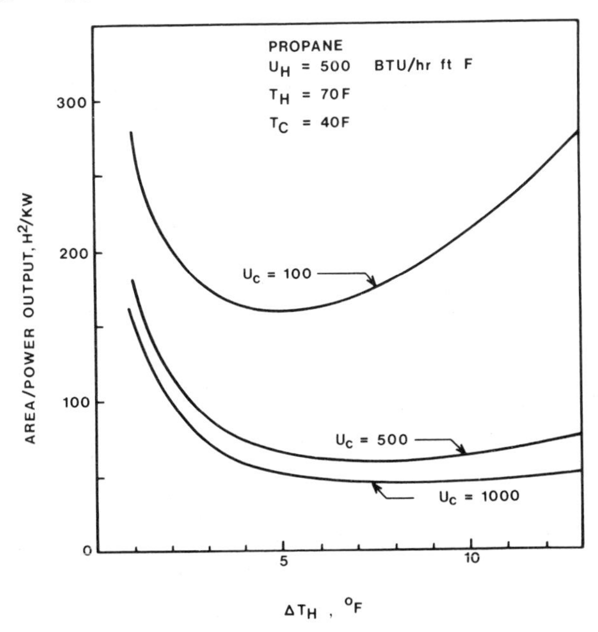

Fig. 7.11 Optimum area (based on T_c) as a function of ΔT_H.[41]

AMMONIA	(25 MW)
NOMINAL SPEED	1800 RPM
NOMINAL DIAMETER	85 IN
TURBINE EFFICIENCY	90%

SINGLE STAGE UNITS WITH DIFFUSERS

PROPANE	(35 MW)
NOMINAL SPEED	600 RPM
NOMINAL DIAMETER	95 IN
TURBINE EFFICIENCY	90%

SINGLE STAGE UNITS WITH DIFFUSERS

Fig. 7.12 Turbine operating parameters.[41]

The two most critical problems[41] in the engineering are the cold water pipe design and installation concept and also how many square feet will be required for the heat exchanger per unit of power.

A deep site near to shore is superior to a site far from the shore. Heronemus and McGowan[41] suggest the possibility of moving the power plant to the shore and bringing cold water up through a tunnel. On the other hand, the amount of water flow needed for a substantial plant is enormous, for instance, 30 million gallons per minute for 400 MW. In spite of this, the pumping work would be less than 10% of the energy produced.

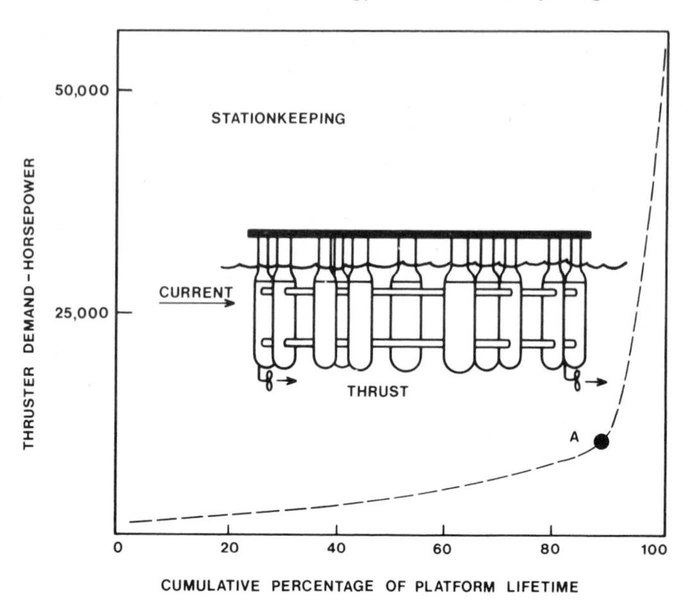

Fig. 7.13 Power demands for a dynamic positioning system.[43] (Calculations performed by G. Rothwell, Oceanic Foundation.)

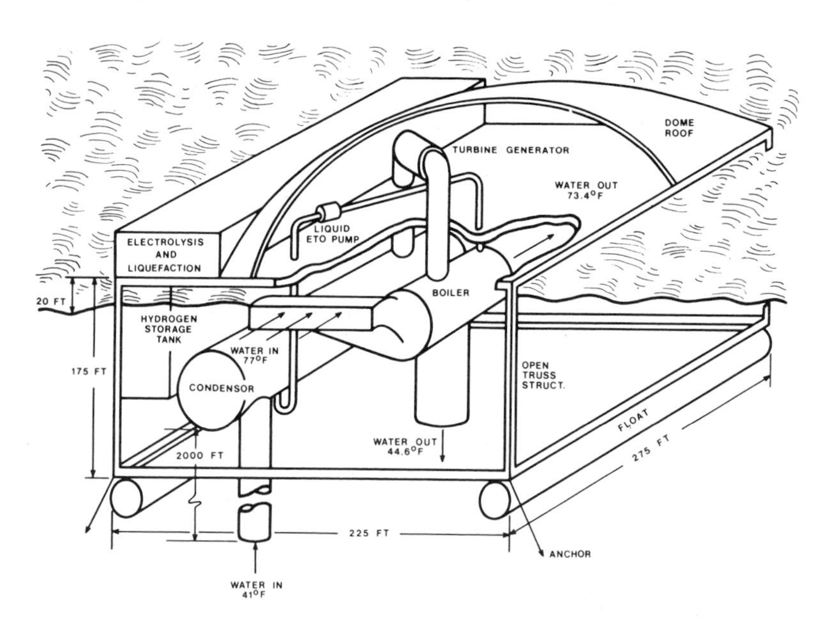

Fig. 7.14 Solar sea power plant schematic.[44]

Suitable ocean platforms for these offshore systems have been discussed by Hanson.[43] The theoretical hydrodynamics of such systems, how to keep them stationary in the presence of currents, tides, etc., has been researched.[43] There are trade-offs between passive mooring systems and dynamic positioning systems. Since passive systems would have to be designed for the worst possible conditions, and would require a high safety factor, they would be expensive.

An active station keeping system offers advantages (Fig. 7.13[43]). It would not require design for most extreme conditions with a large safety factor, for the station could 'give ground' and return to its original position when conditions were less severe. (See also Figs. 7.14[44] and 7.15[45].)

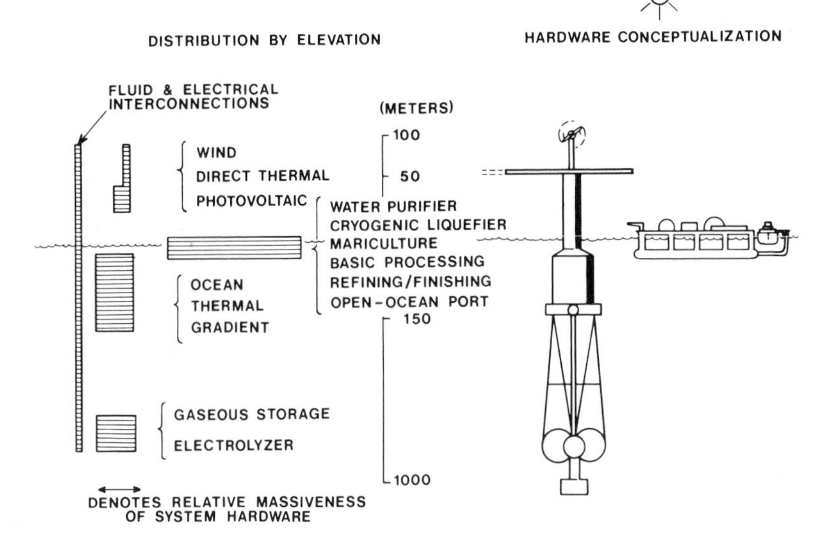

Fig. 7.15 Typical systems location. Wind generators could be included.[45]

TABLE 7.12[40]

COST SUMMARY OF SOLAR SEA POWER

Main plant		\$9,560,000
Auxiliaries		443,200
Structure and assembly		4,210,000
Assembly of cold pipe		262,000
Engineering and supervision		724,000
Contingency		1,448,000
	Total	\$16,647,000

Cost per kW = \$166.00
Yearly owning and operating cost = \$1,870,000
Rated yearly capacity = 876 million kWH
Estimated yearly output = 650 million kWH
Cost per kWH = \$0.00285

Much attention has been paid to the working fluid.[45] One may have[44] ammonia, several hydrocarbons, freons, carbon dioxide, methyl chloride, ethylene oxide, etc.. Ammonia has an advantage over propane from

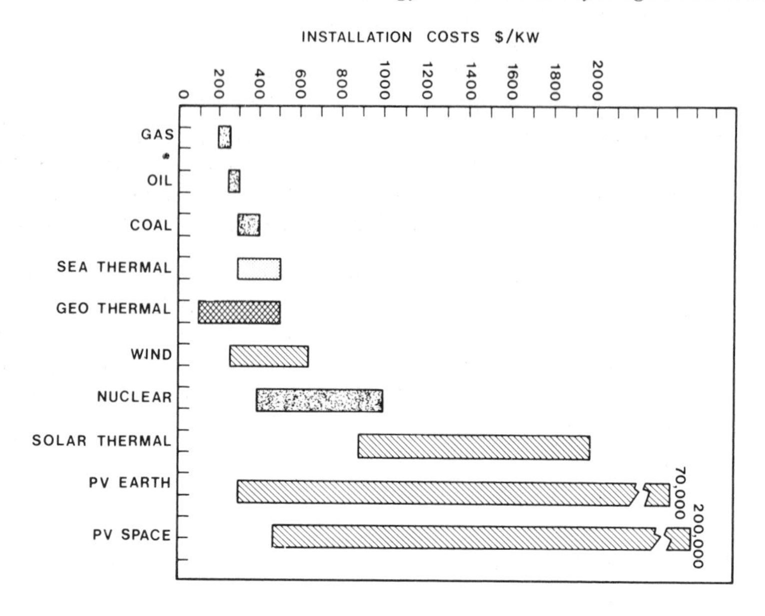

Fig. 7.16 Estimated installation costs for electric generating plants.[46]

the point of view of its thermal properties, though it has a greater toxicity, a disadvantage in view of the fact that some leaks must be expected.

Economics of Sea Solar Power

The economic system in detail is given by Anderson and Anderson.[40] The dollars are 1966 and the cost summary is given in Table 7.12.[40] Power could be produced at less than 5 mils (1974) at the plant.

In Fig. 7.16[46] are shown cost comparisons for various methods of solar conversion. According to a projection made by R. Cohn[40a], solar sea power should become commercial in the 1980s.

HYDROGEN

If the distance between plant site and points of use is more than about 500 miles, the transfer of hydrogen will be cheaper than the passage of the same amount of energy in the form of electricity through wires (Chapter 8). The possibility of the passage of hydrogen over long distances makes it reasonable to consider the high insolation areas of the world (North Africa for Europe, Australia for South-East Asia and the Caribbean region for the Eastern U.S.A.) as potential world energy depots of the future. Si photovoltaic arrays in such areas would fit in with the production of hydrogen by electrolysis at the solar array sites, although the problem of sufficient water for electrolysis may provide difficulties (desalination?). The direct production of hydrogen by photosynthetic (or photoelectrochemical) means is an attractive prospect for highly insolated areas.

APPENDIX

THE EDGE-DEFINED-FILM-FED-GROWTH (EFG) OF SINGLE CRYSTAL RIBBONS

The edge-defined-film-fed growth (EFG) was carried out by LaBelle and Mlavsky[11] for sapphire growth, i.e. the formation of single crystal alumina.[47] The basic apparatus uses the principle of capillary rise, the molten alumina rising in a molybdenum capillary or between parallel molybdenum plates, see Fig. 7.5.[11]

The use of the capillary gives an increased element to mechanical stability compared to other techniques. The temperature at the top of the capillary is close to the melting point and an α-Al_2O_3 seed crystal of the desired orientation is dipped into the melt. As the seed is pulled out, the curvature at the solid-liquid interface is increased and there is a reduction of pressure in the liquid by:

$$\Delta p = \gamma \left\{ \frac{1}{R_1} - \frac{1}{R_2} \right\} \tag{7.2}$$

If the T conditions are adjusted properly, the filament solidifies at a rate at which it is pulled and Δp will remain approximately constant, and the supply of liquid will continue.[11 48–50]

The procedure for silicon is that the silicon rises after melting and spreads across the top of two parallel plates. This defines the breadth of the ribbon which is pulled out by the seed. Crystals of many feet in length should be growable.

A theoretical account of the process is given by Surek and Chalmers[51] and by Chalmers *et al.*[10] The process has difficulties, the main one of which is that, starting with highly purified silicon, impurities are picked up from the plates between which the liquid moves. To avoid this, the plate sides could be cooled and the liquid move between a layer of silicon; or the plates could be made of material containing silicon, e.g. CSi or $ZrSiO_4$.

REFERENCES

1. A.E. Spakowski & I. Shure, NASA Technical Memo. TM-X-2520, March 1972.
2. P.E. Glaser, *Science,* **162,** 857 (1968).
3. A.L. Hammond, *Science,* **173,** 733 (1972).
4. E.L. Ralph, 'A Plan to Utilise Solar Energy as an Electric-Power Source', Proceedings of the 8th Photovoltaic Specialists Conference, Seattle, Washington, August 1970.
5. P.E. Glaser, *Rotarian Magazine,* September 1973.
6. E.L. Ralph, *Solar Energy,* **14,** 11 (1972).
7. C.G. Currin, K.S. Ling, E.L. Ralph, W.A. Smith & R.J. Stirn, 'Feasibility of Low Cost Silicon Solar Cells', Proceedings of the 9th IEEE Photovoltaic Specialists Conference, Silver Spring, Maryland, May 1972.
8. 'Solar Cells: Outlook for Improved Efficiency', National Academy, Washington, D.C., 1973.
9. M. Wolf, 'Cost Goals for Silicon Solar Arrays for Large Scale Terrestial Applications'. Proceedings of the 9th IEEE Photovoltaic Specialists Conference, Silver Spring, Maryland, May 1972.
10. B. Chalmers, T. Surek, A.I. Mlavsky, R.O. Bell, D.W. Jewett & J.C. Swartz, NSF Report/RANN/SE/GI-30767X/PR/73/2, September 1973.

[11] H.E. LaBelle & A.T. Mlavsky, *Mat. Res. Bull.*, **6,** 571 (1971).
[12] F. Collins, National Science Foundation Memo., May 1974.
[13] K. Boer, 'Direct Solar Energy Conversion for Terrestrial Uses', Proceedings of the 9th IEEE Photovoltaic Specialists Conference, Silver Spring, Maryland, May 1972.
[14] F. Shirland, 3rd Conference on Large Scale Energy Conversion for Terrestrial Use, Delaware, October 1971.
[15] J.G. Haynes, Proceedings of the 8th Photovoltaic Specialists Conference, Seattle, Washington, August 1970, p. 184.
[16] L. Partain & M. Sayed, 'Accelerated Life Test of CdS Cu_xS Solar Cells', NSF Grant No. G134872, May 1973.
[17] H.A. Aaron & S.E. Isakoff, 3rd Conference on Large Scale Solar Energy Conversion for Terrestrial Use, Delaware, October 1971.
[18] J.M. Woodall & H.J. Hovel, *Appl. Phys. Letters*, **21,** 379 (1972).
[19] J.J. Loferski, E.E. Crisman, L.J. Chen & W. Armitage, 'Methods of Improving the Efficiency of Photovoltaic Cells', 7th Semi-Annual Report on NASA Grant No. NGR40-002-093, 1973.
[20] J.R. Benemann, *Hydrogen Production from Water and Sunlight by Photosynthetic Processes*, University of California, San Diego, La Jolla, California, December 1973.
[21] A. Meinel, 'Research Applied to Solar-Thermal Power Conversion', Vol. 1, Executive Summary, NSF/RANN/SE/GI-30022/FR/73/1, 1973.
[22] T.L. Nystrom & J.P. Angello, 'High Power Density Electric Power Sources', Proceedings of the Intersociety Energy Conversion Engineering Conference, 1971.
[23] J.J. Loferski, A. Wold, J. Schewchun, R. Arnott, E.E. Crisman, S.D. Mittleman, G. Adegboyega, R. Beaulieu, P. Russo, H.L. Hwang, A. Loshkajian & C.C. Wu, 'Investigation of Thin Film Solar Cells Based on Cu_2S and Ternary Compounds'. 1st Semi-Annual Progress Report, NSF/RANN/SE/GI-38102X/73/4, January 1974.
[24] P.E. Glaser, *Solar Energy,* **12,** 353 (1969).
[25] Hearing before the U.S. Senate, 31 October 1973, Committee on Aeronautical and Space Sciences.
[26] P.E. Glaser, *Astronautics and Aeronautics*, p. 60 (August 1973).
[27] W.C. Brown, 'Experiments in the Transportation of Energy by Microwave Beams', IEEE Intersociety Conference Record, Vol. 12, Pt. 2, p. 8-17, 1964.
[28] V.J. Falcone, Jr., *Microwave Power*, **5,** No. 4, 269 (December 1970).
[29] W.C. Brown, *Microwave Power*, **5,** No. 4, 245 (December 1970).
[30] Special issue on Active and Adaptive Antennae, IEEE Transactions of Professional Group on Antennae and Propagation, March 1964.
[31] cf. V.A. Baum, R.R. Aparase & B.A. Garf, *Solar Energy*, **1,** 6 (1957).
[32] A.F. Hildebrandt, G.M. Haas, W.R. Jenkins & J.P. Colaco, E. & S. Transactions of American Geophysical Union, **53,** 684 (1972).
[33] R.E. Henderson & D.L. Dresser, 'Solar Concentration Associated with Stirling Engine', in *Space Power Systems, Progress in Aeronautics and Rocketry*, Vol. 4, p. 219, Academic Press, New York, 1961.
[34] F. Bennett, *Solar Energy*, **9,** 32 (1965).
[35] A.F. Hildebrandt & L.L. Vant-Hull, 'A Tower-top Point Focus Solar Energy Collector', presented at the Hydrogen Economy Miami Energy (THEME) Conference. March 1974.
[36] A.P. Bradford & G. Haas, *Solar Energy*, **9,** 32 (1965).
[37] F. Trombe, *Solar Energy,* **1,** 9 (1957).
[38] F. Daniels, *Solar Energy,* **6,** 78 (1962).
[39] L.W. Jones, *Science,* **174,** 367 (1971).
[40] J. Hilbert Anderson & James H. Anderson, *Mech. Engr.*, p. 41 (April 1966).
[40a] R. Cohn, priv. comm., December 1974.
[41] W.E. Heronemus & J.G. McGowan, (University of Massachusetts, Energy Alternatives Programme), in 'Proceedings Solar Sea Power Plant Conference and Workshop', sponsored by NSF, Division of Advanced Technology Applications, Washington, June 1973, p. 21.
[42] C. Zener, A. Lavi & C. Chang Wu, 'Solar Sea Power', First Quarterly Progress Report on NSF/RANN Grant No. 39114, October 1973.
[43] J.A. Hanson, in 'Proceedings of Solar Sea Power Plant Conference and Workshop', sponsored by NSF, Division of Advanced Technology Applications, Washington, June 1973, p. 153.
[44] H.L. Olsen, G.L. Dugger, W.B. Shippen & W.H. Avery, in *ibid.*, p. 185.

[45] W. Escher, in *ibid.*, p. 96.
[46] J. Hilbert Anderson, in *ibid.*, p. 126.
[47] H.E. Bates, F.H. Cocks & A.L. Mlavsky, 'The Edge-Defined Film-Fed Growth (EFG) of Silicon Single Crystal Ribbon for Solar Cell Applications', Proceedings of the 9th Photovoltaic Specialists Conference, Silver Spring, Maryland, May 1972.
[48] H.E. LaBelle, *Mat. Res. Bull.*, **6,** 581 (1971).
[49] B. Chalmers, H.E. LaBelle & A.L. Mlavsky, *Mat. Res. Bull.*, **6,** 681 (1971).
[50] U.S. Patent No. 3, 591, 348, 6 July 1971.
[51] T. Surek & B. Chalmers, presented at the NSF Workshop on Photovoltaic Conversion of Solar Energy for Terrestrial Applications, 23-25 October 1973, Cherry Hill, New Jersey.

CHAPTER 8

Methods for the Transmission of Energy over Long Distances

INTRODUCTION

THE CONCEPT of a Hydrogen Economy (Chapter 2) originated in the realisation that, if hydrogen were the medium of energy, its transmission over long distances would be cheaper than the transmission in wires of an equivalent amount of energy. The advantages of hydrogen as a fuel for cars[1 2], for aircraft[3], for general use in industry[4] and the solution to the air and water pollutional problem[4], were pointed out independently.

The need to have long distance transmission of energy is shown in Fig. 8.1[5] and Table 8.1.[5] Thus, Alaskan coal would be uneconomic because of the cost of transporting it to centres of use. Were the coal gasified, and the gas sent to use centres, the difficulty would be overcome.

The energy sources of the future are likely to originate far from the centres of population. This is so in respect to coal, because the large undeveloped coal deposits (Alaskan, Australian and Siberian) will be far from population centres. For nuclear energy, a large distance from use centres would reduce pollutional difficulties. For solar energy, areas of maximum insolation are in North Africa, Saudi Arabia, Australia, the Kalahari Desert in South Africa and on the tropical seas.

There are three methods to be considered for energy transportation over 500 miles. They are:

Direct electrical transmission

The cost of transmission through cables depends on the transmission voltage. Potentials above 1 million volts are not feasible. Costs are cheaper in unsightly overhead electric cables. Underground passage of power involves a large increase in cost. The possibility of super-conductivity will be discussed.

Space relay of beamed power

It would be possible to beam from, say, Central Australia to parts of the United States, *via* satellites.[5]

Transmission of hydrogen through pipes

In the non-pollutional technology which must be built, electrical energy arriving from a distant region of the world (or directly from space) *via* microwave beams from a satellite, will probably be converted to

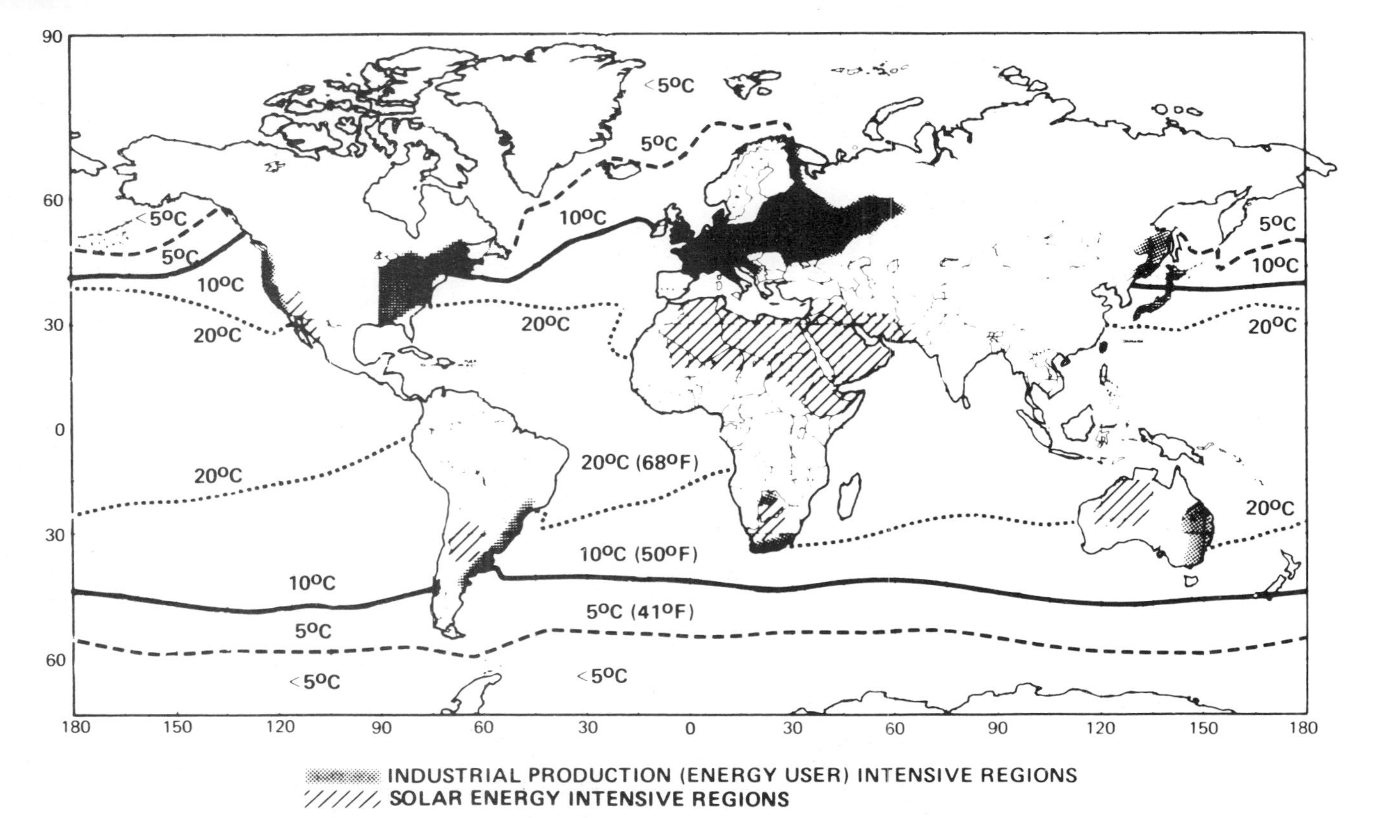

Fig. 8.1 Figure shows the great distances between areas of high insolation; and those of high concentration of affluent groups with manufacture.[5]

TABLE 8.1[5]

SUMMARY OF REASONS FOR REMOVING POWER PLANTS FROM POPULATED AGRICULTURE-INTENSIVE AND INDUSTRY-INTENSIVE AREAS WHICH IMPOSE A HIGH INTRINSIC BURDEN ON THE LOCAL/REGIONAL AIR-WATER-LAND-ENVIRONMENT

	For environmental burden reasons				For hazard reasons	For other reasons
Electric power stations	Air	Water	Land	Thermal		
Low-sulphur coal power stations		*	*		Solid waste	
High-sulphur coal power stations*	*	*	*		Solid waste	Remote geographic location of large bituminous coal reserves
Oil-fired power stations		*				
Nuclear converter power stations		*		*	Nuclear waste	Public fear of power plant failures
Nuclear breeder power stations*		*		*	Nuclear waste; Pu-239 shipping over over highways and by rail	Public fear of power plant failures
Solar power stations*						Remote geographic location of most of world's choice solar energy regions

* Asterisks indicate particularly strong reasons for remote siting.

hydrogen for storage before diffusion to industry, households and for use in transportation. Its *arrival* near the site of use in this form would be consistent with convenience in use, but also as will be shown, under certain conditions, more economic than alternate possibilities.

DIRECT ELECTRICAL TRANSMISSION

A typical loss in Power in a cable 3,000 miles long, consisting of copper, is 55% if the voltage of transmission is 600,000 volts.

As the loss is I^2R, it can be diminished by the use of higher voltage transmissions at lower currents for the same resistance. 700 kV can be used. 1000 kV is discussed and experimental set-ups exist. The difficulty is the corona discharge, which is dangerous in the vicinity of buildings, and electromagnetic effects on mechanisms in the neighbourhood.

One may consider lowering the temperature to reduce the resistance.

Were it economic to reduce the temperature to liquid hydrogen values, one could obtain super-conductivity, and the difficulty of losses in electrical transmission would be resolved.[6]*

In the U.S. there are about 3,000 electricity producing plants. These may evolve to some 300 super-large plants, with a capacity of 2000-3000 MW each.[7] According to Meyerhoff[8], it will not be possible to carry the electrical output from these plants in the normal way. Cables could perhaps be immersed in liquid nitrogen[9] or LNG could be run through a pipe keeping it at low temperatures and the electricity passed through the pipe.[10]

These possibilities are left to the references given and the use of superconductive cables considered. There are two types of superconductors, I and II. Type I superconductors are exemplified by lead, tin and indium. Their very low critical temperature makes it unlikely that they will be practical. Type II superconductors consist of alloys, typified by NbZr, NbTi. These have relatively high critical temperatures, are hence more acceptable. They suffer high losses in an a.c. field. Niobium is little available: however, a cable would consist of only .005 cm of Nb on a 1 cm copper tube.[8]

There are problems connected with the passage of a cable underground, although there is pressure on the utilities to increase the amount of underground transmission because of the poor public reaction to overhead lines. The increase in cost for underground transmission is shown in Fig. 8.2.[11] An underground superconducting power line is indicated in Fig. 8.3.[8]

Difficulties of progressing towards superconducting transmission are various.

Line Termination
When the superconducting line meets up with another conductor, it is carrying very large power per sq cm. The technique of a connection to a non-superconductor is unclear.

Dielectric strength
Materials which have sufficient dielectric strength to be in contact with superconducting cables are difficult to obtain. Pressurised helium or a

---

* The probability that superconductivity will be obtained at temperatures far above that of liquid hydrogen can be judged by a simple consideration of one aspect of the model for superconductivity. Due to their interactions with the lattice, electrons at sufficiently low temperatures *attract* each other and hence move in pairs. The individual electron is a fermion and hence two electrons together will be a boson. When there are many bosons in the same state, there is a large amplitude for the particles to be in the same state. Thus, nearly all the elections will be locked into the *same* state. But resistance is a measure of the difficulty to go from one state to another. Hence, resistance will disappear if all the electrons are in the same state, e.g. if all of them are paired. Now, the probability that the pairs will break up is connected with the factor: $e^{-E_{PAIR}/RT}$, and, in so far as this factor is finite, individual fermions will exist and cause resistance. Suppose $E_{PAIR} = 2$ ev, a change of T from 1° K to 10° K will cause an increase in the equilibrium fraction of pairs by *c*. 10^{39}; and 10° to 100° K about 10^{4} times. Thus the dissociation probability and hence the falling off of superconductivity, increases enormously when one departs from very low T.

vacuum have been suggested[8]: their use over very long distances may not be economic.

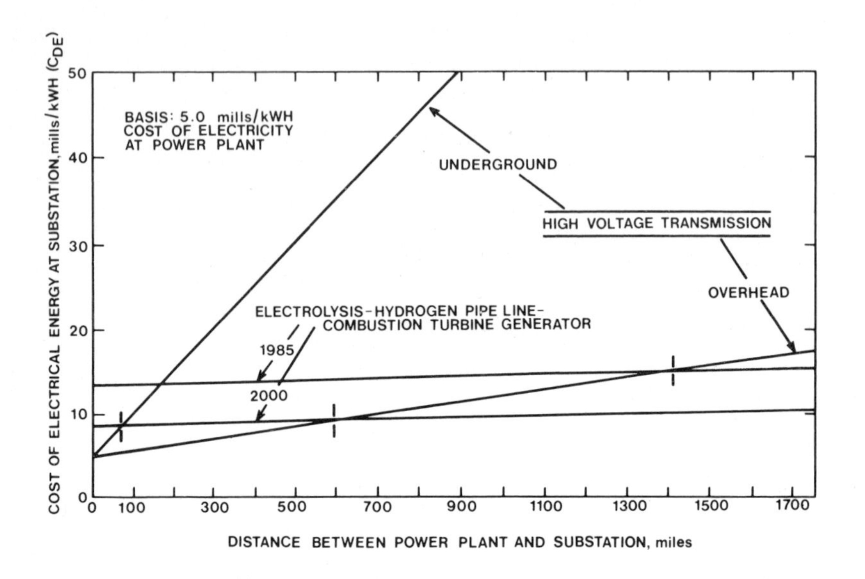

Fig. 8.2 Comparison of cost of electrical energy at distribution substation for electrical transmission v. hydrogen pipeline.[11]

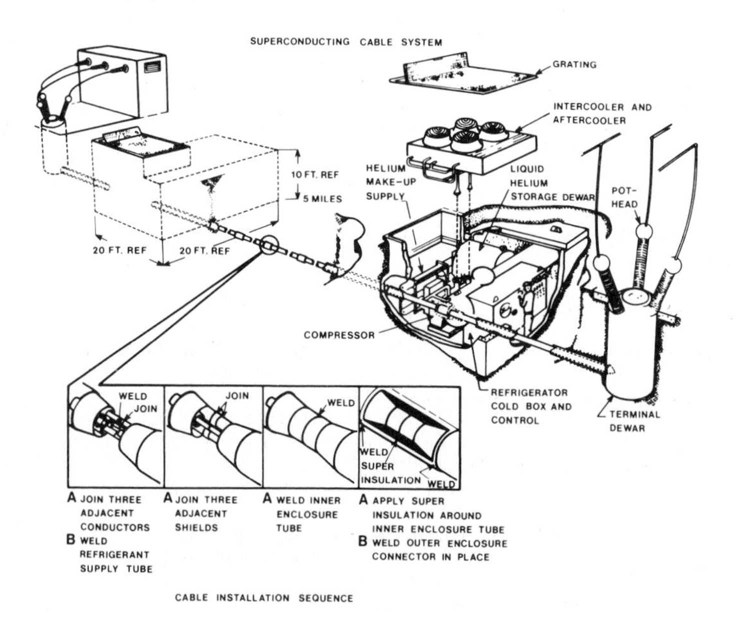

Fig. 8.3 Illustration showing the installation of an underground superconducting ac power line proposed by Union Carbide.[8]

Thermal contraction[8]

A method which allows for thermal contraction between conductors and the cryogenic envelope when the line is cooled has yet to be published.

TABLE 8.2[12]

ELECTRICITY TRANSMISSION COSTS

Data	Value Assumed	Ref. source
Average cost of five 500-kV lines built since 1969 – includes right-of-way, excludes cost of one urban line	$130,000/mile	25
Cost of d.c. line at 0.65 times the cost of comparable a.c. line	$84,500/mile	
Typical power capability of 500-KV line	900,000 kW	25
Typical a.c. terminal cost	$8/kW	25
Typical d.c. terminal cost	$30/kW	25
Average 500-kV overhead line cost	$0.144/kW-mile	
Average 500-kV overhead line cost	$42.32/$10^6$ Btu-hr-mile	
Average d.c. overhead line cost	$0.093/kW-mile	
Average d.c. overhead line cost	$27.50/$10^6$ Btu-hr-mile	
Cost of two terminals for a.c.	$16/kW	
Cost of two terminals for a.c.	$4,687/$10^6$ Btu-hr	
Cost of two terminals for d.c.	$60/kW	
Cost of two terminals for d.c.	$17,580/$10^6$ Btu-hr	
Cost ratio of underground to overhead power transmission	10:1 to 40:1	25
Cost of underground line	$1.44/kW-mile	
Cost of underground line	$423.2/$10^6$ Btu-hr-mile	
Total overhead line plus terminal costs for 200 miles	$0.112/kW-mile	25
Total underground line plus terminal costs for 200 miles	$1.68/kW-mile	25
Projected cost for 138,000-V superconducting line for 10 miles	$0.88/kW-mile $0.60/kW-mile	26 8
Projected cost of 345,000-V superconducting line for 10 miles	$0.20/kW-mile	8

Cryogenics[8]
A cryogenic capability which would cool lines with acceptable economics over very long distances has to be developed.

The cost of introducing superconductors seems prohibitive. They are summarised in Table 8.2 (Fig. 8.4[13]). The case for transmission with superconducting cables, on technical grounds, is weak compared with that of transmission in hydrogen; and the economic aspects prohibitive. Of course, this would change if high temperature superconductivity becomes practical.

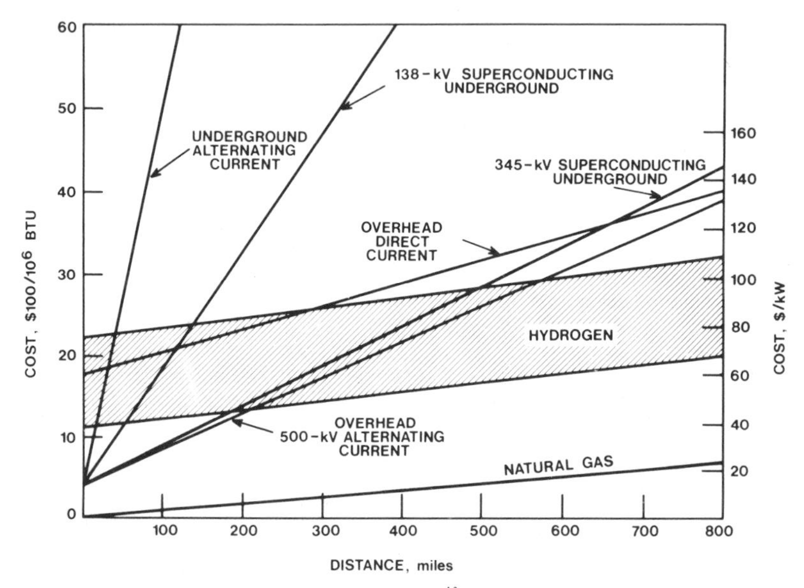

Fig. 8.4 Cost of energy transmission facilities.[13]

TRANSMISSION THROUGH DIRECTED MICROWAVE RADIATION

Glaser[14] has suggested that there should be space platforms which receive energy directly from the sun and beam it to earth (Chapter 7). An analysis of this system leads to a relatively expensive way of obtaining solar energy.[5] Thus, a mid-70s space shuttle can lift about 28 tons into earth orbit, and is planned to fly 60 times per year. To put up a 10,000 MW system, weighing about 10,000 tons, would take, then, about 6 years. To put up a *relay* satellite system which relays power from one continent to another would take less than one year on the same basis. It is desirable to keep the heavy and complex parts of the system on the ground – the lighter parts in space. Such considerations will apply particularly to the cost per kW, for, in the orbiting solar *collector* system, this is overwhelmingly the cost of putting the collector into orbit.

Because of the higher insolation of latitudes between 30°N and 30°S, and the availability of relatively calmer seas and very much flat desert there, it may become economically attractive to transfer massive quantities of solar energy over many thousands of miles. Ehricke[5] suggests that ground-level collected solar energy could be beamed up to a satellite, and then down to a use centre, several thousands of miles distant. The general concept of a world supply based on collectors in Australia, South Africa, Chile, etc., see Figs. 8.1, 8.5[15] and 8.6[15], see also Tables 8.3[16] and 8.4, becomes reasonable to analyse.

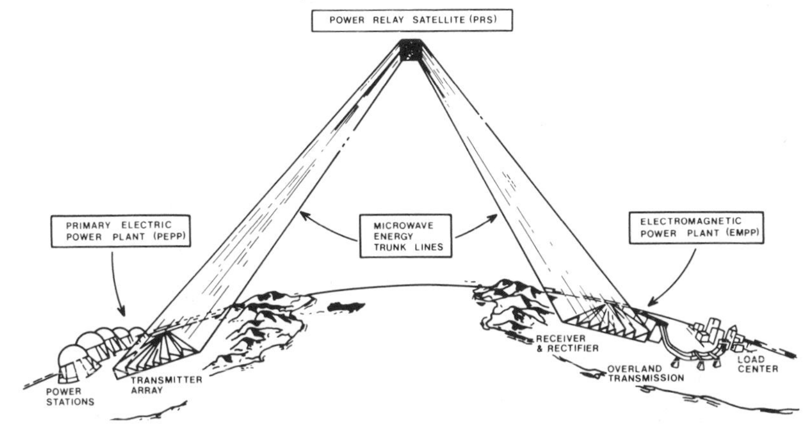

Fig. 8.5 Power relay satellite concept.[15]

TABLE 8.3[16]

POWER RELAY SATELLITE CONCEPT – PERFORMANCE AND COST DATA

Power generated	13,800,000 kwe	Maximum values.
Power delivered	9,000,000 kwe*	Beam power can be
Energy delivered over 30 years	2.1 trillion kwhe*	less.
Transmitter array:		
Construction cost	$3.1 billion	$40/kwe*
Maintenance cost (30 years)	$12 billion	
Overall cost	$15.1 billion	7 mils/kwhe*
Power Relay Satellite (1 kW²; 300 tons)		
Construction/delivery/erection	$0.75 billion	$84/kwe*
Maintenance cost (30 years)	$1.5 billion	
Overall cost	$2.25 billion	1 mil/kwhe*
Electromagnetic power plant (EMPP)		
Construction cost	$1.5 billion	$166/kwe*
Maintenance cost (30 years)	$0.95 billion	
Overall cost	$2.45 billion	1.15 mils/kwe
Grand total		
Construction cost	$5.3 billion	$590/kwe*
Maintenance cost (30 years)	$14.45 billion	
Overall energy transmission cost	$20 billion	9.3 mils/kwhe* †

* Power delivered.

† According to a newer analysis, taking into account the lessened maintenance cost caused by the development of platinum-coated amplitrons – this cost goes down to about 5.2 mils/kWH^{-1} (but this does not include the capital cost, which puts the figure up to 10 mils/kWH^{-1}).[27]

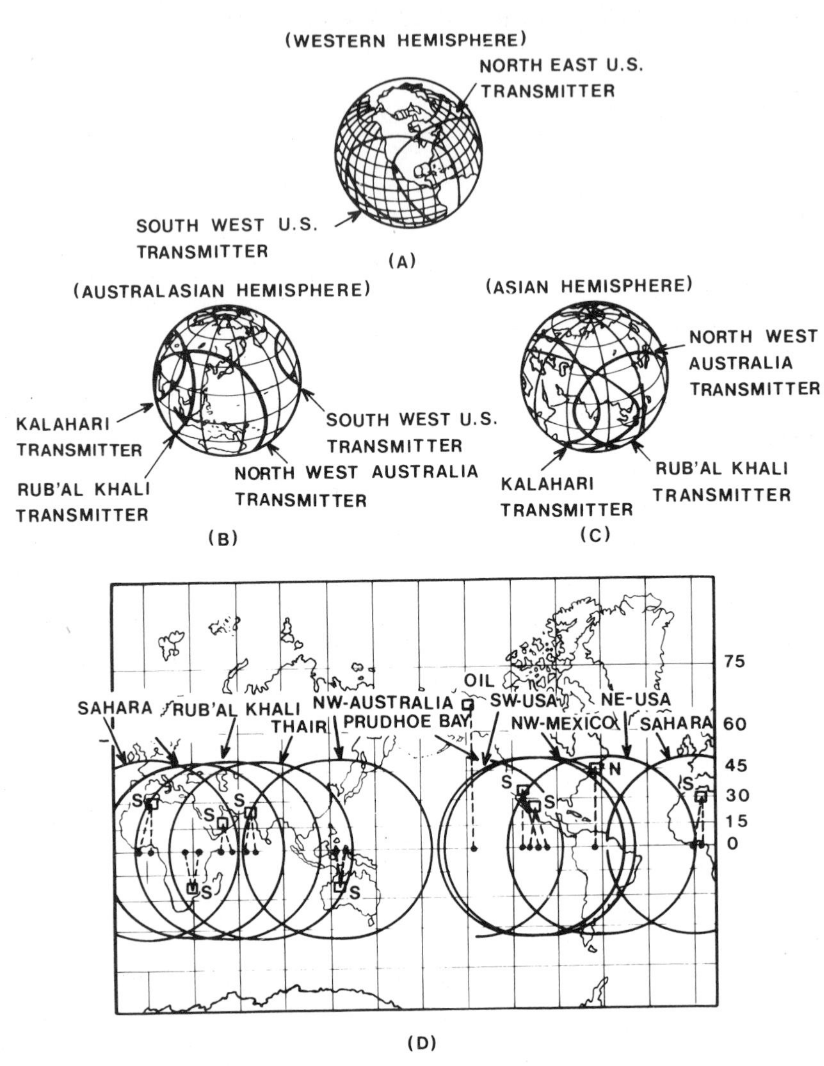

Fig. 8.6 Range of a number of Primary Energy Power Plant Systems.[15]

The transmission of energy by microwave-beam from a terrestrial source to a load centre, *via* a relay satellite, should offer a more economic system of using the transfer of energy through space than that which places the collector in orbit. Ehricke's sketch is shown in Fig. 8.5.[15] The primary electric power plant (PEP) is on the left and would perhaps be a solar collector station in Australia, near Alice Springs. The satellite would be in geosynchronous orbit over the equator, and the receiver

TABLE 8.4*

CHARACTERISTICS OF MICROWAVE-PRS ENERGY SYSTEM

Microwave Energy (point-to-point transmission via power relay satellite).

NOTE:
MW = microwave
PRS = power relay satellite
kwb = kilowatt beam power; as used in (C), kwb refers to the beam power at transmitter aperture
kwe* = dc electric kilowatt as delivered at the bus bar of the receiver power plant.

A *General*
Avoids questions of engineering overseas pipeline; building tanker fleet. Because of high cost of the transmitter, is economically important to use energy sources lasting at least 30 years, i.e. largely for transmission solar energy from high insolation areas or exporting fusion energy.

B *Processing*
Conversion to MW energy at high efficiency ($\gtrsim$90%) in microwave power generators at the expected state of the art at mid-1980s. Transmission efficiency over several thousand miles 50-60%.

C *Loading*
As phase-controlled MW beam in transmitter antenna. Antenna size large, but not prohibitive ($\gtrsim$ 138,000 kwe*/km^2). Relatively high construction cost, due to stringent engineering requirements (~\$3/ft^2; ~\$31M/km^2; ~\$295/kwb; ~\$340/kwe*).

D *Shipping*
Power Relay Satellite (PRS) in geosynchronous orbit needed for beam redirection. Stringent PRS engineering requirements. But moderate construction cost (~\$84/kwe*) due to small size (~1 km^2) and to shuttle which keeps maintenance low. Environment effects of beam on atmosphere are small.

E *Unloading*
Receiver-rectifier system (rectenna) of 85% conversion efficiency to dc current. Therefore, little thermal load. Environmental burden? Receiver size is larger, but not prohibitive ($\gtrsim$90,000 kwe*/km^2). Receiver power plant construction cost is small (~\$166/kwe*). Operating costs low. Operation is simple, making system suitable for export and use in developing countries.

F *Storage*
Has no storage capacity, would need to convert to hydrogen, anyway.

G *Conversion to electricity*
In rectenna, conversion to ac after electric transmission close to load centre.

H *Transmission to user*
Receiver power plant can be located near load centres, due to its low socio-environmental burden quotient. Dc-power transmission yields superior efficiency to ac-power transmission.

* Table originates in ref. 5, but modified.

would be a few tens of miles out to sea from, e.g., Osaka. The transmission 'line' would be a microwave beam.

In Ehricke's plan, transmitting and receiving antennae would consist of very many individual elements. He suggests a helix antenna 1.4" in diameter, 14" in length (Fig. 8.7[16]). Modules would measure 8-12 metres and hold 12,000 helices. A square antenna with a side of 10 km and an aperature of 100 sq km would contain 1.25 modules or 1.6 helices. Each module would be powered by a microwave power generator and have a phase shifter, to keep the beam focussed on the power relay satellite (PRS).[16]

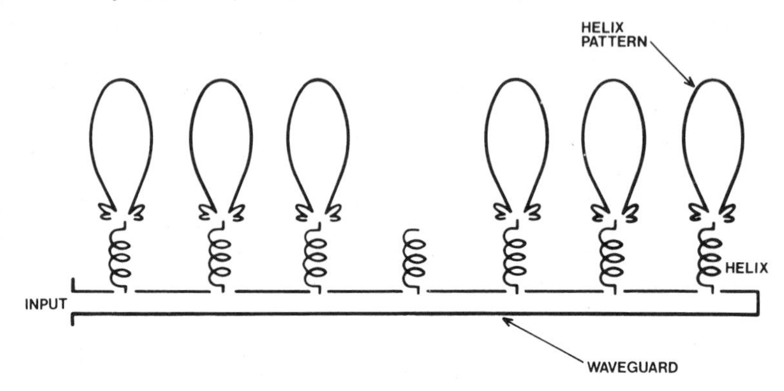

Fig. 8.7 Linear array of waveguide-fed helix elements.[16]

TABLE 8.5*

AUSTRALIAN ENERGY EXPORT POTENTIAL

* Promising
** Conditionally promising
† not promising

Alternatives	Domestic energy supply	Energy Export Suitability
Coal – direct use	** Envir.	† Transp/Environment
– gasification	*	*
Oil Shale/Tar	** Envir.	**
Local nuclear power – direct	** Envir. safety	† Complexity/Pu-239 profile
Remote nuclear power – direct	*	
– hydrogen	*	*
– microwave	*	* If distance > 4,000 miles
Solar power – direct	*	* Power plant equip. export
– hydrogen	*	*
– microwave	*	* If distance > 4,000 miles
– artif. hydrocarbons	† Energy requirement	† cost

* Originated in ref. 5, modified.

A number of influences tend to defocus the beam, so that the outgoing microwave beam would bypass the PRS. One is roughness on the surface of the antenna. Irregularities must be measured and compensated for by phase shifts, whose value would be computer determined. Clouds would decrease the efficiency of transmission only a few per cent, but hail and snow could offer greater attenuation. However, feedback signals between the satellite and the transmitter could be used to compensate negative effects rather completely.[15]

For a small roughness tolerance of 1 mm, required to achieve only 2% scattering loss, the size of the modules in the Power Relay Satellite would have to be reduced, and made adjustable. Precise mechanical pressure could be exerted on the corners of the modules, using electrically powered gear-drive actuators, to bring about phase shifts.[5 16] Several options for surface to control are available.[15]

The overall transmission efficiency would be 60-65%.[5 16] This would correspond to a 600,000 volt line, 4,000 miles long. 9 million kW could be transmitted by a 10 by 10 km transmitter array and a 1 by 1 km PRS.* Calculated, basic performance data and costs are shown in Tables 8.3[16] and 8.4.[5]

The dollars per kW for the PRS plant would be $600. The corresponding figure for the hydrogen synthesis plant would be $100.[16] The cost of the power generation system would be in the same range as that of a non-breeder nuclear plant. The cost is less than that of putting a solar collector into orbit (at present $10,000 per kW with future projections down to $1,000 per kW). Some characteristics are shown in Table 8.4.[5] The export profile is shown in Table 8.5.[5]

TRANSMISSION IN HYDROGEN

The advantages are: low cost, conventional technology and the fact that the fuel arrives in the form in which it is useful in a large variety of ways (Chapter 14). A summary of the relative advantages and disadvantages is made out in Table 8.6.

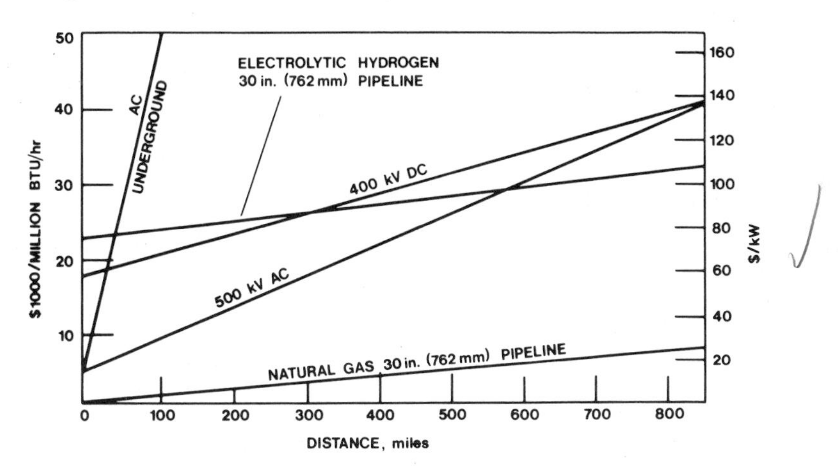

Fig. 8.8 Investment costs for energy transmission facilities.[17]

* Development of amplitrons with platinum-coated cathodes will improve the cost picture.

TABLE 8.6*

SOME CHARACTERISTICS OF HYDROGEN ENERGY SYSTEMS

ENERGY SYSTEM
Hydrogen (from inorganic sources, primarily water; some of (A) and all of (C) through (G), also applied to the use of hydrogen from organic sources, e.g. coal).

A *Extraction*
Electric power source needed. Greater freedom of power plant siting than with electric transmission. Preferred energy sources for electricity: low-cost, low-quality coal (on site); solar.

B *Processing*
Industrial electrolytic conversion efficiency 50-60% at present, possibly 80% in mid-80's. Separation energy of hydrogen from water is 2790 el.kWH per 1,000 cubic metres, or 31,600 el. kWH/ton at 100% conversion.

C *Loading*
Domestically, as gas. For overseas export, as gas in pipelines; or in tankers. Greater hazards than with natural gas.

D *Shipping*
Overland: Gas pipelines. Higher cost, greater risks and higher maintenance standards required than for oil or gas piping.
Overseas: Gas pipelines possible? LH_2 tanker fleet required?

E *Unloading*
As gas, no problems. As liquid, high quality facilities required.

F *Storage*
Lends itself particularly well to storage.

G *Conversion to electricity*
Lends itself particularly to high efficiency fuel cell conversion.

H *Transmission to user*
Hydrogen-electric plants could be central or in home. Zero pollution. Total energy concept.

* The table is from ref. 5, considerably modified.

Some investment costs are in Fig. 8.7.[17] The cost per 1,000 miles is $35 per kW, but this is an incremental cost and the cost for the first 1,000 miles is about $130 per kW (1972 dollars).

A comparison of costs for delivery by electricity with that of delivery through hydrogen clearly depends upon the cost of making hydrogen from water, and this is examined in Chapter 9. Some projected costs are given in Table 8.7.[18]

The costs of transmission in hydrogen are significantly less than those for power relay satellites, excluding the fact that, upon arrival, the electricity from the PRS system would have to be converted to hydrogen for many purposes. The PRS system could be cheaper than H_2 at distances above some 4,000 miles.

Thus, for transmission above about 500, and less than 4,000 miles, hydrogen is the cheapest method which could be attained with known or projected technology. Its use as a medium of energy will be the preferred one, therefore, for many situations of the new energy sources where

(Table 8.1) the distance between source and use area will be in excess of 500 miles. In view of its usefulness as a fuel it might also be employed where the source-user distance is smaller.

TABLE 8.7[8]

RELATIVE PRICES OF DELIVERED ENERGY – 1972
\$/million Btu

	Electricity	Natural Gas	Electrolytic Hydrogen
Production	2.67*	0.17 †	2.95-3.23**
Transmission	0.61	0.20	0.52††
Distribution	1.61	0.27	0.34
Total	4.89	0.64	3.81-4.09

* Equivalent to 9.1 mils/kWH
† 1975 costs will be about 300% greater than this.
** Assuming power purchased at 9.1 mils/kWH.
†† Assuming pipeline hydrogen, at $3.00/million Btu used for compressor fuel in optimised pipelines compared to natural gas at $0.25/million Btu.

DIFFERENCES IN PIPELINES SUITABLE FOR THE TRANSMISSION OF NATURAL GAS AND HYDROGEN[19]

Because of the different compressability of hydrogen and natural gas, there will be a difference in the capacity of the pipe for either gas, depending upon pressure. The capacity is less for hydrogen than for natural gas at a pressure of 750 psi.[19] It would be necessary to increase compressor capacity if the hydrogen is used rather than natural gas, for the present compressors would not be able to handle the extra volume. Thus, an existing line, if not modified, would carry only 26% of the energy in H_2 which it would carry in natural gas. To obtain the same energy in hydrogen, at the same pressure of 750 psi, the compressor capacity would have to be increased 3.8 times and horsepower 5.5 times. Increased pressure of H_2 would help the situation somewhat.

A design optimisation, carried out by Gregory and his colleagues[19], concludes that the cost of transporting hydrogen at 750 psi is 3-5 cents per million Btu per 100 miles (about 1.6 mils per kWH/1,000 miles), whereas, for natural gas, it is 1-1.5 cents/million Btu per 100 miles (throughput 300 trillion Btu per year, and assuming a $3 per million Btu cost of hydrogen). At 2,000 psi, the cost per 100 miles would drop to 2.5-3 cents per million Btu. Results are summarised in Fig. 8.9.[20]

BASIS OF CALCULATIONS ON THE COST OF THE TRANSPORT OF ENERGY IN GASES

The basis is the cost of transferring natural gas through a pipe. This is well known from practice. Thus, the basis of the calculations concerning a Hydrogen Economy is more sound than those concerning other schemes, which have to be systems-analysed before a practical base is set up.

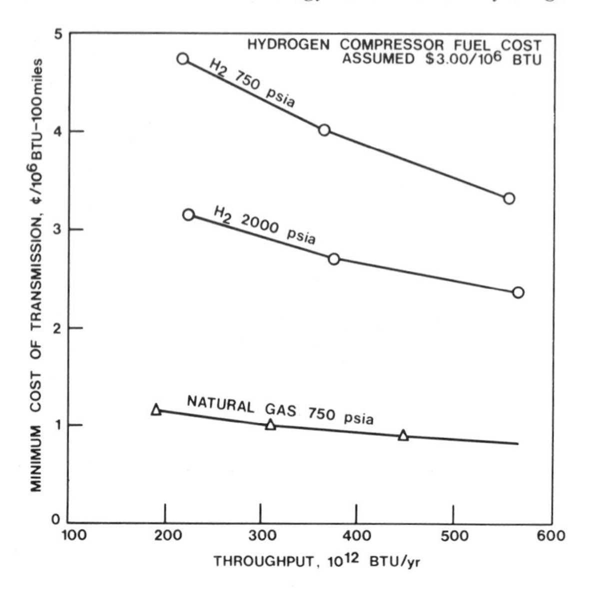

Fig. 8.9 Optimised transmission costs for hydrogen compared to natural gas.[20]

FEATURES OF HYDROGEN PIPELINE DESIGN

Using equations for pipeline design, it is possible to calculate the flow capacity for hydrogen, and then for natural gas. To determine the energy transmission capacity, relative, in Btu hr transmitted, Gregory *et al*[22] considered the relative heating values. These are in a volume ratio of 1:3 for hydrogen to natural gas. The ratio changes as the pressures increase, however, because the compressibilities of the gases differ. The ratio at 750 psi is 1:3.8. The specific heat ratio for hydrogen has to be known over a wide pressure range. Gregory *et al*[22] have extrapolated the value which they used to 100 Ats.

The results of the calculations of Gregory et *al.*[22] are shown in Fig. 8.11[23], which exhibits the energy delivery rates for hydrogen and natural gas of pipelines operating at different pressures. One set of curves shows the effect of varying compressor capacity; the other set shows the effect of varying the compressor input horsepower. The values are given relatively: 1.0 is the standard condition for the original pipeline. Thus the energy carrying power for sections from any pressure from 100-6,000 psi can be interpolated.

The results of Gregory *et al* show that *hydrogen transmission through natural gas pipelines would not be satisfactory, because the energy passed at a given pressure would be reduced compared with that for natural gas.* It is not only a matter of having to pump harder. The pipelines have to be redesigned: the diameter of the pipe has to be changed; the type of blades in the compressor should be modified, and the number of pipes in parallel should be increased.

Details of the calculations by Gregory *et al.* are in ref. 24.

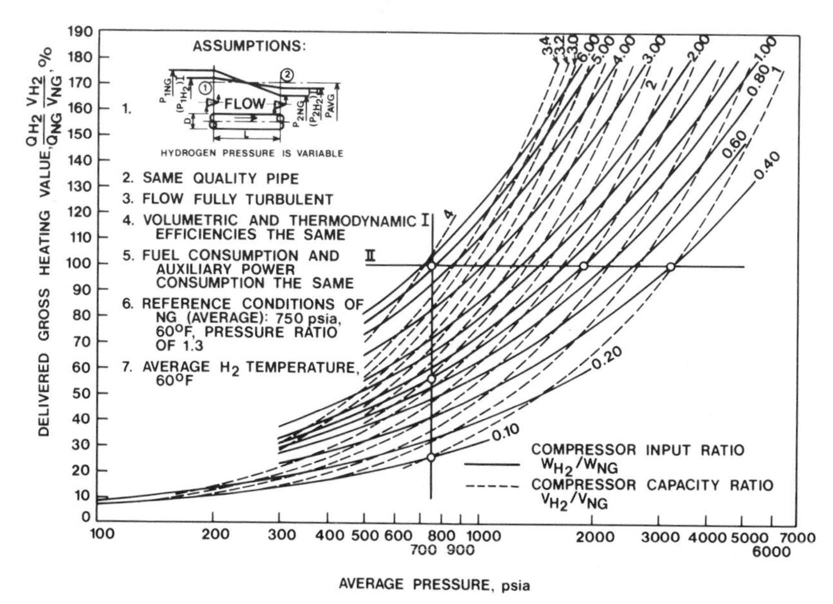

Fig. 8.10 Relative energy transmission capability and required compressor capacity and horsepower of a hydrogen pipeline at various pressures compared with natural gas at 750 psia, 60° F, and compressor spacing corresponding to the pressure ratio $\pi_{NG} = 1.3:1$.

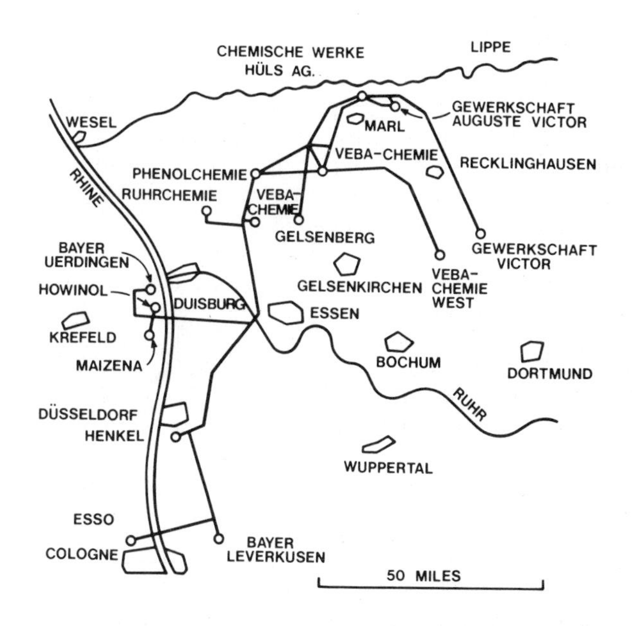

Fig. 8.11 Hydrogen pipeline grid of Chemische Werke Huls AG in Germany, 1970.[21]

HYDROGEN PIPELINES ALREADY WORKING[21]

One of the hydrogen carrying pipelines is in the Ruhr, Germany. It is owned by the Chemische Werke, Huls AG, and is 512 km in length. The line varies in diameter from 80 to 400 mm (Fig. 8.11[21]). The pressure in this pipe is 150 psi. The steel is St35.29, approximately equivalent to SAE1015 or 1016.

The reef area in Johannesburg, South Africa, contains a pipeline 50 miles long, which carries a mixture of hydrogen plus hydrocarbons, with 50% hydrogen.

COULD THE PRESENT NATURAL GAS TRANSMISSION SYSTEM BE USED FOR HYDROGEN?

Much is in favour of using the present natural gas system for hydrogen transmission in the earlier development of a Hydrogen Economy. However, the optimal transmission of energy in hydrogen through pipes needs pipes of a different diameter and pumping power than those which optimise the transmission of natural gas. Hydrogen could only be pushed through natural gas pipes inefficiently. In addition, joints in such lines would have to be examined to guard against leaks. An increased number of pumping stations would have to be added to the system.

Embrittlement of the pipes should not be a difficulty (Chapter 12).

USE OF METAL HYDRIDES TO TRANSPORT HYDROGEN

Metal hydrides have the advantage of volume energy density. There are no liquefaction and cryogenic storage problems. Handling safety is not a problem.

Conversely, hydride storage would have to be limited to very small-scale usage. Thus, there is a poor mass energy density with hydrides, and high cost of the metals. Large scale storage would consume most of the world production of many of the metals produced. Economics would be poor, particularly in the new and otherwise very attractive metal storage hydrides, $LaNi_5H_6$ (Chapter 10).

LIQUID HYDROGEN IN TRANSMISSION

Hitherto, transmission as a gas has been considered in the Hydrogen Economy, and shown to be a cheaper way of transmitting energy than is transmission through wires or satellite over certain longer distances. If the hydrogen is first liquefied, this cost difference would be decreased (with electricity at 7 mils/kWH, liquefaction would cost between 75-95c per 10^6 Btu. Apart from these costs, there is a problem in the insulation of the pipelines. A number of factors involved hydrogen transportation are shown in Table 8.8. However, liquid transmission could have applications.

(1) There may be situations where hydrogen would arrive in a tanker already liquefied. It would then be practical to transfer it in pipes to refuelling stations as a liquid. (See also Table 8.8[5].)

(2) Cryogenic superconductivity. There is the possibility of carrying electricity through pipes which contain liquid hydrogen. There could

be a dual hydrogen and electric economy in which liquid hydrogen was passed through pipes and used for various purposes, but the pipes were also used to pass electricity under superconducting conditions. There may be certain distances in which the dual passage of hydrogen and electricity is economically attractive.

TABLE 8.8[5]

ROUGH ESTIMATES OF COSTS INVOLVED IN LH_2 TRANSPORTATION BY CRYOGENIC TANKERS*

Item	Liquefied Nat. Gas	Liquefied Hydrogen	Comments
Temperature (1 Atm)	−259°F	−423°F	
Gas-to-liquid volume reduction	625	800	
Pounds/cu ft	36.2	4.37	
Volume required for equal heating value	~ 1.64	~ 4.5	Crude oil = 1.0
Refrigerated tanker cost ($/cu ft storage volume)	560-650	800-1,200	Based on $130M per tanker with 2-2.3 10^6 cu ft.
Tanker fleet investment	$4.78B	$21-24B	For 1,800B cu ft. gas equiv/year
		$600-700B	For 9B barrels crude oil equiv/year
Amortization and running operating cost of world-wide energy distribution in hydrogen economy (mils/kWH)		3.8-5.4	30 year operating period. 66% energy utilisation. Annual operative cost of 3% of initial investment.

* Cross-land line gaseous hydrogen transmission through pipes costs in mils/kWH^{-1} (1972):

$$1.3 + 1M,$$

where M is the number of thousand miles transmitted.

CONCLUSIONS

(1) The direct transmission of energy in the medium of electricity through underground cables is extremely expensive at any feasible transmission potential. Overhead transmission at 700 kV is more expensive than the transmission of hydrogen through pipes at distances above about 500 miles. Transmission through high and super-conductors looks both unfeasible; and, if feasible, uneconomic.

(2) Microwave beams up to a synchronous orbiting satellite and the beaming down again to distant cities would lead to the provision of solar energy, collected from highly insolated desert and ocean areas of the world, more cheaply than from (the much heavier) orbiting platforms which would collect solar energy from beyond the earth's shadows in space, and beam it directly to cities. The discrepancies between the costs of these two space-oriented alternatives will remain great until (and if)

the cost of raising a body into orbit drops to the region of $100 per lb.

(3) Gaseous hydrogen passage through pipes is the cheapest way of transmitting large amounts of energy over distances between 500 and 4,000 miles. Transmission by microwave and satellite would be cheaper for the large intercontinental distances, or if, for some reason, the laying of a pipeline became impractical.

(4) There are some situations which will lend themselves to liquefaction and transport by tanker or by liquid hydrogen in a pipe. The latter would then offer super-conducting possibilities.

REFERENCES

[1] R.A. Erren & W.A. Hastings-Campbell, *J. Inst. Fuel*, VI, 277 (1933).
[2] R.O. King, *Can. J. Res.*, **26F**, 264 (1948).
[3] R. Witcofski, paper presented to the American Chemical Society Symposium on Non-Fossil Fuels, Boston, 1972.
[4] J.O'M. Bockris, *Environment*, **13**, 51 (1971).
[5] K.A. Ehricke, 'The Power Relay Satellite (PRS) Concept in the Framework of the Overall Energy Picture', North American Aerospace Group, Rockwell International, December 1973.
[6] R.P. Feynmann, R.B. Leighton & M. Sands, *Lectures in Physics*, Addison Wesley, 1964.
[7] A. Kusko, IEEE *Spectrum*, **5**, 75 (1968).
[8] R.A. Meyerhoff, *Cryogenics*, p. 91 (April 1971).
[9] P. Graneau, IEEE Trans. Power Apparatus and Systems, PAS-89, 1 (1970).
[10] A. Pastuhov & F. Ruccia, Conference on Low Temperatures and Electric Power, IIR, London 1969.
[11] J.H. Russell, L.J. Nuttall & A.P. Fickett, paper presented at the American Chemical Society Meeting Hydrogen Fuel Symposium, Chicago, Illinois, August 1973.
[12] D.P. Gregory, assisted by P.J. Anderson, R.J. Dufour, R.H. Elkins, W.J.D. Escher, R.B. Foster, G.M. Long, J. Wurm & G.G. Yie, 'A Hydrogen-Energy System', prepared for the American Gas Association by the Institute of Gas Technology, Chicago, August 1972, p. X-22.
[13] Reference 12, p. X-23.
[14] P. Glaser, *Astronautics and Aeronautics*, p. 60 (August 1973).
[15] K.A. Ehricke, 'The Power Relay Satellite – A Means of Global Energy Transmission through Space – Part I', North American Space Operations, Rockwell International Corporation, El Segundo, California, March 1974.
[16] K. Ehricke, Statement before the Committee on Aerospace Stations, U.S. Senate, 31 October 1973.
[17] D.P. Gregory, assisted by P.J. Anderson, *et al*, *op. cit.*, p. 6.
[18] *ibid.*, p. 7.
[19] *ibid.*, p. IV-1.
[20] *ibid.*, p. IV-30.
[21] *ibid.*, p. IV-6.
[22] *ibid.*, p. IV-13, 14.
[23] *ibid.*, p. IV-19.
[24] *ibid.*, p. IV-41.
[25] Transmission Technical Advisory Committee for National Power Survey, 'The Transmission of Electric Power', Washington, D.C., Federal Power Commission, February 1971.
[26] D.P. Snowden, 'Superconductors for Power Transmission', *Sci. Am.*, **226**, 84 (April 1972).
[27] K.A. Ehricke, priv. comm., 21 August 1974.

CHAPTER 9

The Large Scale Production of Hydrogen Fuel from Water

INTRODUCTION

THE PRODUCTION OF hydrogen in recent years has been mainly from natural gas. The price had been sinking. Natural gas and oil are unlikely future sources of hydrogen on a large scale, because of the price increases which began in 1973, and will continue as the demand exceeds supply, due to the approaching exhaustion of these supplies (late '80s in the U.S.A.). A Hydrogen Economy must be understood mainly in respect to a coupling of hydrogen with one or more of the inexhaustible sources of energy. Methods of extracting hydrogen from water will be, therefore, the principal relevant methods (though, see Chapter 4).

WATER

The reaction for the production of hydrogen from water is endothermic and hence heat has to be used to make hydrogen.

The enthalpy change in the reaction

$$H_2O_{(g)} \rightarrow H_2 + \tfrac{1}{2}O_2 \tag{9.1}$$

is (150° C):

$$+ 58.14 \text{ kcal mole}^{-1} \tag{9.2}$$

and correspondingly the $\Delta G°$ at 150° C is:

$$+ 52.99 \text{ kcal mole}^{-1}, \tag{9.3}$$

and $\Delta S°$ at this temperature is 12.29 eu. $\Delta G°$ will become less positive as the temperature is increased, and the equilibrium of (1) shifts to the right. ΔG becomes negative at about 4700° K, the temperature, therefore, at which a hypothetical fuel cell containing H_2O at 1 atm, and producing H_2 and O_2 at 1 atm would begin to function. At 2000° C, there is *c.* 1% H_2 in equilibrium with H_2O at 1 atm.

The difficulties of the *direct* thermal dissociation to obtain H_2 are not only the material difficulties arising from the high temperature at which appreciable amounts of hydrogen are in equilibrium with water (refractories over about 1600° C provide difficulties), but is due also to the lack of massive amounts of heat at such temperatures. Nuclear heat could be available only up to about 700°

Irradiation of water should be considered. The strength of the H-O bond is 104 kcals mole^{-1}. Several reactions occur when water is irradiated. Ung and Back[1] have used u.v. radiation at 1849A° (*c.* 6 ev), and found hydrogen. Photo-electrochemical dissociation occurs at room temperature if the electrodes of a cell are subject to energy greater than about 1.2 ev (because the products are H_2 and O_2, i.e. not hydrogen radicals). The photo-electrochemical dissociation[2] occurs when the radiant energy is greater than the band gap for the semiconductor electrode material.

These parameters act as background to photo-oriented methods for the harvesting of hydrogen from water.

Seawater is a likely source of hydrogen in a Hydrogen Economy, because of the advantage of sea-borne collectors of solar and solar-gravitational energy; the cost has to have included in it an allowance for dealing with the Cl_2 which would be evolved (instead of O_2) on the electrolysis of seawater (see later section).

PRODUCTION OF HYDROGEN FROM FOSSIL FUELS

Methods using natural gas and oil are not of interest for innovatory considerations. (The cost of hydrogen from these sources will rise during the next decade with fossil fuel prices reaching more than four times the 1973 values.[3])

The only fossil fuel to be considered for hydrogen production on a large scale is coal. Some methods for obtaining hydrogen from coal have already been given in Chapter 4.[4]

CYCLICAL CHEMICAL SYNTHESIS OF WATER[5]

Basics

The difficulty for the dissociation of water to hydrogen is the large positive value of ΔH. Now $\Delta G = \Delta H - T\Delta S$ and it is not $\Delta G°$, but ΔG, which is to be reduced from a high positive value. We would drive ΔG in a negative direction by provoking a large ratio $p_{H_2O}/p_{H_2} p^{1/2}_{O_2}$. However, this ratio would have to be impractically large to bring ΔG to a negative value at temperatures as low as 1000° C, the temperature at which nuclear heat is expected to be available. Correspondingly, to make ΔG negative by temperature alone, for a one-step dissociation, requires heat at temperatures above those available on a massive scale.

To bring ΔG to a practical value, it is possible to go through a series of reactions, the net result of which is the dissociation of water, but which involve, on the way, several steps, each carried out at temperatures chosen to maximise the effect of a positive ΔS, and minimise those for a negative ΔS , in the individual partial reactions. The initial substance must be regenerated. In this way, the net ΔG can be made less positive for a given sequence, for it is the algebraic sum of the $T\Delta S$ changes in the successive partial reactions which affects the net ΔG for the production of hydrogen. Thus, the sum of the *enthalpies* of various alternate partial reactions which give the overall reaction is independent of the path of the reaction – and enthalpies depend relatively little on temperature. Entropy is also a function of state. However, the algebraic sum of the $T\Delta S$ contribution to the net ΔG depends on various temperatures used in the partial reactions. If those with a high positive Δ are carried out at high temperature, and those with negative ΔS at low temperature, ΔG may be reduced.

As an illustration, suppose that we wish to provoke the reaction AB → A + B. We carry out the reactions:

$$AB + C \rightarrow CB + A \quad (\Delta S \text{ +ve}) \tag{9.4}$$

$$CB + D \rightarrow B + CD \quad (\Delta S \text{ −ve}) \tag{9.5}$$

$$CD \rightarrow C + D \quad (\Delta S \text{ +ve}) \tag{9.6}$$

$$AB \rightarrow A + B \tag{9.7}$$

One assumes that the ΔS values for the different reactions have different signs. Let the ΔS of (9.4) be positive: it will be run at the highest practicable temperature, so that the $T\Delta S$ will be large, and ΔG for the whole reaction will be made decreasingly positive. For (9.5), let ΔS be negative. The reaction would be run at the lowest temperatures permissible from the reaction rate point of view. Suppose that (9.6) has a positive ΔS: a high temperature will again be used.

In this manner,[9] the sum of the $T\Delta S$ terms may be sufficient to bring ΔG for the overall reaction into a favourable value without exceeding a temperature, for any of the individual reactions, of 1000°. For reasons of material stability, an attempt will be made not to exceed 700° C.

Funk and Reinstrom[6–8] attempt to achieve a suitably lessened ΔG by using two-step cycles, but were able to show that it is improbable that any two-step cycles can be found which would give them a value of ΔG° so that water can be firmly dissociated in practice at less than 1000° C. Marchetti[9] organised a large team of the European Atomic Energy in ISPRA between 1970 and 1973 to examine the possibility of a satisfactory result from cycles involving three or four reactions. Within this team, de Beni has been the author of several proposals described by Marchetti.[10, 11] One of these, called Mark I by the author, is:

$$CaBr_2 + 2H_2O \xrightarrow{730^\circ} Ca(OH)_2 + 2HBr$$
$$Hg + 2HBr \xrightarrow{250^\circ} HgBr_2 + H_2$$
$$HgBr_2 + Ca(OH)_2 \xrightarrow{200^\circ} CaBr_2 + H_2O + HgO$$
$$HgO \xrightarrow{600^\circ} Hg + \tfrac{1}{2}O_2$$

Numerous cycles have been proposed by the Italian workers.[10, 11] Examples are:

$2CaBr_2 + 4H_2O \rightarrow 2Ca(OH)_2 + 4HBr$	730° C
$4HBr + Cu_2O \rightarrow 2CuBr_2 + H_2O + H_2$	100° C
$2CuBr_2 + 2Ca(OH)_2 \rightarrow 2CuO + 2CaBr_2 + 2H_2O$	100° C
$2CuO \rightarrow Cu_2O + \tfrac{1}{2}O_2$	900° C

$SrBr_2 + H_2O \rightarrow SrO + 2HBr$	800° C
$2HBr + Hg \rightarrow HgBr_2 + H_2$	200° C
$SrO + HgBr_2 \rightarrow SrBr_2 + Hg + \tfrac{1}{2}O_2$	500° C

$Cl_2 + H_2O \rightarrow 2HCl + \frac{1}{2}O_2$ 800°C
$2HCl + 2VOCl \rightarrow 2VOCl_2 + H_2$ 170°C
$4VOCl_2 \rightarrow 2VOCl + 2VOCl_3$ 600°C
$2VOCl_3 \rightarrow 2VOCl_2 + Cl_2$ 200°C

$H_2O + Cl_2 \rightarrow 2HCl + \frac{1}{2}O_2$ 800°C
$2HCl + S + 2FeCl_2 \rightarrow H_2S + 2FeCl_3$ 100°C
$H_2S \rightarrow H_2 + \frac{1}{2}S_2$ 800°C
$2FeCl_3 \rightarrow 2FeCl_2 + Cl_2$ 420°C

Some of these cycles have not yet been investigated experimentally, and it is therefore not yet known whether all the reactions proposed occur, in comparison with competing possibilities.

The efficiency of cyclical chemical processes
The justification of an attempt at a chemical stepwise synthesis of hydrogen is the avoidance of the overall inefficiency of the three-step electrolytic process. In this process, chemical reactions give heat to produce mechanical work – introducing a Carnot factor. Thereafter, this mechanical work is converted to electricity at very high efficiency and finally the electricity to H_2 *via* the electrolysis process at a rather high efficiency. At the beginning of work on the *chemical* cycle approach, it seems to have been tacitly assumed that a factor analogous to the Carnot factor involved in the conversion of heat to work did not exist, or was so favourable as to be negligible, in successive chemical reactions, carried out at different temperatures.

However[6], in the chemical cycles method, heat is applied to a reaction at the beginning of the cycle, but the gas produced in that reaction is, e.g. cooled and compressed; a second reaction must then occur, gases from it must be heated and expanded again, and so on. The energy needed to drive these happenings is related to the energy used to produce the product in a Carnot-like way. The situation is not excactly Carnot-like because water enters and hydrogen and oxygen escape from the cycle, but nevertheless, there is an *intrinsic* thermal inefficiency about the process which is analogous to that of a heat engine.

In addition, there are practical heat losses. The systems arranged are complex, and consist of many consecutive parts, towers, pumps, condensers, etc.[10] The empirical overall efficiency will be given by:

$$\epsilon = \frac{\text{heat obtained by combusting 1 mole of } H_2}{\text{total heat put in from an inexhaustible source per } H_2 \text{ produced}} \tag{9.8}$$

These real efficiency factors are not known because they depend upon actualisation in pilot plants. Marchetti predicts efficiencies in the region of 45-50%, but the basis of his estimate seems to have been optimistic intuition. One process, examined experimentally by Funk, Conger and Carty,[7] gave 18%.

Advantages of the cyclical chemical reaction method
To use the variation of temperature for various partial reactions, in

combination with the sign of ΔS, is attractive. The use of nuclear heat directly, without having to make it pass through the process of producing mechanical work, with an efficiency of, say, only 38%, originally seemed attractive (because the presence of Carnot-like factors in cyclical chemical methods, *carried out at different temperatures,* was not initially realised).

Difficulties of the cyclical chemical reaction method

(1) The systems to be used necessarily involve water and ionic salts. They give corrosive solutions[12]: water at temperatures above some 500° K is itself highly corrosive. Such effects suggest that larger maintenance costs would have to be borne by the cyclical chemical method than those of electrolytic processes carried out at temperatures of less than 100° C; or even of simple, electrolytic processes at high temperatures (1000° C) if these are one-step processes, instead of the many steps involved in the cyclical chemical method.

(2) It is assumed that the reactions will occur down free energy gradients[13], and not according to the relative kinetic rate constants of the various alternative reactions possible at a given temperature, and increasing in numbers with temperature. At sufficiently high temperatures this is true. It is, however, not true at 200° C, or even, often, at much higher temperatures. The assumed reaction path which actually occurs may not be that which has been assumed to occur in free energy oriented calculations. Any contribution from reactions not in the cycle will decrease the efficiency and give rise to a net use of the substances supposed to be regenerated; lack of 99.99% regeneration would interfere with the economics of the process, because the side product would mount up and have steadily to be removed (thousands of tons daily) from the production site.

(3) The efficiency of the process had not been brought out earlier: The actual efficiencies of tested processes seem to be less than the electrolytic ones.[8]

Funk's analysis of the efficiency of the vandium chloride process

A series of successive reactions has been examined in detail by Funk and co-workers.[6–8] The reactions are shown in Fig. 9.1.[7] The detailed analysis is shown in Table 9.1.[8] The efficiencies are given in the last column. *The calculated overall efficiency of this cyclical chemical process is substantially less than the actual efficiency obtainable from the classical electrolysis process.*

Cost estimates for the cyclical chemical process

A graph showing an estimated cost of a typical cyclical chemical process, devised by Marchetti[11], is given in Fig. 9.2.[11]

If no oxygen credit is taken, and if it is assumed that the process occurs at 50% efficiency, one can estimate a price of about \$1.50 per MBtu (1973 dollars). However, the basis of this efficiency (and hence cost) calculation is not clear and the efficiency assumed may be too high. Furthermore, the fixed costs of the process *are taken to be equal to those of the conventional steam reforming method.* This seems too optimistic in view of the much greater maintenance expenses which would be expected from a high temperature, multi-step method.

TABLE 9.1[8]

VANADIUM CHLORIDE PROCESS DATA TABULATION

Parameter	Units	Systems				Remarks
Maximum helium temperature	°F	2000	2000	1500	1500	
Minimum helium temperature	°F	37	37	67	67	
Helium system pressure	atm	10	1	10	1	He pressure drop equals 10 psi in all cases
Process heat input	kcal	155	155	475	475	
Total heat rejected	kcal	312	382	698	998	Includes unrecovered shaft work
Total heat regenerated	kcal	262	262	872	872	
Helium flow	$\frac{\text{moles}}{\text{mole } H_2}$	113	113	383	383	
Helium pumping power	kcal	3	24	14	105	100% efficiency
Separation work (VCl_4 and He)	kcal	18	18	23	23	
Total separation work	kcal	33	33	38	38	4 stages, plus VCl_4 and He separations
Shaft work input	kcal	69	90	90	18	Separations work at 50%; pumps work at 100% efficiency
Thermal power	kcal	230	300	296	600	30% efficient (heat to work)
Total reactor thermal power	$\frac{\text{kcal}}{\text{mole } H_2}$	385	455	770	1075	Process heat plus shaft work
Figure of Merit, η	%	18	15	9	6	$\Delta H_o / Q_t$

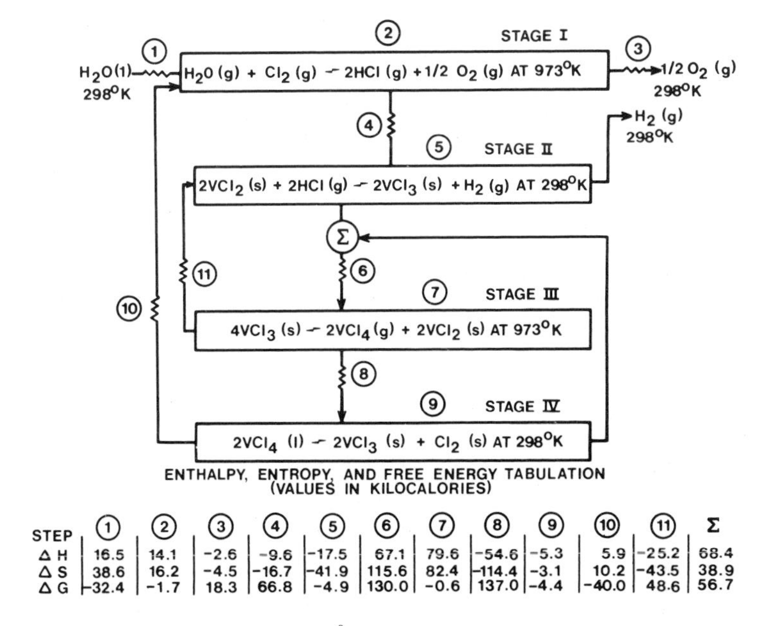

STEP	1	2	3	4	5	6	7	8	9	10	11	Σ
Δ H	16.5	14.1	-2.6	-9.6	-17.5	67.1	79.6	-54.6	-5.3	5.9	-25.2	68.4
Δ S	38.6	16.2	-4.5	-16.7	-41.9	115.6	82.4	-114.4	-3.1	10.2	-43.5	38.9
Δ G	-32.4	-1.7	18.3	66.8	-4.9	130.0	-0.6	137.0	-4.4	-40.0	48.6	56.7

Fig. 9.1 Vanadium chloride process.[8]

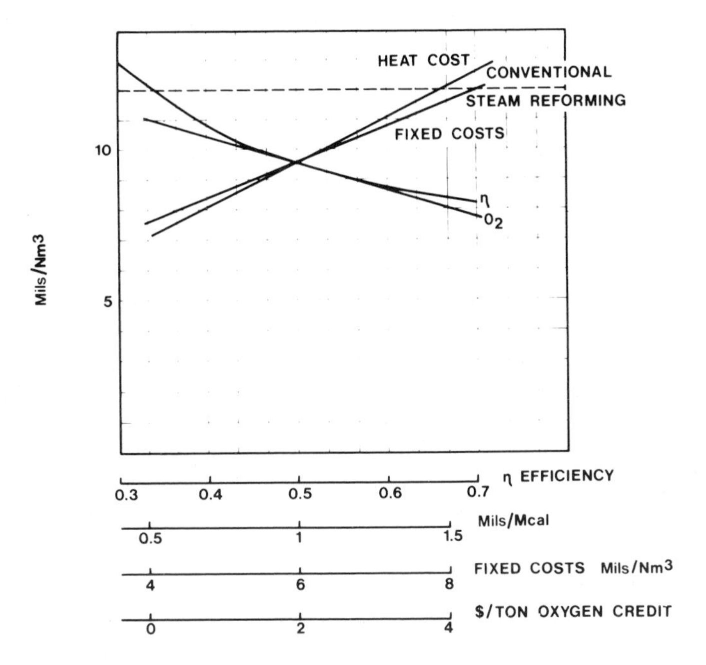

Fig. 9.2 Estimated cost of a typical cyclical chemical process.[11]

ELECTROCHEMICAL PROCESSES

For water dissociation, one can work at room temperatures, and drive the reaction $H_2O \rightarrow H_2 + \frac{1}{2}O_2$, so long as one has available the electrical energy.

The *electrosynthesis* of hydrogen is simple, may be achieved at low temperature; and the products are generated separately. The technology is established, but old; although the fundamentals of electrode kinetics are now understood,[14] and many laboratory improvements have been made (see Fig. 9.9), although none have been incorporated as yet in a large scale plant. There is much to do in the way of new, and modified, methods for the production (and cost lowering) of electrolytic hydrogen.

The effect of temperature upon the electrochemical production of hydrogen

For $H_2O \rightarrow H_2 + \frac{1}{2}O_2$ at 25°C:

$$\Delta G^\circ = 56.69 \text{ kcals mole}^{-1}$$
$$\Delta H = 68.32 \text{ kcals mole}^{-1}$$
$$\Delta S^\circ = 39.13 \text{ eu}$$
$$E_{cell}\ (1 \text{ atm}) = 1.229$$

At the boiling point of water, the parameters change because ΔS and ΔH include a contribution due to the heat of vaporisation.

At 150°, the equivalent quantities have been given in an earlier section. The temperature coefficient of the standard reversible potential is:

$$\frac{\partial E}{\partial t} = 0.25\ mv(°C)^{-1} \tag{9.9}$$

Also[15]:

$$\Delta G = -nFE, \tag{9.10}$$

so that, as the temperature rises, the *voltage* which must be applied to a cell to overcome the thermodynamically necessary energy is reduced. In a hypothetical state, for which $\Delta S = O$, then:

$$\Delta G = \Delta H, \tag{9.11}$$

$$E = \frac{\Delta H}{nF} = \frac{68.32\ .\ 4.18}{2\ .\ 96500} = 1.47 \text{ volts.} \tag{9.12}$$

Under these hypothetical conditions, then, the manufacture of H_2 is carried out entirely by electricity, and there is no heat evolved to the surroundings, and no cooling effect on the cell (or corresponding heat picked up from the surroundings to contribute to the running of the reaction).

The reversible potential of a hydrogen-oxygen cell at 25°C is 1.23v.[15] Hence, under the hypothetical condition that a cell could work thermally with thermodynamic reversibility (no overpotential), 0.24 volts of the energy necessary would come from the $T\Delta S$ term. In this hypothetical situation, heat is withdrawn from the surroundings to be added to that needed for completing the reaction. The electrical energy used is less than the energy needed to decompose water to hydrogen and oxygen. If the potential falls below 1.23 volt, H_2 cannot be produced at 1 At pressure, and 298° K even at extremely low rates.

Correspondingly, if the potential applied exceeds 1.47 volts at 25°, the electrical energy entering the cell will be greater than that needed – thermodynamically – to dissociate water to hydrogen and oxygen at 1 atm pressure, and the excess energy will be given off as heat.[17]

These relations of heat and electricity can be seen in Fig. 9.3.

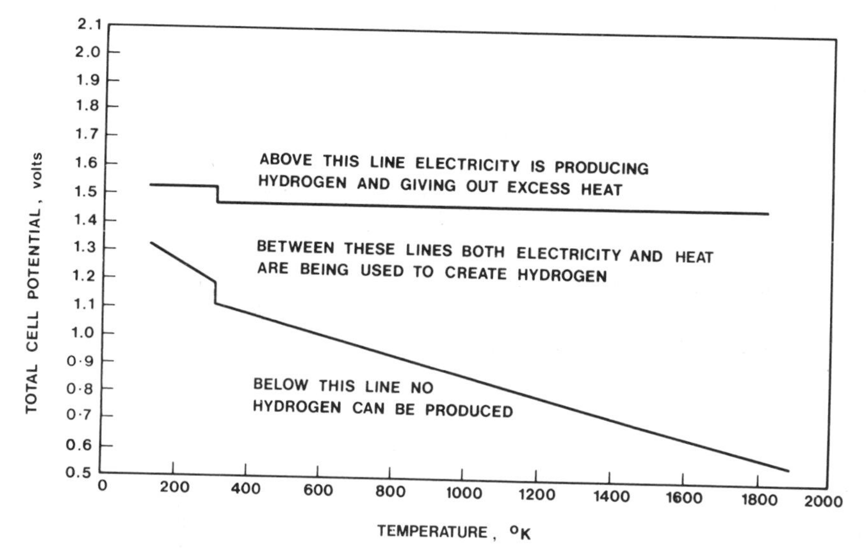

Fig. 9.3 Electricity and heat: gain and loss in the electrosynthesis of hydrogen.

Thermoneutral potential

Gregory[18] called the potential at which heat was neither being taken from the surroundings to create hydrogen, nor being rejected to the surroundings, the 'thermoneutral potential'.

Can ambient heat be converted to hydrogen?

Consideration of the conversion of heat to hydrogen *via* an electrochemical cell can be divided into two ranges. Firstly, if no heat is applied to the cell from outside, then, if electrolysis is occurring below the thermoneutral potential, the cell tends to cool, i.e. extract heat from the surroundings. The temperature could be restored by keeping the surroundings mobile, i.e. blowing air over the cell, but a practical device for extracing energy from the surroundings could only be made with difficulty in this way for kinetic reasons. Thus, at the thermoneutral potential, 1.47 volts, the overpotential of the cell would be 1.47 – 1.23 at 25° = 0.24 volt. Assuming that this overpotential is entirely on the oxygen anode, and that this is working on a planar electrode which has an i_0 (exchange current density) of 10^{-8} amp cm^{-2}, the current density would be given (at 25°C) by:

$$0.25 = \frac{RT}{\alpha F} \log \frac{i}{i_o}, \qquad (9.13)$$

or, with $\alpha = \frac{1}{2}$, $i \simeq 2.\ 10^{-6}$ amp cm^{-2}, i.e., 10^5–10^6 times slower rate than what is desirable. By the use of porous systems, this situation could perhaps be improved by only *one to two* orders of magnitude. A hypothetical solution, and perhaps a goal for research, would be a super-catalyst for the oxygen anode.* Were such a catalyst to give a real i_o of 10^{-4} amp cm^{-2}, and area effects (porosity) to contribute a further ten to one hundred times to the rate of production per apparent unit area[19], then, working at a reasonable 10^{-1} amp cm^{-2}, the overpotential would be:

$$\eta = \frac{RT}{\alpha F} \log \frac{10^{-1}}{10^{-3}} = 0.24, \qquad (9.14)$$

i.e. the loss of energy due to overpotential would compensate the gain in energy from the surroundings, neglecting I^2R loss, because the I would be very low. It follows that, by building a plant for the electrosynthesis of hydrogen which covered a cell area 10-100 times larger than that used hitherto (building upwards?), one might begin to gain some energy from the atmosphere, for at, say, 10^{-2} amps cm^{-2}, the activation overpotential would leave the cell potential at 0.1 volt less than the thermoneutral potential.

However, any hope of an economically meaningful heat to electricity extraction to a cooled cell from normally warm surroundings should be abandoned for a variety of reasons: (a) Increased fixed costs for the land area in building extra large plants for working at low current densities; (b) Neglect of IR drop tends to make the above statements optimistic, even if, at low current densities, the ohmic drop is low; (c) Only gold in alkaline solutions has an i_o as high as 10^{-8} amp cm^{-2} for O_2 evolution, at $T \simeq 25°$C, has been reported. Seeking one remains a good research goal, but it would be foolish to rely upon its attainment in a relatively short time.

Analogous remarks apply to concepts of working in electrolytes below $T = 273°$K. The i_o value would drop with temperature and the energy losses due to polarisation would increase, wiping out the likelihood of a usuable heat transfer to a cooled cell.

The conversion of high temperature heat to electricity and hydrogen

If the current density is increased, a hydrogen-oxygen reactor will experience an increased potential between the electrodes, due to the ohmic drop, IR.[20] Such a potential difference produces heat, at a rate I^2R, and the temperature of the cell tends to rise above that of the surroundings, so that no heat can flow to it from them.

However, due to the $T\Delta S$ term in the equation, $\Delta G = \Delta H - T\Delta S$, the electrical potential thermodynamically needed for the decomposition

---

* This catalyst would not necessarily be the same as that for oxygen cathodes in fuel cells.[19] Oxygen anodes usually have oxide film surfaces; cathodes work on the metal.

of water to hydrogen, is reduced as the temperature is increased. Further, activation overpotential is effectively the amount by which the heat of activation is diminished to achieve a certain rate. Increased temperature leads to an increased rate, and hence the activation overpotential, η, necessary to attain a given rate, is diminished at an increased T. If the temperature of a cell is raised several hundred degrees above room temperature, i.e. the electrolyte is a solid electrolyte with mobile O^{--} ions, and the fuel is steam, there is a possibility of supplying the cell with heat from the surroundings (e.g. 1000° heat from an atomic reactor), and converting such heat directly to hydrogen.

To investigate this, consider the $T\Delta S$ term during the electrolysis of water to H_2 at above 1000°C. The value is for the conversion of 1 mole. The rate of production of hydrogen at a current density i is i/nF moles sec^{-1} cm^{-2}. Hence, the cell loses heat at the rate of:

$$4.18\ T\Delta S \frac{i}{nF} \text{ joules sec}^{-1}\text{ cm}^{-2}. \tag{9.15}$$

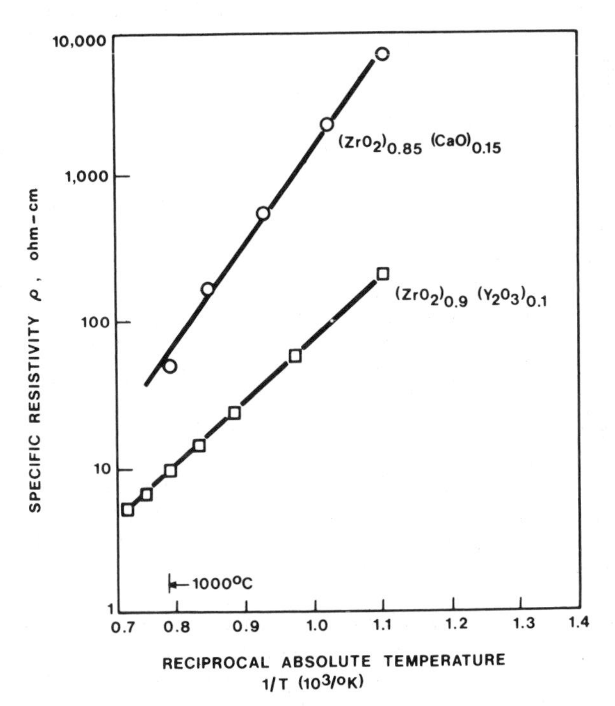

Fig. 9.4 The specific resistivity of solid electrolytes as a function of temperature.[21]

Thus, if at, say, 1000°C, $i \simeq 1$ amp cm^{-2}, and the resistance of the membrane (Fig. 9.4[21]) is <0.01 ohms cm^2, then the iR drop is less than 0.01v (i.e. negligible), although the *potential* needed for the production of hydrogen will be about 0.8 volt (Fig. 9.3). As the thermoneutral potential at 1273°K is about 1.49v, 0.69v or 46% of the energy for the decomposition of water would be coming from the external heat source, which would then be compensating for the cooling tendency of the cell.

The cell has become, therefore, a one-step heat to hydrogen converter which operates near to 1000°C. Such a converter of heat to hydrogen would be superior to a purely electric, electrochemical process, because nuclear heat is cheaper than the same quantity of energy after conversion to electricity. But it is to be preferred to cyclical *chemical* processes using heat, because these are thermally inefficient, complex to build, and would therefore involve high maintenance costs, and may not be 100% cyclical.

The approximations in the above argument are: negligible activation overpotential, a reasonable assumption at 1000°C; and negligible IR drop. The latter condition has not been achieved; but it has also not been desired because the concept of high temperature cells for electrolysing water is usually that they provide their own heat from the I^2R of the passage of current through the solid electrolyte. In so far as it were economic to utilise atomic or solar-derived heat to convert to electricity and hydrogen, then solid electrolytes for function at $T > 1000°C$, with sufficiently small R's, are probably feasible. Their long-term chemical or thermal stability is not explored.

A cost estimate for the hydrogen produced in this way can be made by taking equation 9.30 i.e.:

$$\text{Cost of MBtu of hydrogen} = 229\ Ec + 40, \tag{9.16}$$

using E from the reduced value at 1000°C (about 0.8 volt in the absence of IR and η), and adding to the value obtained from equation (9.16) the necessary heat cost. Atomic heat can be taken at 50c per MBtu. With $E = 0.8$, and 7 mil electricity (very large scale electricity, used at the site of production)[9], the electricity cost would be \$1.28 per MBtu. If 46% of the energy is supplied from an external independent heat source, this adds 46c (assuming 50% efficiency of use), i.e. the cost estimate is \$2.14 per MBtu of H_2. This would make the cost of gaseous H_2 less than that of 1975 petrol.

The electrosynthesis of hydrogen at high temperatures without an external heat source

Let it be assumed an electrochemical cell for the production of hydrogen is working at a sufficiently high temperature so that the activation overpotential is negligible, and IR is the source of the heat which is equal to $T\Delta S$ cooling. As a first approximation, let it be supposed that there are zero radiational and convectional heat losses, then, just to compensate the cooling effects of the endothermic reaction:

$$i^2R = \frac{4.18\ T\Delta S\ i}{nF} \tag{9.17}$$

$$iR = \frac{T\Delta S}{nF}\ 4.18 \tag{9.18}$$

Hence, the total applied potential must be:

$$E_{\text{thermo}} + iR = E_{\text{thermo}} + \frac{T\Delta S}{nF}\ 4.18. \tag{9.19}$$

But $T\Delta S/nF$ 4.8 is the difference between the reversible thermodynamic potential and the thermoneutral potential. Thus, in the absence of an external heat source, one cannot, in practice, electrolyse water to produce H_2 at below the thermoneutral potential and above the temperature of the surroundings. It has sometimes been implied that this can be done with consequent decrease of E, and, from equation 9.31, lowering of costs.[22]

Electrolysis at high temperatures gives, then, no net *thermodynamic* advantage. The lowering of the thermodynamically reversible potential is nugatory, because the IR drop which has to be added to the reversible potential to provide the heat necessary to prevent the cell cooling (in the absence of an external source of heat), is equal to the voltage lowering gained by the increase in temperature. However, a real gain may exist by electrolysis at medium temperatures, because of the reduction of the anodic activation overpotential. A considerable gain *is* available if external heat is used to maintain the cell at constant temperature, whilst electrolysing steam at high temperature.

Heat to hydrogen and net electricity production

If one raised the temperature of a hydrogen-oxygen reactor to that at which $T\Delta S > \Delta H$, the cell would begin spontaneously to produce electricity as well as hydrogen, i.e. it would function as a hydrogen and oxygen *producing* fuel cell. This would be of interest as a method of converting heat to electricity, but it is easy to show that the temperature at which the process would begin, even at low partial pressures of hydrogen and oxygen, is above 2000°C, and hence impractical.

However, a net production of electricity from heat could be achieved by utilising a high temperature hydrogen-oxygen reactor, with negligible membrane IR, to produce H_2, the $T\Delta S$ cooling being compensated by an external heat source, and then using this H_2 to produce electricity in a fuel cell at room temperature. The maximum available potential would be (driving cell at 25; driven cell at 1000°C), about 1.2-0.8, or 0.4 volt, a practical voltage. In effect, the room temperature cell will not function at significant rates with $\eta < 0.2\text{-}0.3$, so that a very poor conversion efficiency would result.

Conclusion concerning an optimal path for heat to hydrogen conversion

There is much to be said for using atomic heat to keep an electrolysis cell at 1000°C, and to convert part of it to hydrogen (at an *estimated* cost of $2.14 per MBtu). Such a use of heat to convert water to hydrogen economically in one step is preferable to the use of multi-step purely chemical processes with their predicted low net efficiencies and high maintenance costs.

The electrode-kinetics of the electrochemical production of hydrogen from water

Although the first textbook on the kinetics of electrode processes was published in 1955[23], most electrode kinetical equations are developed to apply to individual electrodes, and there has been little development of equations which are applicable to the overall cell. If the cathode and anode are of the same area, it is easy to show[24] that:

$$I = (i_o)_{cell} e^{(V-Ve)/q}, \tag{9.20}$$

where V = total applied potential to drive cell at I; V_e = equilibrium potential; and $q = \lambda_1 + \lambda_2$, where $\lambda = RT/\alpha F$, 1 and 2 refer to the two electrode reactions, respectively; α is the so-called transfer coefficient of electrode kinetics. This coefficient is related[25] in a complex way to the shape of the barrier at the electrode-solution interface, and to its stoichiometry in practice, and is often near to $^1/_2$. Detailed values are given elsewhere.[25]

Thus, as the value of the applied potential departs significantly from the equilibrium potential, i.e. as the overpotential, η, begins to be significant, the *rate* of production of the H_2 rises exponentially with the overpotential (Fig. 9.5 and 9.6[20]).

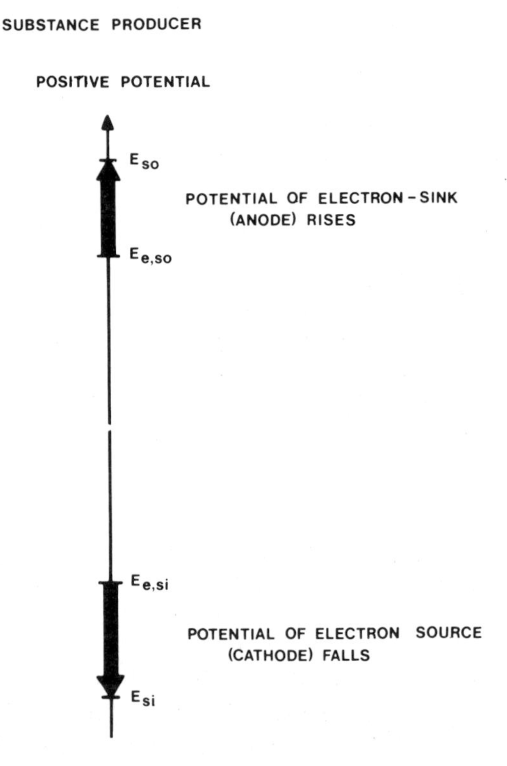

Fig. 9.5 When a substance producer is driven at an appreciable rate, the anode potential becomes more positive and the cathode potential becomes more negative. The net result is an increase in the cell potential compared with that at an open circuit.[20]

Overpotential is like inflation: it always leads one to spend more than one thought would be necessary (i.e. more than the price necessary assuming electrolysis at V_e, i.e. electrolysis at zero overpotential). On the other hand, like inflation, the results are not all bad, because both give rise to an accelaration of the action; currents and economies, respec-

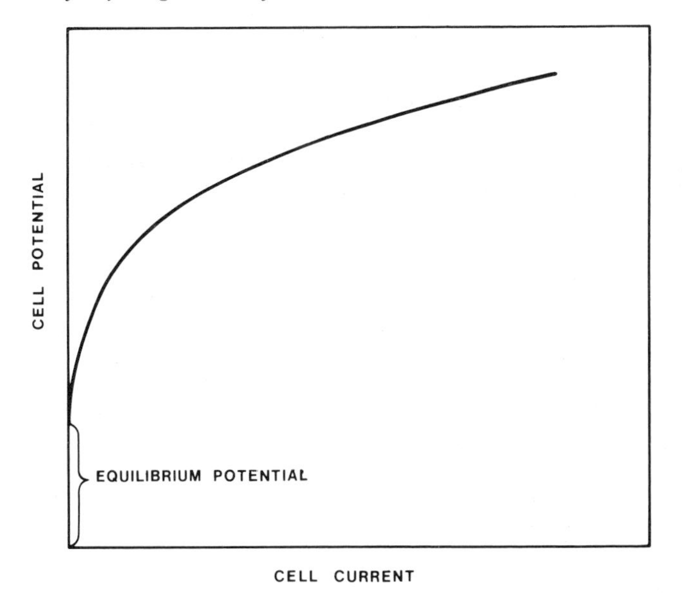

Fig. 9.6 The higher the current driven through a substance producer is, the larger will be the cell potential opposing the driving cell potential.[20]

tively, go faster. However, the overpotential of cells increases the price of the product just as inflation affects the price of goods. One has to pay more (compared with the price hypothetically available at very low rates when overpotential would be nearer to zero), per mole of product, for the electricity to drive the cell by the ratio:

$$1 + \frac{\eta}{V_e} \tag{9.21}$$

Conversely, cells occupy space and space costs. The *faster* the reaction (arising from a greater overpotential), the *less* the space is covered per unit mass and time of production. Hence, there will be an optimal value of overpotential to bring about a high current density depending on the local price of electricity and land. However, $\frac{1}{i_o}$ (see equation 9.20) should always be maximised, for the reaction rate, for a *given* overpotential (i.e. cost of fuel per mole of H_2), is proportional to it.

The overpotential needed to give a certain value of i or rate of production of hydrogen, is proportional to $\log\frac{1}{i_o}$. The overpotential can be broken down to:[1]

$$V = V_e + \eta_{\substack{\text{anodic}\\\text{activation}}} + \eta_{\substack{\text{cathodic}\\\text{activation}}} + \eta_{\substack{\text{anodic}\\\text{concentration}}} + \eta_{\substack{\text{cathodic}\\\text{concentration}}} + IR, \tag{9.22}$$

where the activation η's refer to the overpotential associated with electrocatalysis, and the concentration η's to transport difficulties. A fuller development of equations for electrochemical cells is given in the book by Bockris and Reddy.[26]

Thus, equation (9.22) can be written:

$$V_{\text{working}} = V_e + \Sigma\eta, \tag{9.23}$$

where $\Sigma\eta$ comprises the sum of the overpotentials at the two electrodes.

In the electrolysis of water, concentration overpotentials are negligible and the overpotentials are both activational and ohmic. Oxygen overpotential is a major problem for electrolysis at temperatures in aqueous solutions. Several tenths of a volt deviation from the reversible potential can be observed at current densities in the order of 100 milliamps per sq cm. Where $c_{\text{elec}} \simeq 7$ mils $(\text{kWh})^{-1}$, 0.4v overpotential would increase the cost of an MBtu of hydrogen by *c.* \$1, i.e. 33-50%.

The reduction of these overpotentials can be achieved by going to higher temperatures: above 500°C, the oxygen overvoltage problem is more or less eliminated. Two approaches for reducing overpotential at low temperatures exist:

(1) The use of platinum[27] as an electrode catalyst reduces the overpotential, but is too expensive, and there would not be enough.[28] However, *very* small amounts of platinum may be used[29], the Pt being confined to the active zone within the pore. The cost of the Pt needed is about \$10$(\text{kW})^{-1}$.

(2) Small traces of platinum on tungsten bronze facilitate the evolution of oxygen and make the electrode per unit area have the same properties as Pt, although their costs are small compared with that needed if Pt electrodes were used. This work[30-32] is worthy of pursuit in respect to low temperature electrolysers. Analogous work to this in which the tungsten bronzes are replaced by more complex, but similar, materials, is of interest. Thus, Tseung[33] has reported electrodes which in oxygen evolution, at room temperatures, have overpotential of less than 10 mv at rates of hydrogen production equivalent to 100 milliamps cm$^{-}\Delta$.

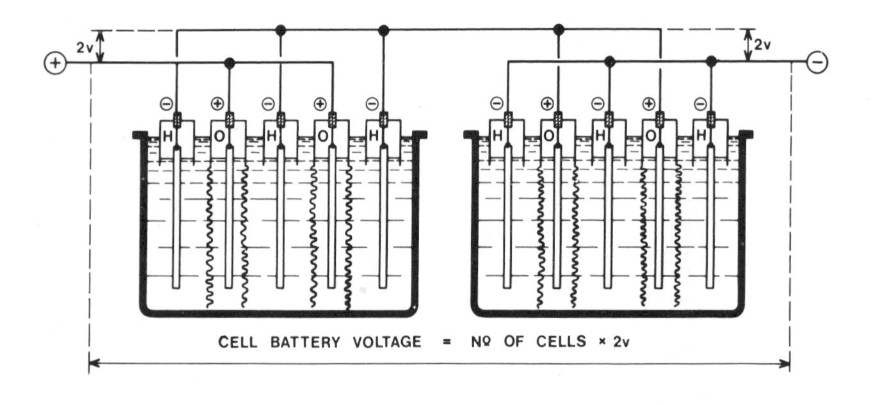

Fig. 9.7 Unipolar cell construction.[34]

CLASSICAL ELECTROLYSERS

Classical electrolysers have been little changed in design for many years. Designs are shown in Fig. 9.7 and 9.8.[34] In Fig. 9.7 the electrodes are in parallel; in Fig. 9.8 they are in series. The series arrangement has an advantage because there is difficulty in obtaining the 2 volts per cell from the rectified AC. Conversely, large total potentials are used in the series arrangements and, when one cell fails, the whole bank ceases to operate. Hence, both types of electrolysers have pros and cons, and both are used.

In classical electrolysers, the potential applied is about 2 volts, and the current density about 100 amps ft^{-2}, i.e. 0.1 amps cm^{-2}.

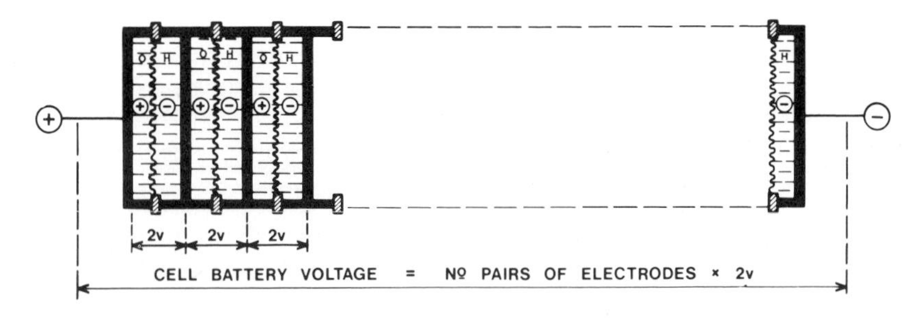

Fig. 9.8 Bipolar cell construction.[34]

The cost of H_2 produced in classical electrolysers is about $5 per MBtu. H_2 can be produced from coal at 1974 prices ($12-14-16 per ton), at about $3 per MBtu.* On this basis, the production of H_2 by electrolysis is sometimes described as too expensive. There are reasons why (particularly with future prices for fossil fuels) this opinion should be revised if low-use period electricity were used, or there were sufficient research funding to independent organisations for the necessary research and development work on new kinds of electrolysers, or breeder reactors become available.

Factors which make probable cheap hydrogen by electrolysis

(1) When the hydrogen which is usually described as produciable from coal is compared with hydrogen produced from electrolysers, it is not the same product. That from coal is some 97-98% pure, and contains CO, together with some H_2S. These impurities may at a cost be reduced to negligible concentrations. However, the cost has not yet been added to the quoted price of H_2 from coal. Hydrogen from electrolysers is completely clean, without further treatments.

(2) The price of H_2 from an electrolyser is sometimes compared 'as a fuel' with the price of crude oil. A typical comparison would be that oil giving fuel at less than $2 per MBtu with H_2 from chemical

*The price of hydrogen from natural gas is lower than this, but not, relevant because natural gas in the U.S.A. will exhaust in about one decade, and this exhaustion will be preceded by large price rises which will make it more expensive to obtain hydrogen from natural gas than from coal.

electrolysers at $5 per MBtu. This is deceptive because the major use of oil is as a refined motor spirit. If the ex-refinery price of petrol is 32 cents per (U.S.) gallon (Sept. 1974 U.S. price), the equivalent is $2.40 per MBtu.* But the $5 price for the electrolytic H_2 prepared by classical electrolysers is the final price of the pure gaseous fuel, which needs no further refining. In view of the further extra-inflational price rises for gasoline now in train, electrolytic hydrogen even from old electrolysers running on coal-based electricity is likely to be a cheaper fuel than gasoline within a decade.

(3) The classical electrolysers – for which the price of $5 per MBtu has been calculated – have not been cost minimised. They have been engineered without use of the insight which modern electrode kinetics has given to the engineering of fuel cells, and hence of electrolysers. Thus, the present technology electrolysers:

(a) Do not use an electrocatalyst to minimise overpotential;

(b) Have no devices to reduce the effect of bubbles on *IR* drop between electrodes;

(c) Do not use flow devices to reduce possible super-saturation hold up;

(d) In fuel cell technology, current densities per external unit area are increased at least 10-100 times by the use of porous electrodes; but the use of porous electrodes has not been introduced into practical electrolyser technology (but see ref. 35).

That a reduction can be achieved in the cost of hydrogen, as a result of appropriate research and development of the above concepts, follows not only from the technical obsolescence of present electrolysers, but also from the performances of many new experimental electrolysers which have been demonstrated on the laboratory scale over the last decade. Current-potential curves are shown in Fig. 9.9.[36] They illustrate that, for the same price of electricity, the potential could be reduced from *c.* 2.2 to *c.* 1.5, i.e. reduced by some 33% for a given electricity cost. There could be a diminution of fixed costs, too, because the higher current densities possible with the emerging technology would give a reduction of area occupied by an electrolyser plant. With electricity at 10 mils per kWh, the established performance of some emerging electrolysers could reduce the price of electrolytic H_2 to about $3.40 instead of $5 per MBtu, quoted earlier for classical electrolysers.

Thus, the proved performance of emerging small-scale electrolysers, if maintained on scale-up, and using electricity costing c. *10 mils per MBtu, could give pure hydrogen fuel from water at about 40% above the ex-refinery price of U.S. gasoline for 1974.*†

(4) The conclusion that electrolytic hydrogen would be 'too expensive' does not account for the results of the application of known principles to the electrolytic process. Some of these results, described below indicate

---

* There are 1.2-1.3 10^5 Btu per U.S. gallon.

† Thus, hydrogen will become economically the preferable fuel were it possible to make large scale electrolysers which have the performance indicated by small scale ones. The cost of electricity should not rise with the cost of oil and natural gas for electricity can be made from coal which is present in many countries in quantities which could form an energy base for 30-50 years if the material can be mined at the rate which would be required.

a probability of lowering the cost to *c.* \$2.30 per MBtu.

(5) Lastly, several concepts under research indicate the possibility that the cost of hydrogen from water could be reduced to *c.* \$2/per MBtu.

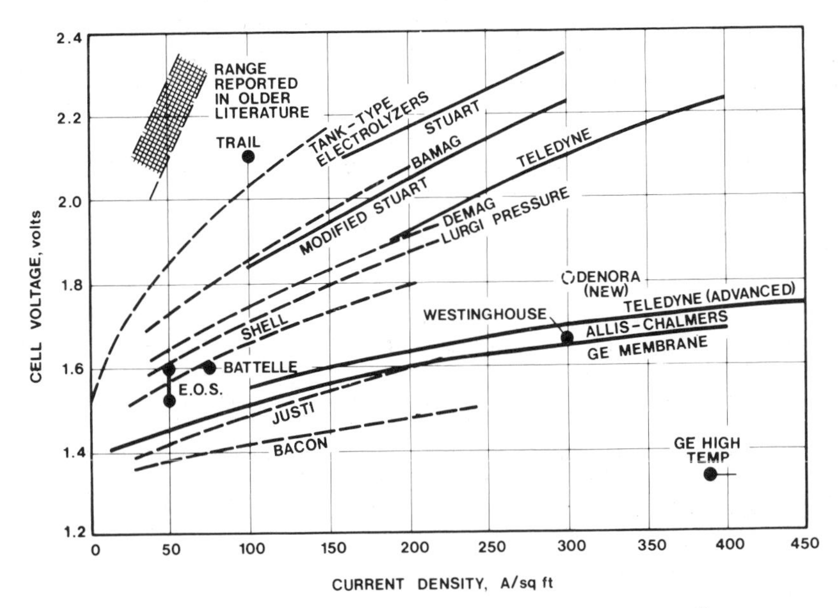

Fig. 9.9 Cell operating performances of various advanced electrolysers.[36]

ON THE COST OF ELECTRICITY

Tables giving the cost of coal-based and nuclear-based electricity have been given in Chapter 7 (see also Table 9.2[37]). The cost is in the region of 10 mils (1 cent) per kWh^{-1}.

Inflational factors have become of great importance. A 1974 estimate breaks down as follows:[38]

An important element is the cost of the power plant. Currently power plant costs are in flux owing to the environmental requirements, but a responsible 1974 figure for a coal-burning plant is in the range of \$225 to \$325 per kW of coal-burning capacity. The fuel at the mouth of the mine, again depending on location, would be anything from 25 cents to 60 cents per million Btu. The other costs, labour for operation and maintenance, are small. Properly designed plants can be put together to assure firmness of supply with a reserve requirement of 15%. Thus, one ends up with the following costs:

Fixed charges at 15%, using a mean value of \$275 for the capital cost of the plant	\$41.20 per kWh
Fixed charges at 8,500 hours use	4.86 mils/kWh
Reserve component	0.73 mils/kWh
Coal fuel cost at 40 cents per million Btu	3.60 mils/kWh
Operating and maintenance costs	.30 mils/kWh
	9.49 mils/kWh

of one MBtu of H_2 would be approximately (using emergency technology parameters):

$$(229 \;.\; 1.5 \;.\; 0.3 + 40) \qquad (9.24)$$

cents per MBtu, i.e. *c.* \$1.43, or far cheaper than the ex-refinery cost of 1974 gasoline (\$2.40 per MBtu).

Such cost analyses are one of the principal supports for a Nuclear-Hydrogen Economy.

LOW-USE-PERIOD POWER

The cost of electricity depends upon the cost of money and the degree to which the plant is used in the number of years the money is rented. There are times (a portion of the night hours) when the load on electricity-producing plants is diminished. During this time, the plant is under-used, and this increases the costs of the plant, because it is supposed to have a lifetime in years and not in kWh produced.

If a suitable industry is arranged which works during the low use hours, and which is sufficiently big so that it is worth running the whole plant to supply the electricity, one could regard the cost of this *extra* electricity as being the incremental cost (fuel + maintenance) only and not count it in part of the amortisation economics, for this has been costed on the basis of low use at night. In this situation, the cost of electricity is reduced. According to the eminent American consultant, Philip Sporn[38], the 1974 cost of off-peak power should be between 4 and 6 mils/kWh (i.e. say 5 mils/kWh^{-1}).

The concept of cheap off-peak power may be very interesting in a Hydrogen Economy, particularly in its initial stages. Table 9.3[40] suggests that, by 2000, the entire electric load to run all the vehicles in the country could be supplied from power obtained during the time when the electricity generating plant has excess capacity. With the cost of the power as a conservative 6 mils/kWh^{-1}, the cost of H_2 with E = 1.5 would be about \$2.46 per 10^6 Btu compared with the 1974 ex-refinery petroleum cost of \$2.40. Hence, if cars could be run on *gaseous* H_2 (and this is established, see Chapter 15), this could be done using

TABLE 9.3[40]

PROJECTED ELECTRICAL ENERGY REQUIREMENTS FOR ELECTRIC PASSENGER CARS

Year	Total No. passenger cars, million	No. electric passenger cars, million	Estimated electric energy required for electric cars, billion kWh	Estimated electric power generation U.S.A., billion kWh	Estimated power available by levelling diurnal load (20% increase), billion kWh*
1980	110	1.1	9	2830	566
	110	11.0	88	2830	566
2000	150	15.0	120	8250	
	150	150.0	1200	8250	1650

* This assumes that charging will occur during off-peak hours.

TABLE 9.2[37]

HYPOTHETICAL COMPARISON OF NUCLEAR AND FOSSIL-FUEL BASED-LOAD GENERATING COSTS AT 80 PER CENT (PLANT FACTOR)
(Mils per kWh – Constant 1970 dollars)

Fixed charges	Nuclear at $400/kW*	Fossil Fuel at $300/kW †
Return Taxes at 15.2% Depreciation	8.68	6.51
Operation and Maintenance	0.30	0.28
Insurance	0.15	0.05
Fuel $0.20/MM Btu at 10,500 Btu/kWH for Nuclear and $0.60/MM Btu for Fossil at 10,000 Btu/kWH	2.10	6.00**
Total	11.23	12.84 ‡

* Estimate for a LWR.
† Coal fired generating plant with stack gas desulphurisation equipment at approximately $80 per kW.
** Fuel cost includes cost of limestone or other reactive agent used in SO_2 scrubbing device.
‡ The fossil fuel calculation would imply a break-even fuel cost of 43.9 cents/MM Btu for coal and associated limestone or other reactive agent.

The range of electricity costs at 100% load factor (in 1974 in the U.S.A.) is from 8 mils to 13 mils/kWh^{-1}, depending upon location.

The cost of electricity herewith estimated may be reduced in special arrangements for large quantities, very near to the site of production. 7 mils per kWh *in these circumstances* was an acceptable figure for 1974. The delivered cost depends on the distance from the source. Fuel costs will increase markedly if the fuel is natural gas or oil, but increase somewhat less if it is coal.

'CHEAP NUCLEAR ELECTRICITY'

It has long been stated by the American A.E.C. that if nuclear powered electricity were produced on a large scale, it would be very cheap. Two types of cost estimates must be considered. Firstly, there is an estimate based upon *incremental* costs, i.e. the cost after the equipment has paid for itself, and the only further costs are fuel and maintenance. Then the electricity should cost about 1 mil kWh^{-1}. This low cost (0.1 cents) must be contrasted with the still very low cost predicted by Weinburg and Hammond[39] of 2-3 mils/kWh^{-1}, taking into account the amortisation costs of the plant, i.e. before the plant has been paid for.

A more recent (1973) U.S. A.E.C. statement projects a price of 3 mils/kWh^{-1} in 1973 dollars for 2020, on the supposition that, by then, breeder reactors are in extensive use.

The realisation of these costs depends upon the development of large reactors and is subject to the difficulties of the developments of the atomic technology set down elsewhere (Chapter 3). However, it is noteworthy that, at a cost of 3 mils/kWh^{-1}, and in 1974 currency, the cost the U.S. electricity generating capacity with negligible increased costs of fuel compared with those of, say, 1975 (and without building special electricity-producing plants).

It would seem possible – economically – to use coal fuel in the interim to provide the extra electricity by the functioning of the present electricity-generating plant during the low-use period and produce hydrogen by electrolysis to run the transportation system. This would seem economic (a) in terms of 1974 costs; and (b) in terms of emerging electrolyser technology. *Even present electrolyser technology* could give H_2 at about \$3.56 per MBtu on 6 mil electricity. Emissions from coal-burning central electricity producers could be controlled better than the emissions from car exhausts. Correspondingly, the cost of electrcitiy, even that for the low-use hours, will increase as the price of coal increases, but it should increase less than that of exhausting gasoline. It is noteworthy that no plant for synthetic natural gas (or synthetic oil) production would be needed.

These concepts of the use of low-use times are, however, not entirely accepted by the electricity utility companies in the U.S.A., however rational they are.

The concepts discussed here are associated with, but not the same as, those of 'off-peak' power. There are *peaks* of power usage and at one time it was customary to build generating equipment which, when run fully, would cover the peak load. At present, the tendency is to have auxiliary equipment, to be turned on for peaking. Independently of the *peak* delivery, there is a mean delivery rate for several hours of the night which is less than that of the day. This is what is referred to here as 'low use-period power'.

FIXED COSTS IN HYDROGEN PRODUCTION

The cost of hydrogen can be broken down into sections: the electricity cost and fixed costs. What the latter comprise are shown in Table 9.4, adopted from Gregory.[41]

In addition, one has to recall that there are capital costs, i.e. one has to rent money to build a plant. Taking this into account, the cost

TABLE 9.4

COSTS, OTHER THAN THOSE OF ELECTRICITY, FOR PRODUCING 1 MBtu OF HYDROGEN

Maintenance and operating supplies	1.38 cents
Labour, \$5/hr	2.04
Overhead at 60% labour	0.90
Depreciation, 6.7%, local taxes and insurance, 2.3%	25.10
	29.42*

* This value, originated in 1971, will be increased by 33% to correspond to a mid-1974 estimate.

of an MBtu of H_2 (advanced type electrolyser) and without the electricity costs, would be 41.5 + 38.9 = 80.4 cents

COST OF ELECTRICITY TO PRODUCE ONE MBTU OF HYDROGEN BY ELECTROLYSIS

Let it be assumed that the hydrogen fuel will be combusted to produce steam. Then:

$$H_2 + \tfrac{1}{2}O_2 \rightarrow H_2O + 58.14 \text{ kcal.} \tag{9.25}$$

1 Btu is the heat to raise the temperature of a lb of water, 1°F. Hence, it is 450 × 5/9 calories = 250 calories. One needs, for 10^6 Btu, $2.5 \cdot 10^8$ calories and this is got from $\frac{2.5 \times 10^8}{5.814 \times 10^4}$ moles of H_2, from which the coulombs necessary are:

$$\frac{2.5 \cdot 10^8 \cdot 2 \cdot 9.6 \cdot 10^4}{5.814 \cdot 10^4} \tag{9.26}$$

Hence, if the cell voltage is E, the watts seconds are:

$$E. \frac{2.5 \cdot 2 \cdot 9.6}{5.814 \cdot 10^4} 10^{12}, \tag{9.27}$$

so that the kWh are:

$$E. \frac{2.5 \cdot 2 \cdot 9.6 \cdot 10^2}{5.814 \cdot 10^4 \cdot 10^3 \cdot 3.6 \cdot 10^3} \tag{9.28}$$

and the cost:

$$\frac{2.5 \cdot 2 \cdot 9.6 \cdot 10^{12} Ec}{5.814 \cdot 3.6 \cdot 10^{10}} \tag{9.29}$$

$$= 229 Ec, \tag{9.30}$$

where c is the electricity cost in cents $(\text{kWh})^{-1}$.

TOTAL COST OF ELECTROLYTIC HYDROGEN

The last two sections give the cost of an MBtu of H_2 in cents as (1974):

$$\text{COST} = 229 Ec + 80, \tag{9.31}$$

where E is the cell voltage and c the cost of the kWh (on a massive bases and at the site of the electricity-producing plant) in cents.

The larger contribution is from the electricity costs. Gregory[42] has suggested that the fixed costs can be halved in future electrolysers, on the argument that there will be more than a doubling of current density and hence a halving in size of plant per unit of fuel produced. The fixed costs tabulated above are based on a 1966 technology.*

* Thus, 'advanced electrolyser concepts', were largely developed as a spin-off from the significant support of fuel cells given from the space programmes. The support of such work was cut back increasingly from 1968, particularly as the pollutive issue began to be expressed in the press (1969 onwards).

With this assumption, the likely formula for new technology for H_2 is:

$$\text{COST} = 229Ec + 40. \tag{9.32}$$

This cost is shown for various values of E and c in Table 9.5.

TABLE 9.5

ESTIMATED COSTS PER MBtu PURE HYDROGEN AS A FUNCTION OF NEW TECHNOLOGY AND COST OF ELECTRICITY

	Cost of 1 kWh in mills					
Cell potential to give high rate production	2	5	7	10	15	20
1.2	0.95	1.77	2.31	3.12	4.51	5.00
1.4	1.04	2.00	2.04	3.60	5.31	6.80
1.5	1.08	2.11	2.80	3.83	5.54	7.26
1.6	1.13	2.27	2.91	4.04	5.85	7.72
2.0	1.26	2.69	3.60	4.98	7.27	9.56

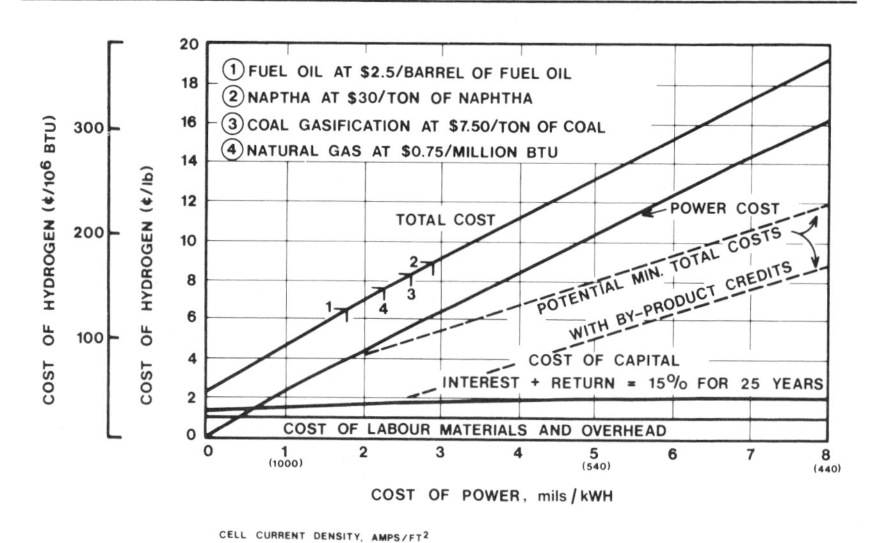

Fig. 9.10 Production cost of hydrogen *via* water electrolysis.[43] A 1973 estimate.

The competitive prices of natural gas, refined petroleum, etc., are likely to be in the range \$1.20-2.00/MBtu (Fig. 9.10[43] and 9.11[44]). Even a cell which works at 2 volts (old technology) will give H_2 in this range at rates often quoted for off-peak power (2-3 mils); or for the costs *predicted* (in 1973) for breeder reactor power in 2020. H_2 prices at near-gasoline prices can be achieved with medium priced electricity (massive quantities at site production, i.e. 7 mils) if the cell potential is 1.5v or less. If the price of lower is 10 mils a cell working at 1.5v would give hydrogen at about 60% more than the ex-refinery price of 1974 (\$2.44). *The electricity price is determinative.* Possible values applicable for massive amounts at source in 1974 vary from 4 to 10 mils, depending on situation. The feasible variation in cell potential is less than twice times. Low use-period power at 5 mil/kWh^{-1} (massive quantities, site of production) would make H_2 a practical fuel for motor vehicles)*c*. 12% greater in cost than gasoline) *even with the present undeveloped old-technology electrolysers.*

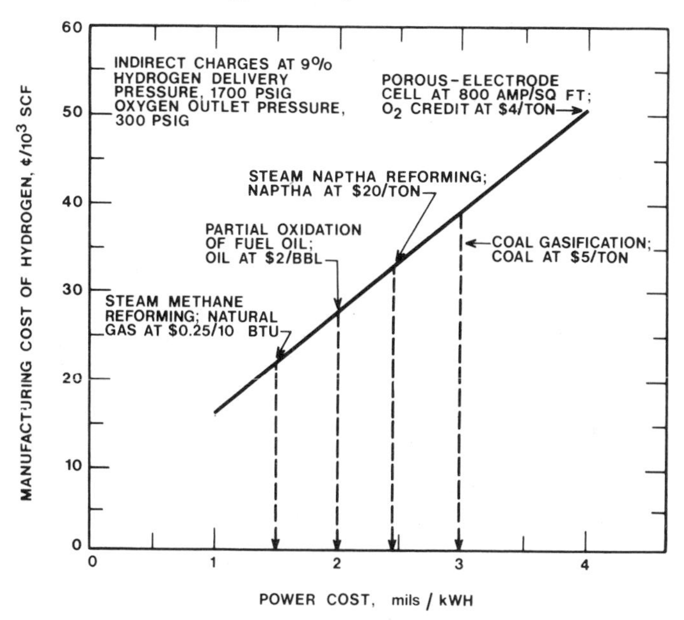

Fig. 9.11 Cost of electricity required for electrolytic hydrogen process to break even with four competing processes with credit assumed for by-product oxygen production; 40 million standard cubic feet of H_2 per day.[44] A 1969 view.

The production of H_2 from electrolysis has the advantages of simplicity, safety, complete purity, etc. To make them comparable, when considered on a practical long-term basis, CH_4 or CH_3OH would have to have added to their costs the price of containing the CO_2 which their burning produces and which cannot be injected into the atmosphere without the possibility of adverse effects after *c*. 2000.

EMERGING TECHNOLOGY

The fact that all the present large-scale electrochemical synthesis plants (Fig. 9.12[45]) producing hydrogen from water were designed without the advantage of modern knowledge of electrode kinetics has been discussed above. No systematic attempt has yet been made to apply the knowledge in this field to large electrolyser construction. However, during the 1960s, support was given to research and development of hydrogen-oxygen fuel cells, for the supply of auxiliary power to space vehicles. The largest investment here was in hydrogen-air fuel cells and two organisations developed the government supported research on fuel cells into the reverse process, the production of hydrogen and oxygen. Most of the discussions of the 'emerging technology' of electrolysers refer to these minor and short-lived spin-offs of

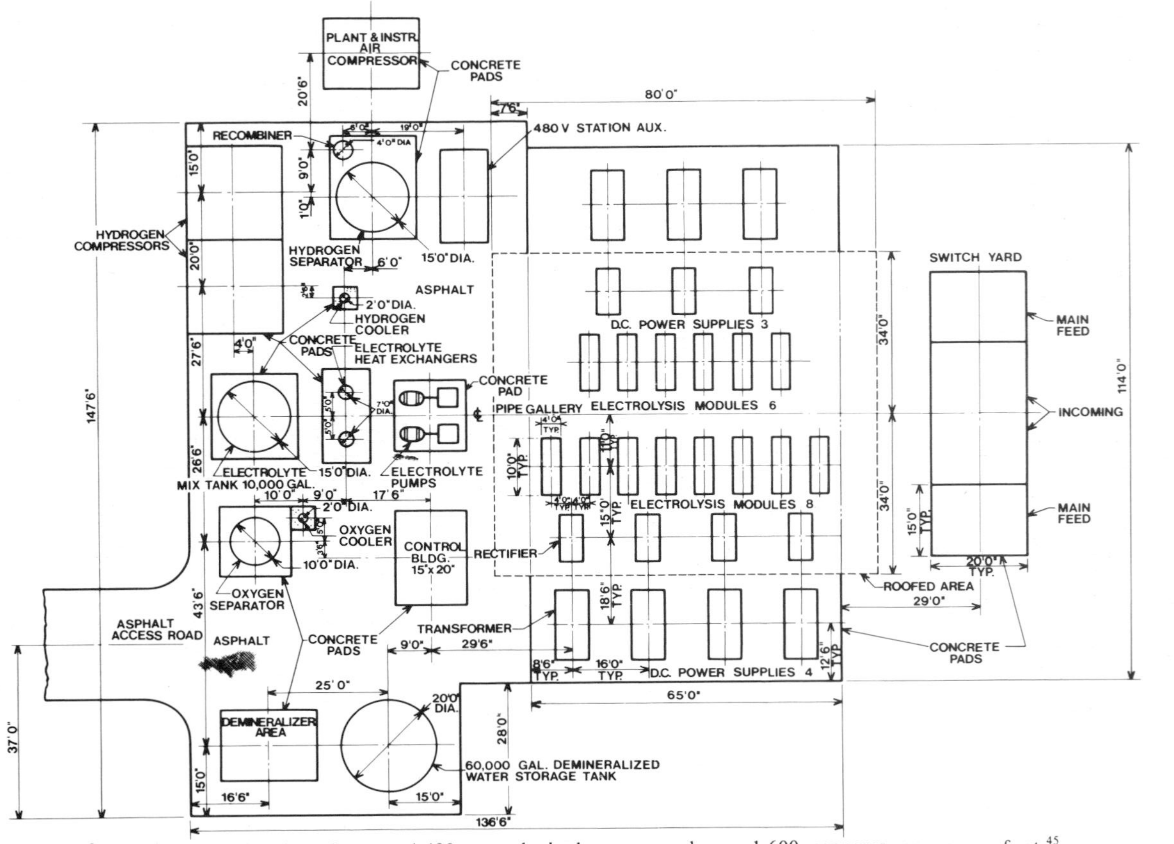

Fig. 9.12 Electrolysis plant layout. 4,400 pounds hydrogen per hour; 1,600 amperes per square foot.[45]

space research, and it is important, in judging the probability of the success of further investment in this direction, to note the almost negligible quantity (particularly in the U.S.A.) of research on modern hydrogen synthesisers.*

The two contributions (but they are 10 years old) to 'new' technology sprang from the laboratories of the Allis-Chalmers Company[46] and those of the General Electric Company.[47]

(1) The Allis-Chalmers contribution aims at reducing the *IR* drop between electrodes. It uses porous electrodes of Ni and these are separated by a membrane, against which the electrodes are pressed. This configuration (Fig. 9.13[48]) forces the bubbles of H_2 and O_2 back into the pores and they emerge on the opposite side of the electrode, i.e. not *between* the electrodes. Thus, they are prevented from entering the inter-electrode space, and thus do not contribute to the *IR* drop which then depends mainly upon the resistance of the membrane.

The maximum current density in the Allis-Chalmers cells is 1600 amps ft^{-2} at 93°C. This is a significant improvement on older technology, where, partly because of the difficulty of the large *IR* drop, current densities are less than 100 amp ft^{-2}. Because of the higher current density,

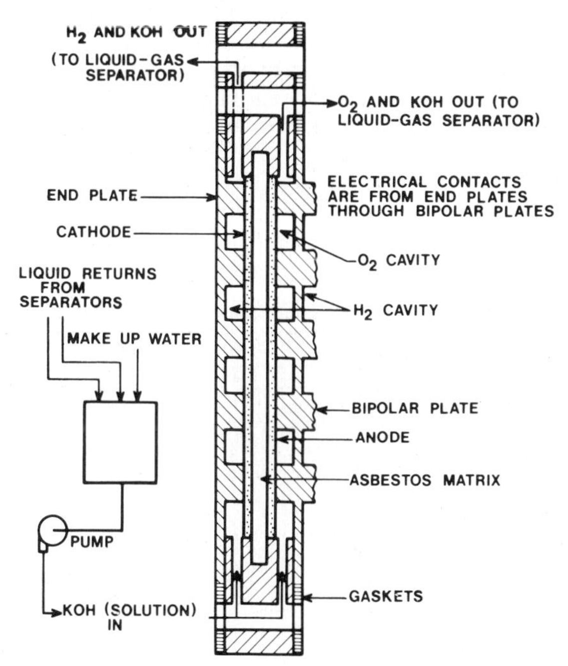

Fig. 9.13 Schematic design of end cell in Allis-Chalmers bipolar water-electrolysis cell.[48]

---

* In the United States, just one company, Teledyne Isotopes, has a small research team occupied with modern research into hydrogen production by electrolysis, whilst many thousands of workers, in dozens of companies, supported by hundreds of millions, attempt to gasify and liquefy exhaustible, polluting coal (needed for other uses) to C-containing products which will end up as CO_2. In particular, at the Institute of Gas Technology, which has contributed seminally to the concepts of a Hydrogen Economy, there was no research on the electrosynthesis of hydrogen in 1974.

the plant can produce 10 to 20 times more hydrogen per unit area occupied by cells, and this, and other size-intensive factors, cuts fixed costs. Conversely, the cell voltage in the Allis-Chalmers cell is high, about 2.2 volts for the current density indicated.

(2) The General Electric cell[47] uses steam at 1000°C. The cell is effectively a high temperature fuel cell with solid-electrolyte membrane, worked backwards. The steam (Fig. 9.14[49]) is injected and on one of the electrodes, the cathode, the water dissociates and leaves H_2 on the side where the contact has first occurred, + O^{--}. The O^{--} diffuses through the electrode, made of a conducting oxide, e.g. ZrO_2 (+ Y_2O_3, etc.). On the anode side, the O^{--} becomes O_2. The entry gas is steam mixed with hydrogen and the exit gas is predominantly hydrogen, mixed with steam.

The principal objective of this cell is to reduce the polarisation which is associated with the anodic reaction. There may have been, in the original objective, the aim of lowering the reversible potential with the thought that there was a greater ease of electrolysis (i.e. a lowering of cost per unit of hydrogen produced). It is not clear from the literature[47] whether the cell in the original experiments had external heating, but it seems probable.

The apparent performance of this high temperature electrolyser is better than any so far reported. The current density recorded is up to 3260 amps ft^{-2}, and the voltage is 1.3. The latter value corresponds to a reversible potential of about 0.95; at 1000°C, the polarisation at the anode can be regarded as negligible. The potential suggests an *IR* drop of about 0.35v.

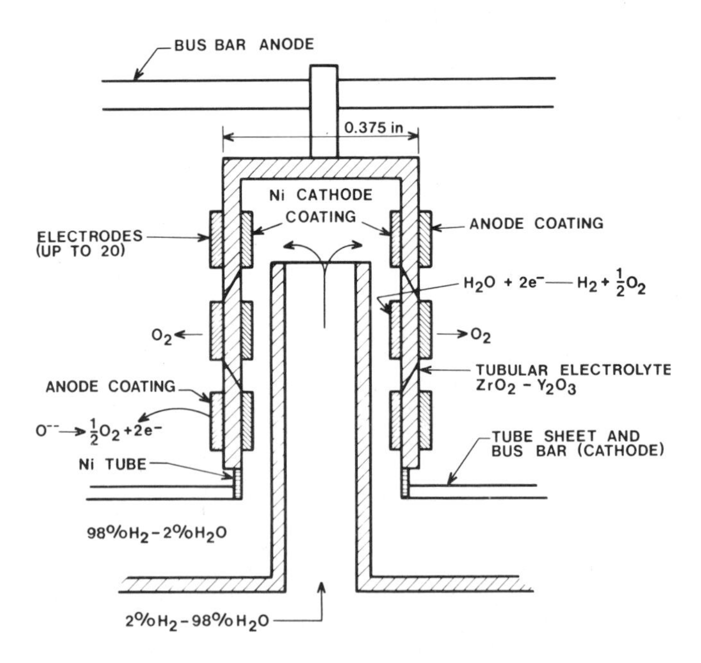

Fig. 9.14 Schematic design of single tube in proposed General Electric steam-hydrogen electrolysis cell.[49]

This electrolyser is working below the thermoneutral potential, and hence heat is used to run the cell, i.e. the I^2R heat contributes to the $T\Delta S$ of the needed energy. The heat balance of the cell, on which the reports were made, remains unclear and this must place in doubt the interpretation of the potential quoted (i.e., was an external heat source used for, if so, the cost for this heat would have to be added to those calculable from the potential alone). More research remains to be done in the manufacture, and in the stability and conduction of the high temperature-stable membranes.[50 51]

There are several questions open in the optimisation of electrolysers. The catalyst needs research. Catalysis reduces overpotential, and hence the total E to work the cell at a given rate. Hence, the cost of the electricity will fall with a good catalyst, and this gain has to be laid against the cost of the catalyst. Studies lack such information. The properties of very small amounts of catalyst, and their apparent specially active properties[31 32], is a possible approach.

Correspondingly, the Allis-Chalmers technique of eliminating the contribution of bubbles to the IR drop and hence to the cell voltage is only partly successful. A membrane has to be used and this introduces an IR drop and extra costs. Studies of the rapid flow of electrolyte between the electrodes (e.g. removal of the gas bubbles before they grow to size) would be of interest.

The advantages of cells which electrolyse water require optimisation studies to be made between catalyst loadings and the temperature. A mild temperature of *c.* 200° C, with a concentrated electrolyte, may make sufficient reduction of IR and η to optimise costs. For cells without an external heat source, the IR drop – which causes the heat to reduce η – increases running costs, and this must be layed against savings from the absence of catalyst.

Many aspects of the engineering and cost optimisation have still to be researched in this central field.

COST ESTIMATES DEVELOPED FROM EMERGING ELECTROLYSERS

Gregory and his colleagues at the Institute of Gas Technology have developed estimates of what could be expected in hydrogen price, as a function of the electricity cost. These are shown in Fig. 9.15.[52]

With 4-7 mil electricity, the predicted costs on an 'advanced' electrolyser are $1.65-$3.00 for one MBtu. If nuclear power at 3 mils became available, the cost would be $1.55 for an MBtu. However, these figures, which involve the assumption that all overpotential is negligible, are probably 25% below the minimum *feasible*.

IS A CATALYST WORTH HAVING?

In the electrochemical water decomposition reaction, the cell exhibits overpotential predominantly at the anode. This polarisation, to which the hydrogen cost is linearly related, is decreased by increasing temperature and reaches a negligible value at temperatures above 250-300° C.

The worth of a catalyst depends on the cost of electricity used if the extra potential, η, is still present in the absence of catalyst, and that of the metal catalyst. In alkaline solution, non-noble metal catalysts are possible; in acid solution, newer work gives rise to uses of tungsten

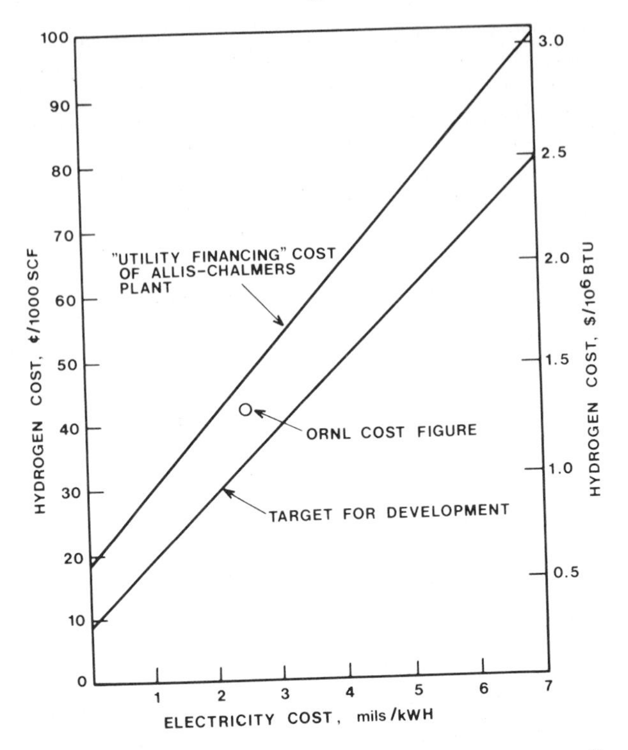

Fig. 9.15 Cost of hydrogen as a function of purchase price of electricity.[52]

bronzes as cathodes.[31] [32] The usefulness depends on the type of cell, e.g. whether the electrodes are porous, and the area within the pores which must be occupied by catalyst. Nickel sulphide coatings have proved useful on de-Nora's recent cells.[53]

The position is under-researched, although porous electrode theory and a knowledge of the exchange current density of the catalyst would allow calculation of effects (but data on the temperature coefficient of i_o are sparse[54]).

CORROSION

Some difficulties occur in hydrogen electrolysers due to corrosion, even at 100° C. The corrosion is caused by inner stray currents and is little studied.

LOW POTENTIAL ELECTROLYSIS COMBINED WITH THERMAL ASSISTANCE

Whilst the cost of bulk electricity at the site of production remains nearer to 10 mils $(kWh)^{-1}$ than to 3 mils $(kWh)^{-1}$, it may be advantageous to electrolyse another substance, for the decomposition of which to $H_2 + X_2$ the standard reversible cell potential is less than that for the decomposition of water. Then, one has to get back to O_2 by the reduction of X_2 with H_2O (which will cost heat).

Electrolysis of HI
The standard potential of the I_2 electrode at 25°C is 0.535. The standard entropy change in the reaction $2HI \rightarrow H_2 + I_2$ has a low negative value and hence the potential needed for decomposition of HI to the $H_2 + I_2$ will not be decreased with increase of temperature. Conversely, there is little disadvantage in increasing the temperature > 144°C, the m.p. of I_2. The kinetics of the $I^- \rightarrow \frac{1}{2}I_2 + e$ reaction are fast, i.e., the anodic reaction would contribute little overpotential, and this would be an advantage compared with the H_2O decomposition (overpotential of O_2 evolution is large at room temperatures).

The electrolysis part of the cost, assuming 0.65 volts for the real cell potential at, say, 1 amp cm^{-2} is \$1.27 for 7 mils electricity.

The I_2 produced would now be taken to O_2 by an analogue of the Deacon reaction:

$$I_2 + H_2O \rightarrow 2HI + \tfrac{1}{2}O_2. \tag{9.33}$$

The equilibrium is less favourable for I_2 to O_2 than Cl_2 to O_2. However, at 1400°C, HI exists at *c.* $10^{-1.5}$atm in equilibrium with H_2O, I_2 and O_2 at 1 atm. The chemical heat necessary to generate the amount of O_2 associated with this indirect synthesis of an MBtu of H_2 would cost (50% efficiency) *c.*81c $(MBtu)^{-1}$. If one (by comparison with the detailed costing of other processes) adds 50c per MBtu for fixed costs, one arrives at an estimated cost of H_2 by this method of \$2.58 per MBtu (7 mils electricity).

With HCl electrolysis, the potential is 1.41 volts – the electricity costs would be considerably larger. Chlorine evolution overpotential is negligible. The costs of the heat would be about 44c on the same basis as for the I_2-involved case. With 50 cents fixed costs, the hydrogen price per MBtu would be about \$3.61 (7 mils electricity).

Electrolysis of cuprous chloride
If one electrolyses CuCl in a cell containing HCl:

$$2Cu^{+} \rightarrow 2Cu^{2+} + 2e \tag{9.34}$$

$$2H^{+} + 2e \rightarrow H_2, \tag{9.35}$$

the cell potential is 0.356v under appropriate conditions.

The cupric chloride is decomposed by heat back to cuprous chloride. For this reaction, the standard enthalpy change is: + 39.6 kcal $mole^{-1}$.

Finally, the Deacon process would be used to produce $\frac{1}{2}O_2$ from Cl_2. The enthalpy change in this process is: + 13.68 kcal $mole^{-1}$.

The basic costs consist in the equivalents of these enthalpies, recalculated for the proportionate amounts for an MBtu of H_2. An efficiency factor must be taken into account in the thermal amounts and this depends on the temperature for the thermal equilibrium process of Cl_2 to O_2. It is arbitrarily taken here as 40%. The heat cost is assumed to be 50c per MBtu. Then:

$$\begin{array}{ll} \text{Heat for } Cl_2 \rightarrow O_2\text{:} & \left.\begin{array}{r}12c\\ 33c\end{array}\right] \times \dfrac{1}{0.40} = 1.12 \\ \text{Heat for } CuCl_2 \rightarrow CuCl + \tfrac{1}{2}Cl_2\text{:} & \end{array}$$

Electrolysis to give H_2: (7 mil electricity)	57c
Fixed costs, analogy to other processes, *c*:	50c
	$2.19

Electrolysis of Ferrous and Ferric Chloride
The hydrogen producing cell would be:

$$\begin{array}{l} 2Fe^{2+} \rightarrow 2Fe^{3+} + 2e \\ 2H^{+} + 2e \rightarrow H_2 \\ \hline 2Fe^{2+} + 2H^{+} \rightarrow 2Fe^{3} + H_2 \end{array} \quad (9.36)$$

The E_o for the ferrous to ferric equilibrium is 0.77 ($Fe^{2+} \rightarrow Fe^{3+} + e$). Hence, the electricity costs per MBtu are: $2.46.
Then, $FeCl_3$ is heated to produce $FeCl_2$ and chlorine.

$$2FeCl_3 \rightarrow 2FeCl_2 + Cl_2. \quad (9.37)$$

The enthalpy change in this direction is 30.60 kcal $mole^{-1}$, equivalent to 27 cents per MBtu.
The Cl_2 is taken to O_2 by the Deacon process at 12 cents per MBtu.
Taking the heat changes at 40% efficiency, and the fixed costs at 50c, one has:

$$\begin{array}{ll} \text{Electricity costs:} & 2.46 \\ FeCl_3 \rightarrow FeCl_2\text{:} & 0.27 \\ Cl_2 \rightarrow O_2\text{:} & 0.12 \\ \text{Fixed costs:} & 0.50 \end{array} \quad \left.\begin{array}{l} 0.27 \\ 0.12 \end{array}\right] \times \frac{1}{0.40}$$

The total is $3.93.
The process is deceiving. The potential quoted for the redox reaction is for one electron and the actual electrical costs are greater than for straight electrochemical decomposition of water.

Stannous to Stannic Electrolysis
Similarly to the foregoing examples, one anodically oxidises Sn^{2+} to Sn^{4+}; decomposes the stannic salt to a stannous salt thermally and brings the Cl_2 to O_2 *via* Deacon. The attraction here is that the stannous to stannic potential (a two electron change) is only 0.14 volts.

$$\begin{array}{ll} \text{Electricity costs:} & 0.22 \\ SnCl_4 \rightarrow SnCl_2\text{:} & 0.41 \\ Cl_2 \rightarrow O_2\text{:} & 0.12 \\ \text{Fixed costs:} & 0.50 \end{array} \quad \left.\begin{array}{l} 0.41 \\ 0.12 \end{array}\right] \times \frac{1}{0.40}$$

The total is $2.05.
There could be difficulties with the hydrolysis of $SnCl_4$.

ANODE DEPOLARISATION

The potential necessary to produce hydrogen electrolytically is contributed to by the O_2 electrode potential of 1.23v (standard value). If one could avoid this reaction by substituting a material which is cheap, or a pollutant which can be made inert by oxidation at the anode, it would be possible to achieve the pollutant removal. Correspondingly, if the anodic reaction of the pollutant removal occurs thermodynamically at a lesser potential than that of O_2 evolution, the removal of the pollutant will be accompanied by a drop in the price of H_2 (see equation 9.30).

Juda and Moulton[55] experimented on these lines with SO_2. They introduced this in the form of 6% H_2SO_3 into 30% H_2SO_4. The voltage for the electrolysis of water before SO_2 addition was 1.6-2.4 volts for the current density 1-500 ma cm^{-2}, and with SO_2, 0.74 to 1.62v for the same current density (Fig. 9.16[55]).

H_2SO_4 production is possible; NO from stack gas could be converted to HNO_3; sewage could be oxidised to CO_2. Catalyst costs could be involved.

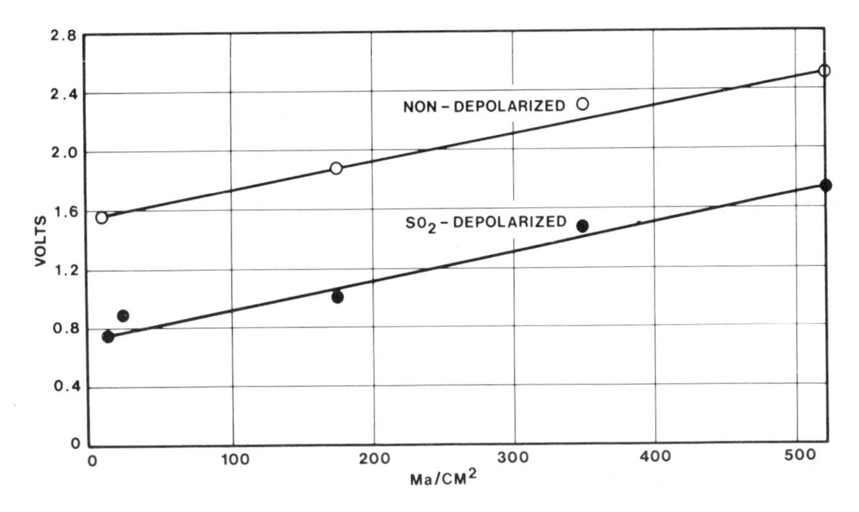

Fig. 9.16 The effect of SO_2 on electrolysis.[55]

PHOTO-ELECTROCHEMICAL METHODS

Introduction

The effect of light on electrode processes has been little studied. Early work[56, 57] gave effects which were ill-defined, i.e. were carried out on electrodes which had oxide surfaces and it was not clear to what extent they were photogalvanic or photoelectric. The first theoretical formulation of photogalvanic kinetics was due to Hilson and Rideal.[58] Bockris[56] formulated the first charge-transfer theory. Matthews[59] has formulated the effect in quantum mechanical terms. Several effects may be distinguished:

Photochemical effects

Here the light has effects on the solution and these have secondary effects on the interface. For example, certain organic materials may be raised to a higher valency state by light and then this state discharged at an electrode with charge transfer.[60]

Photogenerative fuel cells

The light produces the fuel for the fuel cell by photo-dissociating a product which is re-made in the fuel cell. For example, one might use:

$$2HI \rightarrow H_2 + I_2. \tag{9.38}$$

The hydrogen-bromine fuel cell is well known[61], and the hydrogen-iodine fuel cell[62] should give potentials to about 0.4 volts. Several solar-driven cells in series should drive one cell in the production of hydrogen. The costs here would be interest charges, plus maintenance. No expensive materials (e.g. no single crystal semi-conductors) would be required in the photo-dissociation. Allowing $100 per kW for the irradiation equipment and $200 per kW for the fuel cell, the interest costs at 15% would be $1.40 $(MBtu)^{-1}$ and, allowing 10c for maintenance, one comes to $1.50 $(MBtu)^{-1}$. However, efficiency would be very low because of the low degree of overlap between the solar spectrum and the range in which HI absorbs.

Among methods of energy conversion, the conversion of light at electrodes is the least investigated and one of the more attractive. Thus[63], the efficiency of conversion in *casual experiments* is about one-third that at cadmium sulphide cells, but the cost of the electrode materials would be much less than those for thick photovoltaics.

Direct photoelectrochemical generation of hydrogen

Interesting work has been carried out on the effects of light on hydrogen-oxygen cells which are shifted from equilibrium by light to produce hydrogen and oxygen. Reports have been scanty. Thus, Casey[64] reported in 1960 that a hydrogen-oxygen cell of unstated metal electrode material produced hydrogen at a few milliamps cm^{-2} when the oxygen electrode was irradiated with γ emission. A ferrous-ferric couple was present.

In recent work, Fujishima[2] reported the irradiation of the oxygen electrode in a hydrogen-oxygen cell, using acid electrolytes in the presence of salts. The current potential curve between light and dark is given in Fig. 6.15. The quantum efficiency was about 10% at 4150A°. The hydrogen-oxygen cell potential was about 0.5v and the maximum current was about 3 milliamps cm^{-2}. The oxygen electrode was a single crystal of TiO_2. The author did not simultaneously irradiate the hydrogen electrode. These (preliminary) results of Fujishima[2] indicate a power density of their cell as a solar converter in the milliwatt cm^{-2} range. (For solar cells, the power density is in the tens of milliwatts cm^{-2}.)

It is assumed that the entire manufacturing cost is represented by $C per lb.

Then, if the efficiency is ϵ in percent, one needs $100/\epsilon$ square metres at the equator at midday to gather a kW. Diurnal variation will make this about $300/\epsilon$ and weather effects, say, $600/\epsilon$ m^2. If the average thickness of the material is $t_{(cm)}$, the volume in cc is $6.10^6/\epsilon$ cc, and the weight $6.10^6\, t\, \rho/450\epsilon$ lbs. Hence, if the cost of the manufactured article is C$ per lb, the cost of the plant per kW is $6.10^6\, t\rho C/450\epsilon$\$ per kW, so that the $ per kWh are:

$$\frac{6 \,.\, 10^6 \, t\rho Cp}{450.365.24\epsilon.100}$$

(the diurnal and cloud effects are allowed for above), where p is the cost of money in %. Then, 291 kWh are equivalent to a MBtu so that the cost of this would be:

$$\frac{291.6.10^6 t\rho Cp}{450.365.24.\epsilon.100} = \frac{4.43}{\epsilon} t\rho Cp \text{ \$ per MBtu} \tag{9.39}$$

Not enough is known about the requirements of the photoelectrochemical method to make an assessment of the cost which is very meaningful. For the worst case, take $t = 0.01$ cm, $\rho = 7$, $C = 1$, $p = 15$, $\epsilon = 1$. Then, H_2 would cost \$4.65 per MBtu. However, if $t = 0.01$ cm, $\pi = 7$, $C = 0.25$, $p = 10$, and $\epsilon = 5$, hydrogen could cost \$0.16 per MBtu. It would be reasonable to say: $\epsilon = 3$, $t = 0.01$, $\rho = 3$, $C = 0.5$, and $p = 10$. Then, the cost would be \$0.26. One should double this cost to allow for a storage unit and add 50c per MBtu for maintenance. One could still look to a cost per MBtu of ~ \$1.

PHOTOSYNTHETIC

In normal photosynthesis photons in the region of 0.5μ in wavelength are absorbed by chlorophyl (a), and transferred over some distance, to a photon trap, which provides the activation energy for the reaction.

$$2H_2O \rightarrow 4H^+ + O_2 + 4e, \tag{9.40}$$

at the one site, whilst at the other site:

$$CO_2 + 4H^+ + 4e \rightarrow CH_2O + H_2O. \tag{9.41}$$

The formaldehyde polymerises to carbohydrates. The overall reaction:

$$H_2O + CO_2 + h\nu \rightarrow CH_2O + O_2 \tag{9.42}$$

is *endothermic* by 4.8 ev, and has to be driven. The quantum efficiency is 0.125, 8 quanta per unit of reaction. As the quanta are each about 1.8 ev, then about 2.6 quanta would be needed as a minimum – in practice 8 are needed to make carbohydrates.

The electron-doning reaction for photosynthetic H_2 evolution can be thought of as associated with the enzyme hydrogenase, known to produce hydrogen from water.[65–67] What has been done here occurs with hydrogenase and ferrodoxin (an electron carrier). But the oxygen interferes with the reaction, as was proved by adding sodium dithionate which reacts with oxygen, whereupon the reaction is immediately stimulated.[65–67] Probably oxygen reacts with hydrogen to produce water again.

Much more hopeful effects come from blue/green algae which produce hydrogen – *the most important advantage being that the reaction can be maintained in oxygen.*[67] The blue/green algae is grown in long chains of cells and contains heterocysts. These contain nitrogenase, which catalyses nitrogen to ammonia in protein and aids the nucleaic acid synthesis. Nitrogenase is labile in oxygen and the heterocysts have evolved a different kind of cell which excludes oxygen. But we know that nitrogenase in the absence of nitrogen gives hydrogen as well, without being poisoned by O_2, so that, presumably, the same mechanism for the exclusion of oxygen is occurring here (see Fig. 9.17[65]).

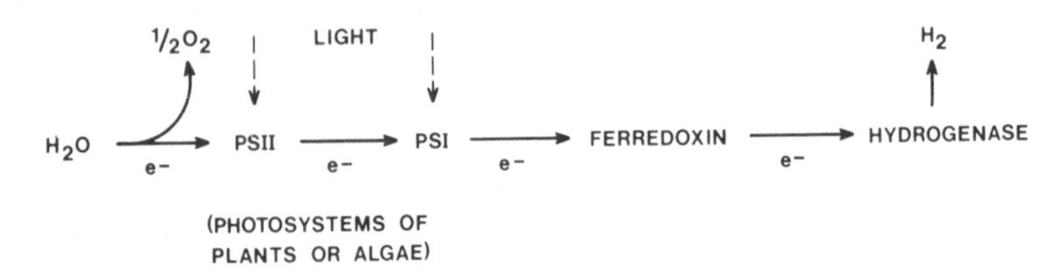

Fig. 9.17 Benemann's representation of the photosynthesis of hydrogen.[65]

The situation could be occurring electrochemically at two sets of sites in the cell. Speculatively:

$$2H_2O \rightarrow 4H^+ + O_2 + 4e. \qquad (9.43)$$

The e and perhaps H^+ then travel along the system to the heterocysts which contain the nitrogenase and thereupon evolution of hydrogen occurs by means of:

$$4H^+ + 4e \rightarrow H_2. \qquad (9.44)$$

Thus the electrochemical hydrogen-oxygen cell is being driven by light.

In a practical arrangement, a thin, but dense, layer of algae, or eventually chloroplast preparation, would be placed under the sunlight in a sealed container with a transparent top. The hydrogen and oxygen produced would be led to a combustion engine or fuel cell. Roof-top production of electrical energy might thus be possible. However, the engineering possibilities can only be very tenuously sketched. What of heat effects on the algae?

Efficiency
One could not at first assume a higher efficiency than those observed in natural photosynthesis (1-3%). Perhaps evolution of bacteria which could give a modification of nitrogenase could be reached. Photosynthetic efficiencies of 9% have been observed in the laboratory.[68]

Production rate from a given area
No quantitative measurements for chloroplasts of the production rate in the above situation exist. One could use the 6 grams m^{-2} day^{-1} (averaged over the whole year), which is a mean-rate of dry matter production observed in vegetation.[69] This corresponds to a 0.4% efficiency. If one increases this figure to 1.5% (the mean of the range for photosynthesis), the production is *c.* 23 gm^{-2} day^{-1}. One may obtain an order of magnitude result by taking this as equivalent to 23/18 moles of water or *c.* 1.3 moles of H_2 m^{-2} day^{-1} (averaged over the whole year).

Suppose we take an Si photocell collecting at 10% efficiency, this is equivalent to about 1 kW per 10 sq m at high noon, say, 0.01

kW m^{-2} averaged through the year. This is equivalent to about 4 moles H_2 m^{-2} day^{-1}.

Thus, there is some indirect evidence that photosynthetic H_2 could be of the same order as that equivalent to hydrogen from Si photocells, the electrolysis being taken as 100% efficient. (About 80% efficiency is possible at this time.)

Cost

Apparatus for receiving light on water and collecting evolved gases would be simpler than equipment implied in the other methods for obtaining hydrogen in this book and it seems reasonable to equate this part of the cost to the lower limit of fixed cost used above, i.e. 25c $(MBtu)^{-1}$. The major unknown cost is that of the enzyme, nitrogenase. It seems reasonable to assume, as it is made in nature, that it could be manufactured on a large scale at $1 per lb. Spread over an aqueous surface at the equivalent of 1μ thickness, and assuming a density of 1, this is equivalent to 0.2 cents m^{-2}. Supposing the enzyme lasts one month, it produces, at the above rate $365 \times (1.3/12) = 40$ moles of H_2 per month m^{-2} or $(475/4.3\ 10^3)$ of the MBtu unit, i.e. about 0.1 of an MBtu. Thus, enzyme costs may be about 2 cents per MBtu. This extremely speculative estimate suggests a price as low as 30c $(MBtu)^{-1}$ for H_2 for photosynthesis, using nitrogenase.

A major doubt here is whether the production rate assumed, based on those of other photosynthetic reactions, can be attained. The efficiency assumed is very low. Unless high efficiencies (5-10%) can be obtained, the method would scarcely be applicable on land because the areas needed would be so great.

HYDROGEN FROM SEA-BORNE AERO GENERATORS

The existence of constant wind belts, and areas of the world where average wind speeds are > 30 km per hour, indicates wind as an important and massive energy source. The fuel is free. The difficulty of large distances between the source and sink of this energy would be diminished by transmission in hydrogen (see Chapter 5).

Here, the cost of hydrogen at source is effectively the fixed costs, i.e. the cost of the money, and maintenance and insurance. The cost of wind generators of great size is subject to large variation in estimate. For large generators, the cost is likely to have a lower limit of $250 per kW (including the electrolysis device) and the corresponding price of H_2 at 10% on this money is $1.60 per MBtu and, with about 40 cents per MBtu for maintenance, insurance, etc., this suggests $2.00 per MBtu. Alternatively, one can take the cost of an aero generator as $250 per kW and this gives (15% money), 4.2 mils per kWh for the electricity. Using the equation already given (eq. 9.31), one obtains $1.78 per MBtu. Other authors suggest $2.17 (1973 dollars) for the year 2000[70]. $2.10 seems a reasonable estimate (1974 dollars), but this does depend on relatively cheap aero generators.*

---

* The NASA-AEC report[70] quotes lower cost limits similar to those taken here, but suggests an upper limit at *c.* $5 for $ MBtu of H_2 at 2000 in 1973 dollars.

PLASMA TORCH PHOTOLYSIS

The solar spectrum is shown in Fig. 6.1. Water absorbs at several regions between 0.5 and 2e.v. Thus, the frequencies in solar radiation which would photolyse water have been partly absorbed when solar radiation passes through the atmosphere and photolysis of water from sunlight on earth is not an attractive proposition.

Eastlund and Gough[71] have suggested that fusion reactors will make available photons from leakage plasma and these could be used to dissociate water on a large scale. If certain elements are injected into the hydrogen plasma, optical and u.v. of various wavelengths can be produced[72], e.g. if we add aluminum to the plasma, we obtain light in the region of 1800-1950A°, in which range water photolysis to hydrogen occurs.[1]

The leakage plasma, though very hot, is dilute (10^{10} atoms per cc). It would seem better to inject cool hydrogen and then give a stream with a high density ($10^{13}-10^{14}$ per cc) at a lower temperature.

Using the data of Ung *et al.*[1], 222,500 kWh of plasma energy would produce 1 ton of hydrogen. The main gain proposed in the method is that the thermal energy absorbed in the water cell could be converted to electricity at 30% efficiency in a heat engine. Then, 53,000 kWh of electricity could be produced per ton of hydrogen. The electricity could be used to provide hydrogen and then only 123,000 kWh of plasma energy would be required to produce 1 ton of hydrogen. As electrolysis requires 71,000 kWh to produce 1 ton of hydrogen, then, with a 40% efficiency reactor, 177,000 kWh is the thermal energy requirement, compared with 123,000 kWh of plasma energy.

Thus, the authors suggest that only about 70% of the energy needed to obtain H_2 by electrolysis would be needed in photolysis, but this is rough indeed, because it neglects both losses before the plasma reaches the cell and also gains from the use of the waste heat. The economic status of the method is unclear until the cost of the fusion heat is known. The recombination rate of H and OH radicals, and the absorption efficiency of the radiation seems to have been neglected.

SEAWATER AS A SOURCE OF ELECTROCHEMICAL HYDROGEN

Large breeder reactors placed on the sea to reduce thermal pollution, solar collectors working from ocean thermal gradients and the use of constant wind belts to drive aero-generators on the sea, all indicate seawater, and its direct electrolysis, as the most convenient source of hydrogen on a very large scale.

There has been little consideration of the degree to which O_2 or Cl_2 respectively will be evolved, although Cl_2 can be prepared by the electrolysis of brine. There is a tacit assumption underlying some discussions of the Hydrogen Economy that O_2 alone is evolved in seawater electrolysis. An elucidation is necessary.

Thermodynamic properties of H_2-O_2 cells

The thermodynamic cell potential is[74]:

$$E = \left[1.23 + \frac{RT}{F} \log\left(a_{H^+} \right)_{an} + \frac{RT}{4F} \log p_{O_2} \right]$$

$$-\left[\frac{RT}{F}\log\left(a_{H^+}\right)_{cath} - \frac{RT}{2F}\log p_{H_2}\right]$$

$$= 1.23 + \frac{RT}{4F}\log p_{O_2}\, p^2_{H_2}. \quad (9.45)$$

Hence, if both electrodes are at the same pH, the cell potential is independent of the value of that pH. The pH of sea water will be near to seven. As electrolysis proceeds, and, if it occurs in a confined space, the regions near the cathode become alkaline, and sodium hydroxide is locally produced, whereas the areas near the anode will have an excess of H^+ and HCl will be produced.

Thus, in an extreme case in which the anodic environment reaches a pH of O and the cathodic one a pH of 14, one has (for $P_{H_2} = P_{O_2} = 1$ atm):

$$E = 1.23 + \frac{RT}{F}\ln 1 - \frac{RT}{F}\ln 10^{-14} \quad (9.46)$$

$$= 2.04 \text{ volt.} \quad (9.47)$$

$$\therefore E_{\substack{cell\\thermo}} = 1.23 - 2.04. \quad (9.48)$$

Thermodynamic properties of H_2–Cl_2 cells

The standard potential of the Cl^-–Cl_2 couple is 1.31 volt and at a concentration of Cl^- of 0.47, and 1 atm Cl_2, the thermodynamic potential is 1.35v. Thus, on the standard hydrogen scale:

$$e_{REV/Cl_2\,(sea)} = 1.35. \quad (9.49)$$

The thermodynamic potential at which H_2 will be evolved is dependent on pH. Assuming pH = 7, the reversible hydrogen potential is 0.058 log $a_{H^+} = -0.406$. Thus, under the condition given, the thermodynamic potential of the H_2–Cl_2 cell would be 1.76.

The potential near a cathode evolving H_2 will become alkaline. The pH attained will depend on agitation and the geometry of the situation. As an extreme let it be assumed that it reaches 14. Hence:

$$e_{REV/H_2\ 1\,atm} = -0.058 \times 14 = -0.812. \quad (9.50)$$

Hence, under circumstances which would lead to the high pH near the cathode,

$$E_{cell\ H_2\ Cl_2} = 2.16 \text{ volt.} \quad (9.51)$$

The *thermodynamic* potential for an H_2–Cl_2 cell in sea water will therefore be between:

$$1.8–2.2 \text{ volt}, \tag{9.52}$$

depending on current density, for, the higher this is, the larger the necessary applied potential, due to the increased pH.

Effect of kinetics
As an approximation, one can assume the hydrogen electrode in water electrolysis to have a negligible polarisation, so that the polarisation is to be attributed to the anodic reaction. This is a good approximation for the hydrogen-oxygen cell, poor for H_2–Cl_2. Using as order of magnitude parameters:

$$(i_o)_{H_2O \rightarrow O_2} = 10^{-10} \text{ amp cm}^{-2} \tag{9.53}$$

$$(i_o)_{Cl^- \rightarrow Cl_2} = 10^{-3} \text{ amp cm}^{-2}, \tag{9.54}$$

the overpotential which must be added to the approximate thermodynamic potentials given is[75]:

$$\eta = \frac{RT}{\alpha F} \log \frac{i}{i_o}, \tag{9.55}$$

for a current density, i, with $\alpha \simeq \frac{1}{2}$.

Thus, the actual cell potential needed to evolve H_2 from seawater at a current density, i, neglecting the effects of transport (which will become important only at higher current densities), will be:

$$E_{H_2\text{-}Cl_2} = (1.8 - 2.2) + \frac{2RT}{F} \log \frac{i}{10^{-3}} \tag{9.56}$$

$$E_{H_2\text{-}O_2} = 1.23 + \frac{2RT}{F} \log \frac{i}{10^{-10}} . \tag{9.57}$$

Thus, the H_2–Cl_2 cell begins to function at a higher applied potential difference than the H_2–O_2 cell. However, once the thermodynamic conditions are reached whereby H_2 and Cl_2 can be evolved in seawater, the Cl_2 evolution rapidly becomes more favourable. This will occur for an applied potential cell potential between 1.8 and 2.2v. *IR* drop is neglected in the above calculation: the *R* can be reduced greatly by engineering design, but in any case at:

$$E_{\text{cell, real}} - IR > 1.8, \tag{9.58}$$

the evolution of Cl_2 will start to displace the evolution of O_2 in the production of H_2 from seawater.

The course of the likely current-potential relations is shown in Fig. 9.18. The figure is approximate, though more than schematic. The values plotted are inexact because of the ambiguity of the reversible cell potential,

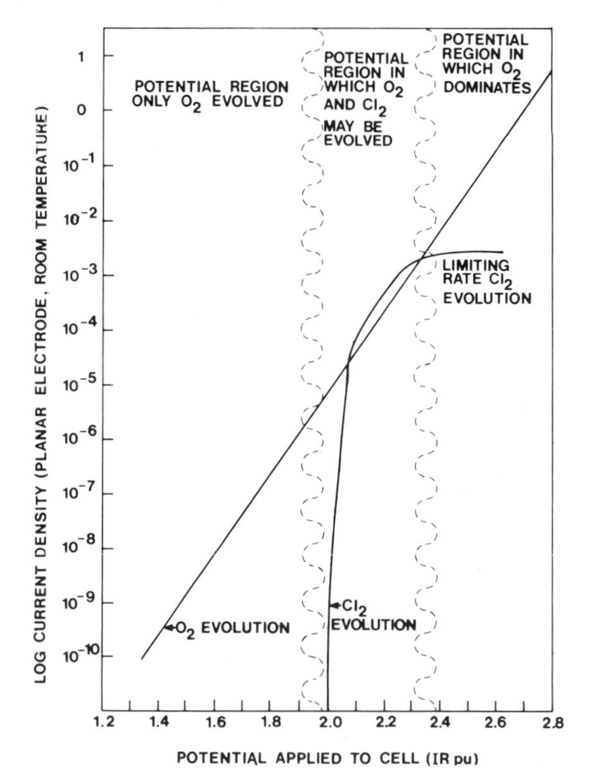

Fig. 9.18 Rough computations for planar electrode configurations suggest chlorine evolution from seawater exceeds O_2 evolution in a cell potential range (excluding *IR*) of about 2.0-2.2 volt (with possibility of competitive evolution beginning at 1.8 v).

which depends on the pH near to the electrodes. This latter quantity depending upon the degree to which free access of seawater is allowed.

Transport effects

Cl_2 evolution will replace O_2 evolution between 1.8 and 2.2v. However, the transport of Cl^- ions to the electrode is limited, whereas there is no corresponding problem with the availability of water. Hence, there will be a potential sufficiently high at which there will be no further increase in the evolution rate of Cl_2, whereas that of O_2 will increase with potential. If ΔE is the increase in cell potential over that at which $i_{O_2} = (i_L)_{Cl_2}$, then:

$$\frac{2RT}{F} \ln \frac{i_{O_2}}{(i_L)_{Cl_2}} \quad \text{at } \Delta E \tag{9.59}$$

Thus, at sufficiently high cell potentials ($\Delta E \simeq 0.2$ volt above that at which the limiting current for Cl^- sets in), the Cl_2 content of the evolved gas will be less than 1% (Fig. 9.18). A precise calculation of the transport control of Cl^- will depend (as with the pH at the interface) upon a knowledge of the transport conditions, e.g. natural and forced convection.[76] This, in turn, depends not only upon the flow of water in a macro-sense, but in the diffusion and normal convection conditions in porous systems, apart, of course, from temperature, forced convection and etc. A calculation (which can be made[76]) implies knowledge of the system geometry. It is enough to use here a rough and simple calculation of the limiting current density,

$$i_L = 0.02\, nC, \tag{9.60}$$

in which n is the number of electrons involved in the reaction, C the ionic concentration in moles l^{-1}. The expression[56] implies zero agitation and gives the *minimum* limiting current, that due to diffusion and natural convection in the steady state at a planar electrode without pores. As agitation increases, the i_L increases until it is 10-100 times greater than the 'still' value. We arbitrarily take the factor as 25, whereupon, with $n = 2$, $C_{Cl^-} \simeq 0.4M$,

$$i_L = 0.4 \text{ amp cm}^{-2}. \tag{9.61}$$

Assuming this as the limiting current for Cl_2 evolution from seawater in the considered geometry, there will develop, as the real current density approximates 0.04 amp cm^{-1} and above, a rapidly increasing extra overpotential (the concentration overpotential), according to[77]:

$$\eta_c = \frac{RT}{F} \ln \left[1 - \frac{i}{i_L}\right]. \tag{9.62}$$

As $i \rightarrow i_L$, η_c becomes very large and there will be no further increase in Cl_2 evolution as the applied potential increases.

Fig. 9.18 schematizes these predictions. Reality would differ from Fig. 9.18 due to approximations made in respect to pH at the interface (it has been assumed that it is such as to displace the reversible potentials 0.2v); neglect of H_2 overpotential; approximate values for i, η, Cl_2 and O_2 evolution; assumption of planar conditions for Cl_2 evolution; and etc. One concludes that, up to about 1.8 v., Cl_2 evolution can be neglected, that after 2.2, it will predominate, but, that after 2.6 volts, it will become a rapidly decreasing component of the evolved gas.

However, the *higher* range of potentials in which Cl_2 evolution rates increase no more, but H_2 and O_2 evolution increase exponentially with the applied potential, is not a practical range. Thus, at > 1 amp cm^{-2}, *IR* drop and heating will be too large, and the applied potential uneconomic (see equation 9.31).

The aim will be, then, to achieve evolution at potentials < 1.8 volts. It is not possible, without more detailed transport calculation, and without knowledge of the degree of lowering of the anode potential (catalysis, depolarisation), to know the practicality of such a potential. Thus, the use of such measures, and special electrode structures, may achieve, or partly achieve, the desirable situation of negligible Cl_2 evolution (Fig. 9.18 has been calculated for planar electrodes, 25° C, and $i_o = 10^{-10}$ for O_2 evolution). Although the calculations on which Fig. 9.18 is based are limited and crude, it seems likely that, under practical conditions for seawater, a mixture of Cl_2 and O_2 would be evolved. The fact that electrode area is of less importance on or under the sea, makes a lower current density (and hence a cell potential nearer that at which Cl_2 does not evolve) acceptable.

Methods of dealing with chlorine evolution

Several methods may be available for preventing Cl_2 evolution into the atmosphere, which is, of course, unacceptable. Reactors on the sea are likely to be in deep water, because of the desirability of having cold water for cooling (ocean thermal gradient collectors would need deep water, too). Suppose the electricity manufactured on the surface was transmitted to a depth of 1 km, the pressure available would be *c.* 100 atm. H_2 could be evolved at this pressure for an extra potential (for H_2 and O_2) of about 0.1 volt, with a corresponding advantage for storage at depth.

(a) Assuming that the evolution is at depth, Cl_2 could be released into the sea. The reaction:

$$H_2O + Cl_2 \rightarrow HCl + HOCl \tag{9.63}$$

is likely. The HCl would react with the NaOH produced and re-establish the original constituents of seawater. The hydrogen could be trapped and stored at pressure (see Chapter 10).

(b) The Cl_2 could be collected at the surface platform and the reaction:

$$Cl_2 + H_2O \rightarrow 2HCl + \tfrac{1}{2}O_2 \tag{9.64}$$

performed, at 650°C. The process would add to the cost because heat would be needed to drive the reaction to the right. O_2 would then be vented and HCl redissolved. The HCl-NaOH reaction heat might be recoverable and used to provide fresh water.

(c) If these approaches are not economic, desalination would be necessary. In the presence of a nuclear reactor, on a floating platform, this would be the most acceptable method, because of the nuclear waste heat. The electrolysis might better be carried out in the sea depth for reasons of high pressure storage. KOH would be the inexhaustible electrolyte. Solar platforms might have difficulty with economic large scale desalination: the costs of this hitherto have been placed 5-10 times above that of nuclear.

Untoward effects of seawater

A number of material difficulties will arise in sea-borne plants and have been referred to in connection with sea-borne platforms for solar sea power. Aluminum is the preferred material for anti-corrosion properties. Barnacles, etc. can be eliminated by means of a minute concentration of Cl_2 in the water, easily produced by making the plant materials anodic at very low current densities in sea water (An auxiliary cathode would be necessary). Storms can be faced either by careful anchoring or more likely by providing the platforms with a form of mobility.

CREDITS

In the estimates made for the cost of H_2 given here, no credits have been assumed for the use of O_2. This is produced as a by-product of water electrolysis and, in the initial stages of a hydrogen economy, could be sold, i.e. an O_2 credit could be allowed for in reducing the price of H_2. In the longer term, the cost of piping O_2 over long distances; and the saturation of the market, may mean that the O_2 credit is diminished towards zero. Conversely, the uses of O_2 may be greatly increased (see Chapter 14). At a plant site, O_2 would be extremely cheap.

It is not possible, at present, therefore, to give a significant O_2 credit for the long term. Selling O_2 in 1975 would yield a credit of *c.* \$0.3 per MBtu of hydrogen[78].

SUMMARY OF COST ESTIMATES

These are given in Table 9.6[79] and Fig. 9.19.[80]

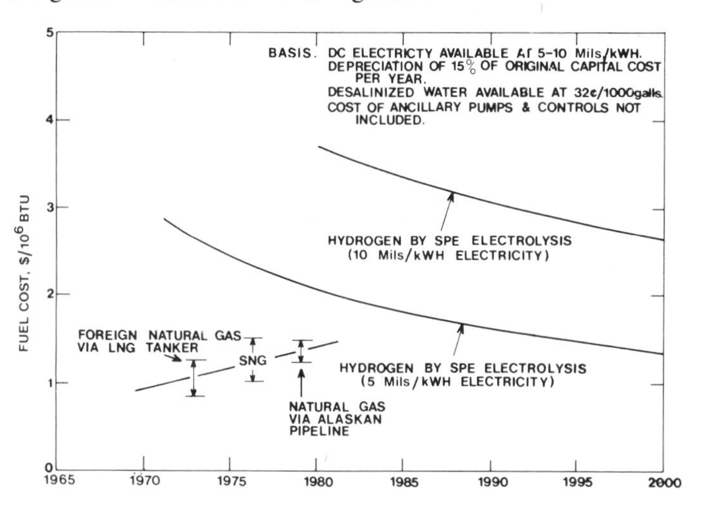

Fig. 9.19 Projected cost of hydrogen production by SPE electrolysis.[80]

DISCUSSION

The first thing is the overwhelming influence on the cost of electrolytic hydrogen of the price of electricity. In the presence of 3 mils electricity, projected as available from breeder reactors,[39] the classical electrolyser, without further development, could produce gaseous hydrogen cheaper (\$2.17-2.31 per MBtu) than gas from cheap coal (*c.* \$2.65-3.10 per MBtu) and at less than the 1974 price of gasoline. Thus *when cheap electricity from breeder reactors is available,* it will be impossible to compete with hydrogen as a fuel and at prices (1975 dollars) of $<$ \$2.50 per MBtu.

The degree of availability of low-use-time power at this time is of consequence for the commencement of test areas for the Hydrogen Economy. According to Hietbrink *et al.*[40], about one-quarter of the automotive energy need could have been supplied by batteries charged through low-use power, even with the utility plant of 1972. The same fraction could be suppliable for H_2-driven cars. In 2000, the entire automotive

TABLE 9.6[79]

COST ESTIMATES COMPARED (SEE FIG. 9.19[80])

Method	Estimate for MBtu, 1974	Remarks
Coal	2.65-3.10	Coal assumed $14 per ton. 50c added purify 96% → 99.9%. Process not yet realised. $1.50 quoted June 1973 for 96% H_2 from coal at $8 per ton^{-1}.
Cyclical thermal decomposition	1.40-2.10	50% efficiency *assumed.* $1.40 accepts author's estimate fixed costs = steam reforming.[10] $2.10 takes fixed costs as doubled. Process not realised.
Electrolytic, classical	1.76-2.72-5.66	Range is for 2-4-10 mils electricity. Technology obsolete.
Emerging technology	1.09-1.77-3.83	Range is for 2-4-10 mils electricity. Good experimental basis.
Solid electrolyte, 1000°C + atomic heat	2.14	Assumes no *IR* heat, negligible overpotential.
Anode depolarised aqueous	1.51	Assumes available zero cost CO, SO_2, NO, sewage in large quantities used in reaction at anode. Experimental basis.
Electrolysis of HI	2.35	At 7 mils. Some experimental basis.
Electrolysis of CuCl	2.19	At 7 mils. Outlook for feasibility excellent.
Electrolysis of $FeCl_2$	3.93	At 7 mils.
Electrolysis of $SnCl_2$	2.05	At 7 mils. Side products?
Photodecomposition of HI	1.50	Some experimental basis.
Photoelectrochemical direct to H_2	1.00	Assumes negligible use noble metals. Experimental basis at 3% efficiency.
Photosynthesis, nitrogenase enzyme	0.30	Speculative estimate. Assumes unknown reaction rate is middle range other photosynthetic reactions; assumes nitrogenase cheaply mass-producable.
Aero generators	2.10	Cost of aero generators main contribution. As yet speculative.
Plasma torch photolysis	See remarks	Depends on price of fusion produced electricity. 70% cost of electrolysis using that electricity. Hypothetical.

load in the U.S.A. could be supplied. If 6 mils kWh^{-1} is the price at the electricity producing plant of this electricity, the corresponding hydrogen fuel would cost between \$2.50 and \$3.00 (Ex-refinery price of U.S. gasoline July '74 $\simeq$ \$2.44 per MBtu).

The second point is that steam electrolysis at high temperatures is more attractive than the modern form of low temperature electrolysis, and the development of the high temperature situation is to be preferred, so long as nuclear heat, the needed auxiliary, is available at relatively cheap rates. (In 1975 currency, $<$ 75c per MBtu; and as long as reasonable progress is made in O^{--} ion conducting solid electrolyte).

Thirdly, one of the intermediate solutions to cheap hydrogen, one which would be viable during the mixed economy of the first years of the introduction of a Hydrogen Economy, is anode depolarised electrolysis, from which pure hydrogen could now compete with dirty hydrogen from cheap coal, using 7 mils electricity.

Fourthly, of the thermally assisted two-step processes, CuCl electrolysis seems the most attractive. It would operate with seawater to give hydrogen at *c.* \$1 $(MBtu)^{-1}$ less than that of emerging technology water electrolysis.

The photo-oriented methods look attractive, although the research is just beginning. The cost estimates given here are tenuously based, and some of the extrapolations may turn out to have been too optimistic. It depends principally upon whether an efficiency of at least 3% can be attained. But the main cost of breaking the H-O bond in all the other methods is the cost of the fuel. Light is free.

RECOMMENDATIONS

(a) The light-oriented methods of going directly to hydrogen are the most hopeful-looking and the strongest recommendation is that wide-ranging support for investigations of the theme 'photo-production of hydrogen' should be given.

(b) *There seems no rational cause for the absence of considerable quantities of gaseous hydrogen at prices marginally more than the 1974 price of U.S. gasoline using low use-period power.* The storage qualities of hydrogen pipelines should make such a development dependent only on building the electrolysis plants. No *new* technology is needed were the low use power to be sold at 6 mils kWh^{-1} or less, although, to attain this result, the emergent new technology would have to be scaled up.

(c) The most important ideas for the new electrolysis of water as yet non-emergent are: anode depolarised; and CuCl electrolysis with thermal conversion of Cl_2 to O_2.

(d) As seawater seems the eventual source of H_2, the reconversion of Cl_2 to O_2 should be examined.

(e) Atomic heat must be cheaper than the corresponding energy in electricity. Its best application is in the production of hydrogen in aiding the one-step (Carnot-efficiency-loss-free) electrosynthesis of hydrogen from steam by supplying the entropic heat and thus reducing electricity costs. This one-step simple way of using atomic heat seems a more efficient use than in multi-step chemical processes because of difficulties expected in attaining practical versions of multi-step high temperature processes which are only acceptable if very highly cyclical.

REFERENCES

[1] A.Y.-M. Ung & R.A. Back, *Can. J. of Chem.*, **42**, 753 (1964).
[2] A. Fujishima & G. Honda, *Nature*, **238**, 38 (1972).
[3] G.H. Gough, paper presented at the Cornell National Symposium on the Hydrogen Economy, August 1973.
[4] A.L. Hammond, W. D. Metz & T.H. Maugh II, 'Energy and the Future', American Association for the Advancement of Science, Washington, D.C., 1973.
[5] Memo. on Project S-128 to Sponsors Committee, Chicago, Institute of Gas Technology, 30 October 1959.
[6] J.E. Funk & R. Reinstrom, *I. & E.C. Process Design Develop.*, **5**, p. 336 (July 1966).
[7] J.E. Funk, W.L. Conger & R.H. Carty, paper presented at the Hydrogen Economy Miami Energy (THEME) Conference, March 1974.
[8] J.E. Funk, paper presented at the Am. Chem. Soc. Div. Fuel Chem., p. 79 (April 1972).
[9] C. Marchetti, *Chemical Economy & Engineering Review*, **5**, 7 (1973).
[10] 'Hydrogen Production from Water Using Nuclear Heat', Progress Report No. 3, EURATOM Joint Nuclear Research Centre, Ispra, Italy, December 1972.
[11] G. deBeni & C. Marchetti, 'Mark-1, A Chemical Process to Decompose Water Using Nuclear Heat', paper presented to the Am. Chem. Soc. Meeting, Boston, 9 April 1972.
[12] J.O'M. Bockris, J.D. Mackenzie & J.L. White, *Physicochemical Measurements at High Temperatures*, Butterworths, London, 1960, p. 95-97.
[13] *ibid.*, Appendix 5.
[14] J.O'M. Bockris & A.K.N. Reddy, *Modern Electrochemistry*, Rosetta Edition, Plenum, New York 1973.
[15] J.O'M. Bockris, N. Bonciocat & F. Gutmann, *An Introduction to Electrochemical Science*, Wykeham press, London, 1974, p. 78.
[16] J.O'M. Bockris & D. Drazic, *Electrochemical Science*, Taylor and Francis, London, 1972, p. 7.
[17] D. Gregory, assisted by P.J. Anderson, R.J. Dufour, R.H. Elkins, W.J.D. Escher, R.B. Foster, G.M. Long, J. Wurm & G.G. Yie, "The Hydrogen-Energy System", prepared for American Gas Association by I.G.T., 1973, p. III-3.
[18] *ibid.*, p. III-4.
[19] J.O'M Bockris & S. Srinivasan, *Fuel Cells: Their Electrochemistry*, McGraw-Hill, New York, 1970.
[20] J.O'M Bockris & A.K.N. Reddy, *Modern Electrochemistry*, Rosetta Edition, Plenum, New York, 1973, p. 1132.
[21] J.O'M Bockris & S. Srinivasan, *op cit.*, p. 541.
[22] D. Gregory *et al.*, *op. cit.*, p. III-16.
[23] K.J. Vetter, *Electrode Kinetics*, Springer, 1955.
[24] J.O'M Bockris & A.K.N. Reddy, *op. cit.*, p. 1135.
[25] *ibid.*, p. 1007.
[26] *ibid.*, p. 1369.
[27] J.O'M. Bockris & S. Srinivasan, *op. cit.*, p. 289.
[28] *ibid.*, p. 633.
[29] B. Cahan, Thesis, University of Pennsylvania, 1968.
[30] A. Damjanovic, D. Sepa & J.O'M Bockris, *J. Res. Inst. Catalysis*, Hokkaido Univ., **16**, 1 (1968).
[31] J. McHardy & J.O'M. Bockris, *J. Electrochem. Soc.*, **120**, (1973).
[32] *ibid.*, **120**, 53 (1973).
[33] A. Tseung & H. Bevan, *J. Electroanalytical Chem.*, **45**, 429 (1973).
[34] A.K. Stuart, paper presented at the American Chem. Soc. Symposium on Non-Fossil Fuels, Boston, 13 April 1972.
[35] J.O'M Bockris & D. Drazic, *Electrochemical Science*, Taylor and Francis, London, 1972, p. 222.
[36] D. Gregory *et al.*, *op. cit.*, p. III-18.
[37] Table supplied by U.S. National Petroleum Council, 1970.
[38] P. Sporn, priv. comm., 25 January 1974.
[39] A.M. Weinburg & R.P. Hammond, *Am. Sci.*, **58**, 1970, July/August, p. 412.
[40] E.H. Hietbrink, J. McBreen, S.M. Selis, S.B. Tricklebank & R.R. Witherspoon,

in *The Electrochemistry of Cleaner Environments*, ed. J.O'M. Bockris, Plenum, New York, 1972.
[41] D. Gregory *et al.*, *op.*, *cit.*, p. III-29.
[42] *ibid.*, p. III-37.
[43] J.W. Michel, paper presented at the American Chem. Soc. Div. of Fuel Chemistry, 116th National Meeting, August 1973.
[44] J.E. Mrochek, 'The Economics of Hydrogen and Oxygen Production by Water Electrolysis and Competitive Processes', in W.W. Grigorieff, ed., 'Abundant Nuclear Energy', p. 107, Washington, D.C., U.S. Atomic Energy Commission, 1969.
[45] R.L. Costa & P.G. Grimes, Chem. Engr. Progr. Symp. Ser. No. 71, **63**, 45 (1967).
[46] D. Gregory, *op. cit.*, p. III-16.
[47] *ibid.*, p. III-17.
[48] Figure from the Allis-Chalmers Manufacturing Co..
[49] Figure from the General Electric Co..
[50] T.H. Estell & S.N. Flengas, *J. Electrochem. Soc.*, **118**, 1890 (1971).
[51] E.S. Volchenkova & Y.M. Nedopekin, *Inorg. Materials*, **9**, No. 6, 960 (1973).
[52] D. Gregory *et al.*, *op. cit.*, p. III-32.
[53] P. Gallone, priv. comm., 11 February 1974.
[54] J.O'M. Bockris, A. Damjanovic & R. Mannan, *J. Electroanal. Chem.*, **18**, 349 (1968).
[55] W. Juda & D. McL. Moulton, Chem. Eng. Symp. Series, p. 59 (1972).
[56] J.O'M. Bockris, in *Modern Aspects of Electrochemistry*, Vol. 1, ed. J.O'M. Bockris, Butterworths, London, 1954, p. 243.
[57] Copeland, Black & Garrett, *Chem. Rev.*, **31**, 177 (1942).
[58] R. Hilson & E.K. Rideal, Proc. Roy. Soc., **199A**, 295 (1949).
[59] D.B. Matthews, *Aust. J. Chem.*, 24, 1 (1971), *idem.*, **25**, 2061 (1972).
[60] B. Quickenden, priv. comm.
[61] J.O'M. Bockris & S. Srinivasan, *op. cit.*, p. 515.
[62] H. Wroblowa & G. Razumney, priv. comm.
[63] J.O'M. Bockris & S. Srinivasan, *op. cit.*, p. 606.
[64] E.J. Casey, paper presented at the Bournemouth Battery Conference, 1960.
[65] J.R. Benemann, *Hydrogen Production from Water and Sunlight by Photosynthetic Processes*, Uni. of Calif., San Diego, La Jolla, December 1973.
[66] J.R. Benemann, J.A. Berenson, N.O. Kaplan & M.D. Kamen, *Proc. Nat. Acad. Sci.*, U.S.A., **70**, 2317 (1973).
[67] J.R. Beneman, Symposium on 'Prokaryotic Photosynthetic Organisms', Freiburg, September 1973.
[68] J.E. Begg, *Nature*, **205**, 1025 (1965).
[69] Australian National Academy's Report on 'Solar Energy Research in Australia', Report No. 17, September 1973, p. 39.
[70] N.A.S.A., A.S.E.C.
[71] B.J. Eastlund & W.C. Gough, paper presented at the 163rd National Meeting of the American Chem. Soc., Boston, Mass., 9-14 April 1972.
[72] R.W.P. McWhirter in *Plasma Diagnostics*, ed. R.H. Huddlestone & S.L. Leonard, Academic Press, New York, 1965.
[73] H.J. Karr, E.A. Knapp & J.E. Osher, *Phys. of Fluids*, 4, 424 (1961).
[74] G. Kortum & J.O'M. Bockris, *Textbook of Electrochemistry*, Elsevier, 1959, Vol. 1, Chapter VIII.
[75] J.O'M. Bockris & A.K.N. Reddy, *Modern Electrochemistry*, Rosetta Edition, Plenum, New York, 1973, p. 845.
[76] V.G. Levich in *Physical Hydrodynamics*, Chapter 6, Prentice Hall, Englewood Cliffs, New Jersey, 1962.
[77] J.O'M. Bockris & A.K.N. Reddy, *op. cit.*, p. 1059.
[78] D. Gregory *et al.*, *op. cit.*, p. VII-40.
[79] J.O'M. Bockris, paper presented at the Hydrogen Economy Miami Energy (THEME) Conference, March 1974 (with revisions).
[80] J.H. Russell, L.J. Nuttall & A.P. Fickett, paper presented at the American Chem. Soc. Meeting Hydrogen Fuel Symposium, Chicago, Illinois, August 1973.

CHAPTER 10

Storage of Massive Amounts of Energy

INTRODUCTION

THERE IS A daily and a seasonal variation in demand for energy. If the post-2000 supply of energy is to come substantially from solar sources, there will be diurnal variations in supply. Further, there will be times at which the sun is obscured by cloud, with known probabilities of this being for several days in succession.

Whereas for atomic energy, a storage system is desirable, for solar energy one is essential. In addition to the storage forced by the out-of-phaseness of supply and demand, there should be attention to storage of very large amounts of hydrogen for possible export (shipping or piping) from, for instance, Australia to Japan.

Storage of solar energy in hydrogen will probably be preferable to other methods of energy storage, because it lends itself easily to transmission and does not need special geographic features, e.g. suitable formations for pumped storage, etc. Nevertheless, other methods, except pumped storage*, will be discussed for comparison.

THERMAL ENERGY STORAGE

Storage in eutectics is a method to be considered, both for the storage of solar energy and nuclear energy; so long as one refers to those energy sources and methods of transduction (e.g. photothermic or fission reactors, respectively), where heat is the form of energy produced or demanded.

The method depends upon the latent heat of the liquid concerned. During freezing, the liquid undergoes solidification at a constant temperature, whilst the latent heat of fusion is given out.

Other phase changes such as the crystallisation of Glaubers salt, Na_2SO_4 $10H_2O$ are used.[1] This stores 1.7 times the amount of energy per unit volume that water does, but near the transition point – not over a large temperature range.

A high latent energy of fusion, coupled with a low price per unit volume, would be the principal scientific criteria to be used. Other

---

* Pumped storage[1] is the oldest method of storing large amounts of energy. However, it depends upon pumping water up to fill a reservoir and thus is limited to a few parts of most countries.

criteria, stability and lack of corrosive properties, in contact with relatively cheap container materials, are important.

Most of the materials used are eutectics of simple inorganic salts, e.g. alkali and alkaline earth metal fluorides, such as NaF-MgF_2. The temperature range at which the storage occurs for various eutectics is between 150-850°C. In Fig. 10.1 are shown systems which have been stressed in the recent work of Schroeder at Philips.[2]

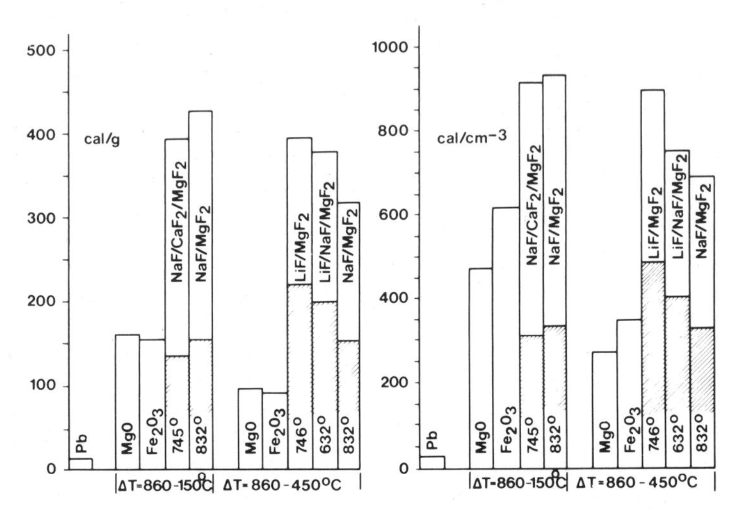

Fig. 10.1 The new heat-storage materials have a much greater storage capacity than those used hitherto. Left: storage capacity per unit weight. Right: storage capacity per unit volume.[2]

Thus, calories per gram are in the region of a few hundred and the calories per cc approach 1,000. This may be compared with the lead-acid battery, where the energy in calories per gram is 20-30. These storers of heat have an energy capacity per unit weight several times greater than that of even modern electricity storers such as Na-S. They store, however, about half the energy per unit weight of a hydrogen-air battery.

The heat storage method has difficulties in the insulation and lifetime of the storer. Corrosion of steel vessels by molten salts at the temperature concerned is a difficulty. Its removal depends upon the removal to a high degree of water and air from the salt. This can be done, it is a matter of a trade-off between the cost of removing water and oxygen, *versus* the cost of a shorter life for the steel vessel. Addition of Al also reduces corrosion.

Among the classical work in this field is that of Telkes[3] and among the more recent that of Schroeder.[2] The cost of these molten salt storers involves a number of factors which are difficult to estimate. It may be as low as \$0.5 per kWh (year).

The method is an easy one. It improves on scale-up, because the heat per unit volume will escape more slowly as the surface area of the storer increases.

The negative aspects of the storage of energy as heat are the difficulty of heat loss; the material problems associated with situations at high temperatures; and the fact that heat is not a good form in which to transport energy over long distances. The method is not one for long-term storage.

Storage in steam seems to have prospects, because of a possibility of transmission of energy in steam-containing pipes.

ELECTROCHEMICAL ENERGY STORAGE

The principles are well known. If the energy source produces electricity directly, storage in electrochemical energy storers – batteries – is particularly acceptable. Thus, batteries do not need any special site; emit no noise; have no pollutants; are modular in nature; and hence have no problems about size. The requirements are 50 watts per kilogram; 200 watt hours per kilogram; 4 years life and a cycle life of at least 1,000. The first two requirements are met by several batteries of Table 10.1.[4] The cycle and total life are not yet clear. A 100 megawatt-hour store would be contained in a space of 8 metres cubed.

The lead-acid battery is suitable for the purpose for which it is used – starting up internal combustion engines. The amount of energy needed is small. The fact that the lead-acid battery has *the poorest energy density of all used batteries* is of little importance. What is necessary is a high power density, and the lead-acid battery does have a good power density, compared with other batteries. The misunderstanding is that the lead-acid battery is typical of other batteries. The facts are shown by the parameters in Table 10.1.[5]

TABLE 10.1[5]

BATTERY PARAMETERS*

	W/kg	Wh/kg
Zinc-air	100	100
$Ni\text{-}H_2$ (compressed)	300	80
Ni-Fe	132	50
$Zn\text{-}O_2$	300	160
Aluminum-air	150	240
Sodium-sulphur	100	100
Sodium-air	200	400
$Li\text{-}Cl_2$ (liquid storage)	300	500
H_2-air	100	2000
Pb Acid	200	20-30

* The figures given are in the middle range in terms of the effect of rate of discharge. As the *rate* of discharge increases, the storage per unit mass decreases.

An attractive possibility is the hydrogen-air battery.

Costs[6] of lead-acid battery storage are \$1-5 per kWh (year).

Firm costs of developing battery systems of lower weight (e.g. zinc-air, and hydrogen-air) are unavailable, because such systems have not been commercialised.

In recent times most development money has been put into the sodium-sulphur battery. Costs of sodium-sulphur battery storage are *c.* \$1 per kWh[7](year).

The sodium-sulphur battery has an energy density of 3 to 5 times greater than that of the lead-acid battery. The working temperature of 350°C is a less desirable feature.[8] It has reached a level of reliability which allows it to be part of a scheme for electrification of British trains. The main difficulty has been the cracking of the β-Al_2O_3 membrane. Development of a cheap iron-air battery, which has characteristics which are 2 to 3 times better than that of lead-acid for energy density, looks promising.[9]

The pros and cons of electrochemical storage are easy to discern. Among the advantages are the consistence with photovoltaic power, or with thermo-electric conversion, for here electricity is produced. Relatively high power densities (100 watts per lb) are obtainable. Although there is a gradual loss of energy in electrochemical batteries when they are not discharged, the problem is negligible compared to the rate of loss of energy to the surroundings in modern heat storers (Table 10.2[10]).

A disadvantage of electrochemical systems for massive storage lies with the amount of material which would be needed. There is not enough lead to supply lead batteries for cars[6], were their use to become widespread. Conversely, iron-air cells, and perhaps the sodium-sulphur cell, would be free from difficulties of material scarcity.

On a smaller scale, for cars and railway engines, electrochemical storage has great (and probably unique) advantages.[11] Battery storage would seem limited to about 1,000 megawatt hours.

One of the more under-developed parts of electrochemical storage seems to be shown by the hydrogen-air fuel-cell battery. The fuel cell is well developed. Hydrogen liquefaction or large gaseous storage, with coupled electrochemical regeneration, is a promising avenue.

TABLE 10.2[10]

SOME APPROXIMATE ESTIMATES ON RATE OF LOSS ON STORAGE

Type of system	Rate of loss % per day	Remarks
Pumped water storage	Negligible	Needs special geographic features.
Thermal	5 to 10	This with conventional insulation (fibre-glass). Losses could be used in space heating.
Battery	0.1-0.2	For lead-acid battery. Varies greatly with type of battery.
Inertial	0.5	About 2 kW continuous dissipation from a 10 MWh store.
Chemicals	zero	
Pipes	zero	
Hydrogen (liquid)	0.3 to 0.8	But boil-off can be used.
Underground (gas)	Not known. Probably negligible.	

INERTIAL ENERGY STORAGE

Introduction

Early fly-wheels had a storage capacity of 3 watt hours per lb (lead-acid battery, 10-15 watt hrs per lb) and were not attractive. There were two difficulties to be overcome in refurbishing the fly-wheel. Firstly, the low energy density; secondly, the danger of the fly-wheel breaking loose.

These difficulties have been reduced by work at the John Hopkins University laboratory under Rabenhorst.[12]

The Super-flywheel

The principal concept in the improvement of the fly-wheel is the replacement of a massive metallic wheel by certain configurations consisting of anisotropic materials, such as glass, graphite and boron. These materials are stronger for their weight than isotropic materials, so long as they are used in certain shapes. This will give rise to an increase in energy density, but, in particular, a diminishing of danger.

One of the new configurations is a thin solid bar of composite material with a concentration of high strength filaments, all running in the same direction. Alternatively, a number of brush-like rotors consisting of thousands of thin rods, bonded to a hub ('whirling spaghetti'). The stress is along the direction of maximum strength (Table 10.3[13] and Fig. 10.2[13]).

TABLE 10.3[13]

STRENGTH-TO-DENSITY RATIOS FOR FLY-WHEEL MATERIALS

Material	Ultimate tensile strength (ksi)	Density (lb/in^3)	Ultimate strength-to-density ratio (10^6 in-lb/lb)
S glass	260	0.072	3.61
E glass	200	0.075	2.67
Maraging steel (18Ni-400)	409	0.289	1.41
Sitka spurce	19	0.015	1.27
Phillipine mahogany	24	0.019	1.26
Hickory	32	0.028	1.14
Redwood	16	0.014	1.14
Titanium (6A1-4V)	150	0.160	0.94
4340 steel	260	0.283	0.92
Aluminum (2024-T851)	66	0.100	0.66
Cast iron	55	0.280	0.19

Energy density of super-flywheels[12]

This is up to 30 watt hours per lb – at the maximum when the rotation begins – but if one takes into account retardation, an average of 12 watt hours per lb is found. This improvement of several hundred per cent on the 1950 flywheels is of great interest. However, although equal to a lead-acid battery in energy density, that of the super flywheel is some 3-5 times less than that of e.g. the sodium-sulphur battery (see Table 10.1[5]).

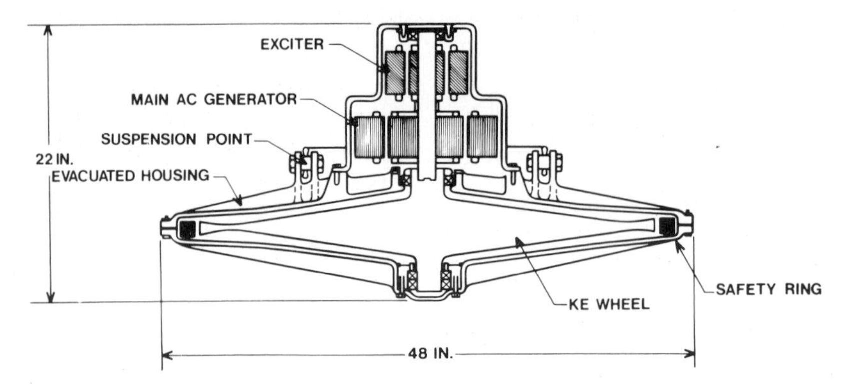

Fig. 10.2 A conical fly-wheel with a constant-radius flare at the tip is being proposed as the energy-storage device for a metropolitan transportation system.[13]

Materials[14,15]
The material factors have been reviewed by Rabenhorst[14,15] and some of the materials which he states would be suitable for the flywheel construction are fibreglass, epoxy, wood, carbon fibres and some special organics, e.g. a du Pont product called Kevlar. Bulk glass may also be suitable. It is unclear where the optimal trade-off would be between price and the properties of these materials.

Advanced engineering concepts[16]
Magnetic suspension giving zero contact between solids, and hence diminished friction, has been proposed.

Pros and cons
The concept of using anisotropic spaghetti-like rods in the fly-wheel has led to an improvement in the energy density for fly-wheels, a diminution in danger of their use. There is no indication that super-flywheels are likely to approach electrochemical power sources in respect to energy density. They may well exceed them in respect to power density. The cost of short-time storage in dollars per kWh-year may well be cheaper than that for batteries.

There may be some cases where the fly-wheel concept has advantages. Indeed, Gorman[17] has discussed the possibilities of fly-wheels in respect to solar energy. One of the difficulties is the necessity of charging the fly-wheel through the medium of mechanical energy.

Thus, in spite of the enthusiasm shown by the originators of the super-flywheels, it is unlikely that they would be important for the large energy storage plants which are needed for the storage of solar and nuclear energy.

In respect to their potential use in transportation, the nearest competitor would be the H_2-O_2 battery with energy densities about 1,000 watt hours per lb. One use of fly-wheels might be in conjunction with fuel cells in cars. Thus, the fuel cell has a lower power density and the fly-wheel has a higher power density than batteries (even the new ones). Fuel cell-energized fly-wheels could provide the accelerative surges needed in starting and overtaking.

STORAGE IN CHEMICALS

Introduction

A compromise between storage in inertial systems and storage in hydrogen is storage in hydrides. This would avoid the expense of liquefaction, and the bulk involved in gaseous storage. In particular, hydrides have been suggested for automotive applications by American workers at Brookhaven[18] and by Dutch workers at Philips.[19]

The attraction of the storage in hydrides is that it gives a higher density than liquid hydrogen without the difficulties of liquefaction. Correspondingly, there is no safety problem.

Intermetallic compounds and their high degree of absorption

Workers at Philips were the first to discover that AB_5 intermetallic compounds (A = a rare earth, and B = Ni or Co) absorbed and desorbed large amounts of hydrogen at small pressures and near to room temperatures.[20] Examples are $LaNi_5$ and $SmCo_5$. The equilibrium pressures are shown in Table 10.4.[20]

In $SmCo_5$–H and $LaNi_5$–H, there is a two-phase region in which hysteresis occurs and the magnitude of this hysteresis is related to the expansion which accompanies formation of the hydride.[21] The La-Ni system shows compounds with lattice constants, which define the compound concerned, e.g. $LaNi_2$.[22]

TABLE 10.4[20]

EQUILIBRIUM PRESSURES (AT ROOM TEMPERATURE) FOR SOME AB_5 COMPOUNDS AT HALFWAY SATURATION WITH HYDROGEN

Compound	Pressure (atm)
$PrCo_5$	0.6
$NdCo_5$	0.8
$SmCo_5$	4.0
$GdCo_5$	20.0
$LaNi_5$	2.5
$PrNi_5$	12.0
$NdNi_5$	20.0
$SmNi_5$	60.0

Prospect of use of hydrides in transportation[23]

The positive aspects are the up-take of hydrogen and its reproduction upon gentle heating. The volume taken up for the large amount of hydrogen stored would be small. The negative aspects are the high price of such rare materials, corresponding to the small amounts available of, e.g., lanthamum and samarium.

Use of cheaper materials for storage as hydrides[24]

Magnesium hydride, MgH_2, is the most promising of these. It contains 7.6% of hydrogen. It decomposes at 1 atm at a temperature of 287°C. It is cheap. There are possibilities with magnesium-nickel alloys. Fig. 10.3[24] shows the dissociation pressure of several hydrides. The process is being developed for storage of roof-collected solar heat.[24a]

Economics of storage in cheap hydrides

These seem to be negative at present. According to Savage *et al.*[25], the price would be between \$6 and \$46 per MBtu per year, compared with about \$3.50 per MBtu per year for liquid hydrogen storage.*

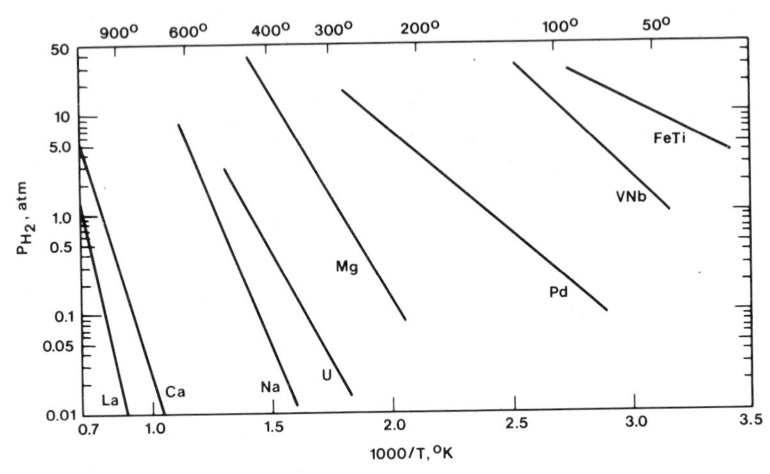

Fig. 10.3 Dissociation pressures of metal hydrides, I. The approximate compositions of the solid phases to which these curves refer are as follows: LaH_2, CaH_2, NaH, UH_3, MgH_2, $PdH_{0.6}$, $VNbZH_3$, FeTiH.[24]

Further information on hydride storage

There are interesting documents by Lynch[26] and Kuijpers.[27]

Summary of hydride storage

In spite of the attractiveness of this method, because of its safety and small volume, the economics, and the scarcity of the really attractive metals, make it unlikely that it will be used on a large scale.

A GENERAL SURVEY OF THE POSSIBILITIES FOR STORAGE IN GASES

Introduction

During the last several decades, natural gas has been stored in mines, caverns in impermeable limestone, and in shale. Cavities of rock salt have also been used.

Several other techniques are possible and have been surveyed by Eakin.[28] Natural gas is soluble, e.g. in liquid propane. At 600 psi, the volume of natural gas soluble in propane is 39 SCF per cubic ft at 104° F. Absorption in solids is a possibility, not a good one.

Reversible chemical combination: this would refer to the formation of a hydride, and can be considered. The possibilities of liquefaction are obvious.

The difficulties with liquefaction are the energy needed to reach the

---

* This is about \$0.01 per kWh year. It is about 10^2 times less than storage in batteries.

low temperatures at which hydrogen condenses and the expense of the container. Chemical storage needs a cheap substance. Water would be the only one cheap enough. The solubility is insufficient and needs a low temperature for absorption, whilst regeneration provides difficulties. One of the difficulties with absorption in a liquid is the huge quantities required.

As to tanks, mines and quarry pits, it is a matter of costs. Mine cavities would seem the cheapest. Too much heat would probably be absorbed from the surroundings and there is much work to be done on the properties of rocks before the parameters are well enough known to be the basis of engineering calculations.

Depleted gas fields[29]

If reservoir size and characteristics are suitable, this is a good method, (a) because the reservoir has already been made, which reduces costs; and (b) because the reservoir is known to be gas-tight.

Aquifers

The relevant considerations are the characteristics of the aquifer, what pressure would be needed to get the gas in, etc. Each case must be evaluated by itself.

Liquefaction

The present concepts envisage liquefaction largely for holding hydrogen energy for peak shaving rather than for dealing with the enormous quantities involved in diurnal swings. Costs are reviewed below.

Salt cavities

These have been, and will be, used for gas storage, see Fig. 10.4.[29] The economics of the salt cavities relate to their location with respect to the transmission centre and the location of the demand centre. Gas storage in salt cavities is cheaper storage than that in liquid form.

Gasometers

Pressure vessels, e.g. 3 to 6 metres in diameter, and 100 metres long, are often used for the storage of gases.

Storage in pipes[30]

One of the advantages of storage in gases is the possibility that, when the system involves a large distance between the transmission centre and the user centre, there is substantial storage in the pipe.

The number of days' supply for a city of 10 million people which could be kept in a pipe 5,000 km long, 2 m in radius and 100 atm in pressure can be calculated thus:

$$n = \frac{PV}{RT} = \frac{100 \,.\, 3.1\, (2 \,.\, 10^2)^2\, 5 \,.\, 10^3 \,.\, 10^2}{0.08 \,.\, 298} \tag{10.1}$$

$$= 3 \,.\, 10^{11} \text{ moles of hydrogen.} \tag{10.2}$$

The amount of energy in this quantity is obtained by recalling that 1 mole of hydrogen gas gives 58.1 kcal per mole upon burning. Hence, the number of moles of hydrogen in the pipe would give $6{\cdot}9 \,.\, 10^{13}$

Btu. New York contains about 10^7 people, each equivalent in overall energy need per unit time to 10 kW, and, on this basis, there would be a total of about one weeks supply of energy for the city in the pipe. Perhaps about half of this would in practice be easily available because as the pressure drops, so the same work of pumping will pump half as much energy per unit of time.

If the pressure of the gas in the pipe is cycled between about 75 and 125 atm, its fatigue life would be about 50 years.

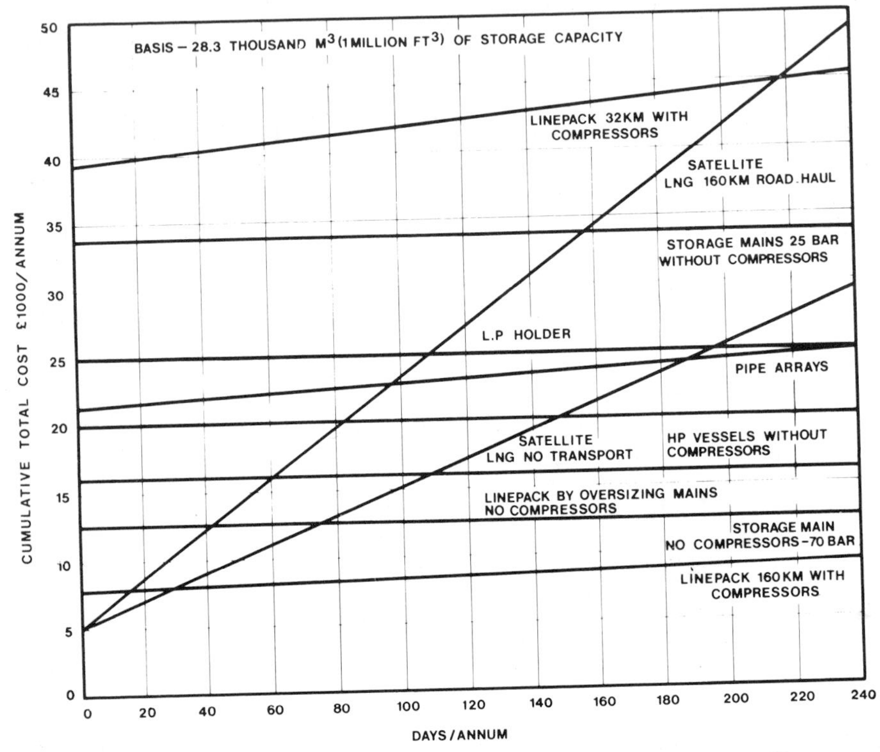

Fig. 10.4 Forms of diurnal storage. Variation of annual costs with load factor.[29]

Storage in air

Oil-driven turbines would compress air and force it into caverns. The air would be compressed to about 40 atm. Little development has been made.[1]

THE EXISTING NATURAL GAS SYSTEM

In the present natural gas system, transmission of gas over distances up to 3,200 km through pipes takes place and the pressures used are 5 to 75 atm. Storage aspects are important, because of the variation in demand, e.g. in the Eastern U.S. there is a 5 times variation between winter and summer.[31]

Local storage in low pressure tanks is possible to about 1 million cubic feet.

Pipeline storage (line-pack storage) is used. Underground storage capacity in the U.S. has some 325 underground pools in 26 States[30] (Table 10.5[30]).

TABLE 10.5[30]

NATURAL GAS SYSTEM DATA (U.S. 1970)

Consumption:	
63.8×10^9 cu ft/day	23.34×10^{15} Btu/yr
Storage capacity:	
Underground gas reservoirs	$5{,}178 \times 10^9$ cu ft
Liquefied natural gas (SCF)	15×10^9 cu ft
Total	$5{,}193 \times 10^9$ cu ft
Length of pipeline:	
Field and gathering main	66,556 mi
Transmission main	252,621 mi
Distribution main	595,653 mi
Total	914,830 mi
Number of customers (meters)	
Residential	38,097,000
Commercial	3,131,000
Industrial	199,000
Total	41,427,000

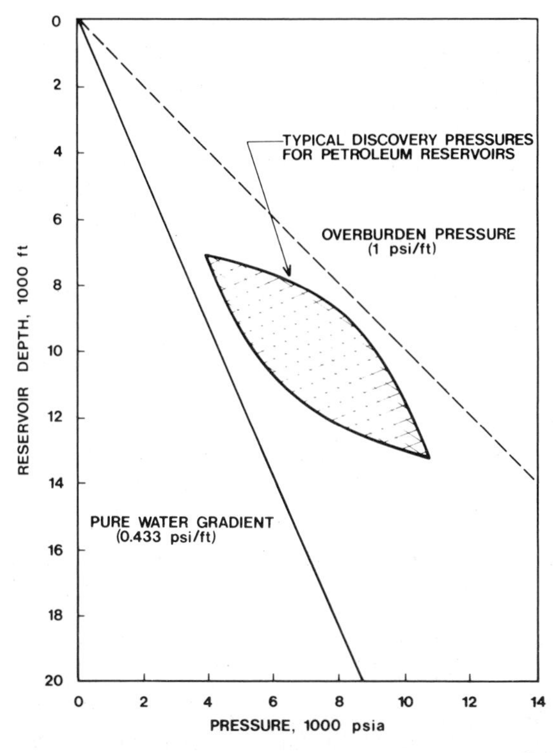

Fig. 10.5 Depth-pressure relationships for underground reservoirs.[30]

STORAGE OF HYDROGEN GAS

Underground storage

One of the attractive ways of storing large amounts of hydrogen is in impermeable rocks and aquifers. Igneous rocks are not suitable. The storage places could be those formerly occupied by natural gas in its original form, or an aquifer could be pumped out and filled with hydrogen.

Depth-pressure relations exist and are shown in Fig. 10.5.[32] There is a specific pressure which must be reached for storage at a certain depth. The sealing of the aquifers, and rock formations, is carried out by cap-rock and the seal is independent of the nature of the gas (natural gas, hydrogen, etc.), because it depends only on the rock-water interfacial tension.

Extensive underground storage of natural gas is carried out at present. The largest store is at Beynes near Paris, where the storage of 7 billion SCF of natural gas takes place.

Mines and other cavities

Mines can be used for the storage of gas, as also underground cavities, natural and created. Here, much more research is needed on the possibilities of forming such cavities by nuclear engineering. Points of doubt are the structural integrity of the cavity; and the leakage of hydrogen stored in the cavity, out through the surrounding earth.

Underwater storage

One requirement of a Solar-Hydrogen Economy is a reserve of energy to guard against abnormalities in climate which may give rise to interruptions in supply. The desirability of having about one month's supply of energy in reserve implies large spaces for the storage of hydrogen. The limitless space beneath the sea has a clear attraction. Thus, it may be advantageous – e.g. in ocean thermal gradients and the wind generator method – to have collectors on the sea and this would increase the attraction of use of the space beneath it. Other advantages of underwater storage would be the availability of high pressures without the need for storage tanks; and the removal of hazards in storage. Such storage would also be less exposed to aerial attack.

The problems are:

(1) There would be an increase in the cost of hydrogen produced at depths. The equation which relates this increase needed to the extra potential of the electrolysis vessel is[32]:

$$\Delta E = \frac{RT}{2F} \ln \; p_{H_2} \tag{10.3}$$

where p_{H_2} is the pressure at the depth considered. If the pressure is 100At. ($\sim$ 1 km depth), use of 10.3 and 9.35, with $c = 1$ cent, indicates a 10 cent per MBtu increase in cost.

(2) Chlorine would be evolved in seawater electrolysis. This would have to be collected and converted to oxygen or allowed to equilibrate with water (Chapter 9).

(3) One method of production and storage would be a membrane which would have to be permeable to ions during electrolysis. The hydrogen

beneath it would have immense lifting power and the engineering difficulties of keeping the membranes down at this depth would be formidable. The life of the membrane might be a limiting factor.

(4) For a week's supply of hydrogen for 10^7 people at 10 kW per person, with a water depth of 1 km and a pressure of 100 atm, about twenty spherical vessels of radius 100m would be needed. Upthrust problems would not be present. The prospect for this storage, at first with smaller vessels, look good (Fig. 10.6).

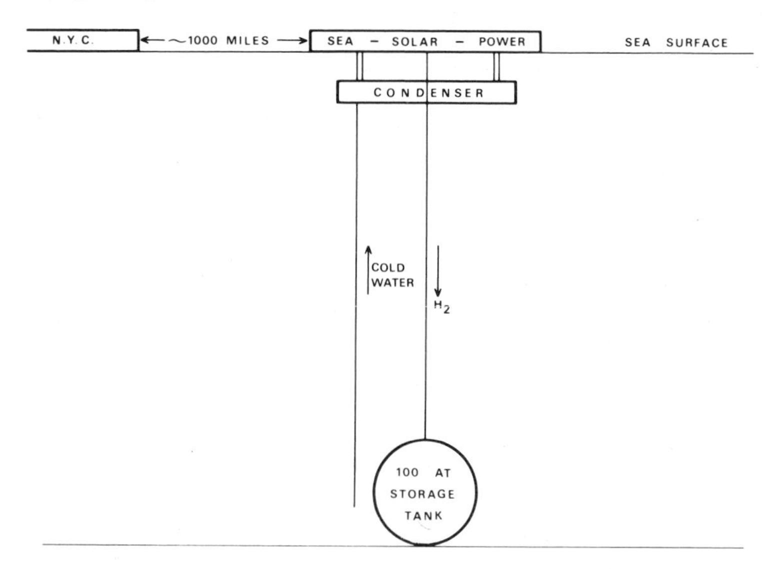

Fig. 10.6 A schematic of part of possible underwater storage for hydrogen.

DIFFERENCES IN THE STORAGE OF HYDROGEN AND NATURAL GAS

(1) Per SCF hydrogen gives 325 Btu and natural gas 1013. Thus, the energy per volume of hydrogen is one-third that of natural gas, a negative factor.

(2) Hydrogen has to be lowered in temperature greatly compared with natural gas if it is to be liquefied. The boiling point of hydrogen is 36° K, of natural gas, 200° K.

(3) Hydrogen has a greater leakiness due to its faster effluent rate through small holes (see Chapter 11). New valves and plugs would have to be constructed, differing from those for natural gas for utensils using hydrogen.

LIQUID HYDROGEN STORAGE AND USE

Most of the considerations in this book have concerned the storage of gaseous hydrogen. It is important to consider whether liquid hydrogen could be used. Various cost factors are given below and in Table 15.4; the liquefaction aspects are reviewed.

Modern dewars lose about 0.8% per day.[34] They can produce refrigeration at 12-13° K. Liquid hydrogen can be shipped by truck in semi-trailers containing about 3.7 tons of hydrogen. Rail cars contain 7.5 tons each, with a slightly smaller loss rate.

The cost of liquid hydrogen[35,36]

Consider a plant of 250-2,500 tons per day, with fixed costs at 12% of the capital investment, and efficiency of 40% Carnot, electricity at 9 mils. Then the costs of liquefying hydrogen are about 5 cents per lb or $0.95 per MBtu. The increase in the price over the original price of obtaining the hydrogen is therefore *c.* 25-50%.

The likely projected price of gaseous hydrogen is $2.00-$3.00 per MBtu. Taking this as $2.50, the projected cost of the liquid hydrogen would be about $3.45 per MBtu compared with an ex-refinery 1974 price of U.S. gasoline without tax of about $2.44.

Hallett[35] considers the use of perlite dewars. A number of advances have been made in the techniques, see below. A typical design of fuel and tank is in Figs. 10.7[35] and 10.8.[35]

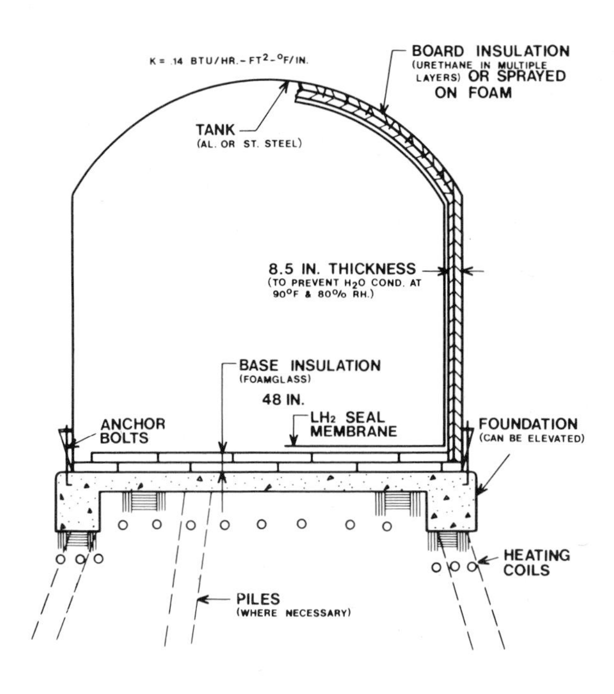

Fig. 10.7 Single-wall internally insulated tank.[35]

The liquefaction process

This is more complicated than for the other gases. The reason is the Joule-Thompson effect becomes negative, i.e. causes cooling, only below 202° K, i.e. far below ambient temperatures. Hence, liquid hydrogen cannot be made by expansion at room temperatures. A liquid hydrogen machine has to start off liquefying nitrogen and the liquid nitrogen causes the hydrogen, below the Joule-Thompson point.

There is an ortho-para conversion problem. 75% of hydrogen is ortho at room temperatures and 25% para. The liquid, however, is all para.

Therefore, the heat of conversion has to be absorbed by the cooling apparatus.[36]

The theoretical work of liquefaction is about 5002 Btu per lb, but in reality is about 3 times more. Including the conversion to the para state, 15000 Btu per lb are needed for liquefaction of hydrogen. This is about 4.5 kWh. At 1c $(kWh)^{-1}$, this amounts to 86c per MBtu.

The storage of liquid hydrogen
The technique of storage of liquid hydrogen was advanced greatly during the last two decades. In 1952, the technique was only highly evacuated dewars with silvered walls. Then, anti-radiation baffles were introduced. Improved insulation – the use of perlite – was started. Recently, a technique described as 'quilted super insulation' has been introduced (Figs. 10.7[35] and 10.8[35]).

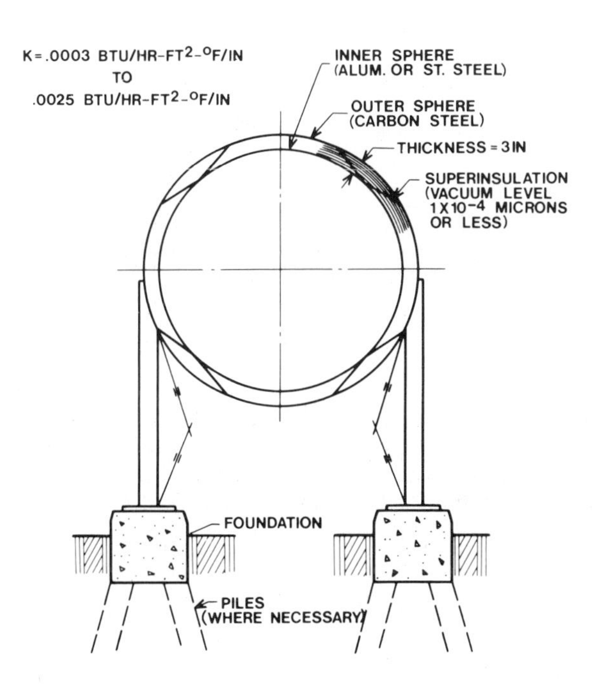

Fig. 10.8 Double-wall evacuated super-insulation storage system.[35]

Uses of liquid hydrogen
It may be that in the Hydrogen Economy, after the hydrogen has been transmitted by pipe, it will be stored as liquid. In this form, it would have a number of direct applications, e.g. in peak shaving, load levelling, transportation (air, road and sea), etc.

SOLID-LIKE FORMS OF HYDROGEN

Some potential may lie in storage in forms of hydrogen at temperatures lower than the melting point (−253° C).

Hydrogen slush may be a method which is of value, a mixture of solid and liquid hydrogen. Metallic hydrogen has been discussed.[37] It is hypothetically formable at a pressure of several megabars. Were it to be stable at room temperatures, it could be a super-conductor and would have metallic strength with a density several times less than that of aluminum. It could possibly become a convenient energy storer.

There are reports which suggest that metallic hydrogen has been formed.[38] The density has been reported to be 1 to 1.3 g. cc^{-1}.

STORAGE IN SUPER-CONDUCTING WIRES[1]

In an electromagnet, the passage of electricity causes a power loss and power must be constantly supplied to maintain the field. If the wire is super-conducting, the magnetic field – once set up – would remain and could give back energy by drawing current from the magnet.

Such a scheme might be feasible for a large store, e.g. 1,000-10,000 megawatt hours. One would need a solenoid with a radius of 50 metres and a height of 50 metres. Nb-Ti would be used.[39] Such technology may be useful before 2000.

CONCLUSION

The best storage medium for large amounts of hydrogen is underground in aquifiers, gas fields or artificial caverns. Hydrogen is difficult to liquefy: the projected cost would be some 40% above 1974 U.S. gasoline. It may be the most convenient form in which to use hydrogen in transportation.

REFERENCES

1. A.L. Robinson, *Science*, **184**, 785 and 884 (1974).
2. Philips Research Press Release No. 724/1230/108E, 1974; 735/1001/122E.
3. M. Telkes, Storage of Solar Heat, Institute of Energy Conversion, 27 June 1974.
4. J.O'M. Bockris, N. Bonciocat & F. Gutman, *An Introduction to Electrochemical Science*, Wykeham Press, London, 1974.
5. 'Technology of Efficient Energy Utilization', Report of a NATO Science Committee Conference, Les Arcs, France, 8-12 October 1973.
6. G. Brown, St. Joe's Mineral Corporation, priv. comm., October 1973; *cf.* report from Ford Scientific Laboratory, 30 June 1974, TNSF Contract C805.
7. K.G. Pankhurst, British Rail Research, priv. comm., October 1974.
8. J.O'M. Bockris & A.K.N. Reddy, *Modern Electrochemistry*, Rosetta Edition, Plenum, New York, 1974, Ch. 11.
9. W.E. Shoupp, Westinghouse Corporation, priv. comm., 8 October 1973.
10. Table constructed largely on data supplied by J. McAllan, C.S.I.R.O., priv. comm., 12 July 1974.
11. J.O'M. Bockris & D. Drazic, *Electrochemical Science*, Taylor and Francis, London, 1973, Ch. 8.
12. D.W. Rabenhorst, Intersociety Energy Conversion Engineering Conference Proceedings, p. 38, 1971.
13. R.T. Dann, *Machine Design*, p. 130 (17 May 1973).
14. D.W. Rabenhorst & R.J. Taylor, John Hopkins University, T.G. 1229, December 1973.
15. D.W. Rabenhorst, presented at the 14th Annual Symposium, New Mexico Sections of the American Society of Mechanical Engineers and the American Society for Metals, University of New Mexico, Albuquerque, 28 February 1974.
16. N.V. Guila & L.D. Linkin, Russian Engineering Journal, **52**, No. 12, p. 3.
17. R. Gorman, 'Solar Cells – Flywheel Storage', Bell Co., Report TM-70-1012-3, 15 September 1970.
18. K.C. Hoffman, W.E. Winsche, R.H. Wiswall, J.J. Reilly, T.V. Sheehan & C.H. Waide, presented at International Automotive Engineering Congress, Detroit, January 1969.

[19] J.H.N. van Vucht, F.A. Kuijpers & H.C.A.M. Bruning, *Philips Research Report* **25**, 113-140 (1970).
[20] *ibid.*, p. 139.
[21] F.A. Kuijpers & H.H. van Mal, *J. of the Less-Common Metals*, **23**, 395 (1971).
[22] K.H.J. Buschow & H.H. van Mal, *ibid.*, **29**, 203 (1972).
[23] P. Hill, *The Engineer*, p. 3 (July 1972).
[24] R.H. Wiswall, Jr. & J.J. Reilly, presented at the 7th Intersociety Energy Conversion Engineering Conference, San Diego, California, September 1972; *Chem. and Eng. News*, 16 September 1974.
[25] R.L. Savage, L. Blank, T. Cady, K. Cox, R. Murray & R.D. Williams, edd. 'A Hydrogen Energy Carrier', Vol. II – Systems Analysis, NASA-ASEE, 1973.
[26] F.E. Lynch, 'Metal Hydrides: The Missing Link in Automotive Hydrogen Technology', Energy Research Corporation, Provo, Utah, 18 August 1973.
[27] F.A. Kuijpers, Thesis, Philips Research Laboratories, Waalre, Eindhoven, 1973.
[28] B.E. Eakin, paper presented at the A.G.A. Operating Section Conference, 1960.
[29] D.J. Clarke, G.S. Cribb & W.J. Walters, presented at the 108th Annual General Meeting, the Institution of Gas Engineers, Solihull, May 1971.
[30] R.A. Reynolds & W.L. Slager, THEME Conference, Miami, March 1974, S2-1.
[31] American Gas Association, '1971 Gas Facts: A Statistical Record of the Gas Utility Industry in 1970', Arlington, Va., 1972.
[32] M.R. Tek *et al*, 'New Concepts in Underground Storage of Natural Gas', Monograph on Project PO-50 at the University of Michigan, 1966.
[33] J.O'M. Bockris & A.K.N. Reddy, *Modern Electrochemistry*, Rosetta Edition, Plenum, New York, 1974, p. 904.
[34] E. McLaughlin, priv. comm. to L.W. Jones.
[35] N.C. Hallett, 'Study, Cost and System Analysis of Liquid Hydrogen Production', NASA Report CR-73226 (1968).
[36] R. Barron, 'Cryogenic Systems', McGraw-Hill, 1966, p. 60-63 and 118-120.
[37] K.W. Bennett, *Iron Age*, 243 (January 1972).
[38] F.V. Grivorev, S.B. Kormer, O.L. Mihailova, A.P. Tolochko & V.D. Urlin, ZhETF Pis. Red., 5, 286 (1972); *Physics Today*, p. 17 (March 1973).
[39] *Science*, 294 (25 February 1974).

CHAPTER 11

Safety Aspects

INTRODUCTION

A DIFFICULTY of the Hydrogen Economy is the image of explosion, arising from a memory of school laboratory experiments. The fire on the air-ship *Hindenburg* in 1938 is well known.* The Hydrogen Bomb is associated with disaster.

Two preliminary remarks:

(1) The safety of a fuel is comparative. The better the fuel, the more dangerous it is in contact with air.

(2) New fuels (e.g. nuclear) are regarded as dangerous until it is shown that safety measures are effective.

PROPERTIES OF HYDROGEN WHICH MAKE IT MORE DANGEROUS TO HANDLE THAN NATURAL GAS[1] (Table 11.1[2])

The reference substances in discussions of hydrogen safety are CH_4 and C_3H_8.

(1) The inflammability limits of hydrogen with oxygen are wider. Thus, hydrogen combines explosively with oxygen when the limits are between 4 and 75% hydrogen. Methane combines explosively with oxygen when the limits are between 5 and 15%.

(2) The ignition energy is lower, hence an explosion is more easily set off.

(3) Leakage is easier. The escape velocity of hydrogen is three times that of methane upon a volume basis. If hydrogen escapes to a closed space, it will exceed the explosion limit at 4% by volume in 0.26 of the time which natural gas needs to exceed its 5% threshold. On the upper side, it takes 1.6 times longer for hydrogen to pass the upper limit of 75% than for methane to pass its upper limit of 15%.

(4) The propogation velocity of a flame is 6 to 100 times more (dependent upon the amount of air present) for hydrogen than for methane.

(5) A hydrogen-oxygen flame is invisible.

(6) There is no Joule-Thompson cooling on expansion of hydrogen

---

* This air-ship, the largest lighter-than-air vehicle built, performed 18 trans-Atlantic crossings before the accident. Its predecessors, the Zeppelins, carried out many hundreds of world-wide journeys over several decades. The mechanism of the fire is understood. The vehicle had passed through an electrical storm just before it approached its landing mast. It projected a conducting contact to the ground, shortly before the beginning of the fire. Such procedures are now recognised as dangerous in the circumstances and would no longer be practised.

TABLE 11.1[2]

HAZARDOUS PROPERTIES OF HYDROGEN AND NATURAL GAS

	H_2	CH_4 (Natural Gas)
Specific gravity air = 1.00	0.0696	0.641
Heating capacity, Btu/SCF	325	1056
Ignition energy, millijoules	0.02 (at 30% H_2 in air) 0.6 (at lower flammable limit)	0.3
Ignition temperature, °F	968 to 1250	1200 to 1310
Ignition velocity, 20% Primary air, ft/s	2.8	<0.1
Lower flammable limit in air, vol%	4	15
Upper flammable limit in air, vol %	75	
Lower detonation limit in air, vol %	18.3	
Upper detonation limit in air, vol %	59.0	
Relative orifice capacity, CF	3.03	1.0
Relative orifice capacity, Btu	0.93	1.0
Viscosity at 0°C, 1 atm, 10^{-6} poise	84.2	100
Molecular speed at 0°C, m/s	1692	600
Quenching distance, cm	0.06	0.25
Colour of flame	Invisible	Blue-yellow

at normal temperature. When hydrogen escapes from an orifice, therefore, it gets hotter and may spontaneously ignite.

WAYS IN WHICH HYDROGEN IS LESS DANGEROUS THAN NATURAL GAS[3]*

There is a better side to hydrogen, in respect of safety, which reduces the weight of some of the negative points made above.

(1) The escape velocity: On a volume basis, the escape velocity of hydrogen is greater than that of methane. However, because hydrogen contains per unit volume three times less energy than does methane, escaping hydrogen brings with it only 0.93 as much *energy* per unit time as does methane.

(2) The energy of a hydrogen-oxygen explosion, which occurs at 4% by volume of a hydrogen-oxygen mixture, is one-quarter that of the

---

* Stanford H. Henry, Union Carbide's Manager for Hydrogen, is quoted in Ref. 4 as stating that handling hydrogen is easier than handling gasoline or propane.

corresponding methane mixture when it reaches its threshold at 5%. A hydrogen-oxygen explosion is hence less violent than that of a hydrogen-methane one.

(3) When hydrogen is released from a container into air, it rises, not only because it diffuses faster than methane, but because of its smaller density.

Propane and petrol are heavier than air. Their vapours remain near the site at which they are accidently vented. In practice, then, the likelihood of an explosion after an accidental venting of hydrogen into an open space is less than for propane. The hydrogen dissipates itself much more quickly than the petroleum.

SAFETY MEASURES NECESSARY FOR THE HANDLING OF HYDROGEN[5]

The situation with the handling of hydrogen gas would be improved were a colorant and/or odorant introducible into the fuel in small quantities. It seems likely that a suitable substance (one effective in very small concentrations) can be found, but research is needed.

Leaks from the joints of pipes passing hydrogen will tend to be greater than when the same pipes contain natural gas.

TABLE 11.2[7]

COMPARATIVE PROPERTIES OF LIQUID HYDROGEN AND LIQUID NATURAL GAS

	LH_2	LNG
Melting point, °F	−434.6	−296.4
Boiling point, °F	−422.9	−258.5
Critical temperature, °F	−400	−117
Critical pressure, atm	12.98	45.8
Specific gravity of liquid, water = 1.0	0.07	0.47
Specific gravity of liquid, lb/cu ft	4.43	26.5
Viscosity, poise	182×10^{-6}	1400×10^{-6}

LIQUID HYDROGEN[6]

Here, the comparison liquid is liquefied natural gas (see Table 11.2[7]).

A major difficulty in respect to the handling of liquid hydrogen compared with LNG arises from the fact that the hydrogen boiling point (-253°C) is lower. Hence, every gas except helium (and perhaps Ne) in contact with vessels containing liquid hydrogen becomes a liquid. For example, air liquefies, and nitrogen tends to distill off preferentially, leaving a layer of oxygen which is more chemically active than air. Foam cannot be used as an insulant because of water within it freezing out. Only helium can be used as a purge gas.

Liquid hydrogen has been used extensively now in the space technology, and references are given below to detailed safety precautions to be used in its handling.*

LEGAL ASPECTS

These are well reviewed in Ref. 3.

ADDITIVES[8]

Natural gas is often odorised to help detection and an odorant would be helpful for hydrogen. It might be a mercaptan.

Illuminants: For hydrogen burners, the flame is invisible. An illuminant would be desirable. A sodium-organic compound might be developable.

Could a colorant be introduced so that a leak with hydrogen escaping from it would be more easily detected? Because the hydrogen would rise rapidly from any leak, it would not concentrate in its vicinity and this would give rise to a difficulty of detection.[9]

APPLIANCES[8]

Appliances using natural gas need different burners than those using hydrogen, so that new appliances would be necessary if the content of the gas were increased to 100% hydrogen.

An improvement would arise in respect to the fact that the only products of burning hydrogen are water and tiny traces of NO. Thus, it would not be necessary to have fuel vents or chimneys for heating or cooking.

INDUSTRIAL USES AND SAFETY

The main contributions here have been from work by NASA in transporting and dealing with hydrogen, both gaseous and liquid, in large quantities.[9] Thus, NASA has developed storage tanks,[10] transport tanks, fittings, etc.

The Linde Corporation has also made many contributions to the technology of handling hydrogen, particularly in liquid form.

SAFETY CODES[11 12]

NASA and the U.S. Air Force have established safety practices and procedures for dealing with hydrogen. A principal document here is NASA Tech. Memo. TMX-5254.[6] This document is a 'Hydrogen Safety Manual' and deals in detail with practical aspects of safety in respect to circumstances in which hydrogen is used. It deals with design principles in the manufacture of apparatus, the elimination of ignition sources, protection of personnel, detailed operating procedures, and etc.

Another report of value is that on 'Hydrogen Leakage and Fire Detection: A Survey', NASA Report SP 5092, 1970.[9]

In Table 11.3[11], there are listed documents which would be helptul to building and designing hydrogen equipment and transporting hydrogen.

---

* Jones[10] relates two incidents of road tankers carrying liquid hydrogen which overturned and spilled liquid. The drivers walked away from the accidents. Neither explosion nor fire occurred.

TABLE 11.3[11]

AMERICAN REGULATORY GUIDELINES FOR DISTRIBUTION OF HYDROGEN

Distribution method	Equipment Specifications	Shipping Regulations	Installation standards
Liquid cylinder	TCG 173.57	TCG 173.316	NFPA 50B
Liquid trailer	ASME/(Ref. CGA 341)	Special permit	—
Liquid tank car	TCG 173.316	TCG 173.316	—
Liquid customer station	ASME	—	NFPA 50B
Gas cylinder	ASME/TCG 178.36.-37	TCG 173.301	NFPA 567
Gas cylinder trailer	ASME/TCG 178.36.-37	TCG 173.301	NFPA 567
Gas pipeline	ANSI B31.8	—	DOT title 9 Part 192

BIBLIOGRAPHY ON HYDROGEN SAFETY

A number of papers on hydrogen safety are given in Ref. 6 and Table 11.3.[11]

SUMMARY

Hydrogen is more difficult to handle than is natural gas, but appropriate handling methods have been worked out.

There are counter aspects of the extra dangers of Hydrogen compared with natural gas: accidentally vented hydrogen rises in air in an unconfined space away from the danger zone and there is less energy per unit volume in a hydrogen explosion than that of methane.

Liquid hydrogen is already handled commercially: in the U.S.A. it circulates routinely in tank cars on railways and on roads.

REFERENCES

[1] D.P. Gregory, assisted by P.J. Anderson, R.J. Dufour, R.H. Elkins, W.J.D. Escher, R.B. Foster, G.M. Long, J. Wurm & G.G. Yie, 'A Hydrogen-Energy System', prepared for the American Gas Association by the Institute of Gas Technology, Chicago, August 1972, p. VIII-1.

[2] *ibid.*, p. VIII-2.

[3] R.L. Savage, L. Blank, T. Cady, K. Cox, R. Murray & R.D. Williams, edd., 'A Hydrogen-Energy Carrier', Vol. II – Systems Analysis, NASA-ASEE, 1973, p. 119.

[4] *Business Week*, 23 September 1972, quoting Union Carbide's Product Manager for Hydrogen, S.H. Henry.

[5] R.L. Savage *et al.*, *op. cit.*, p. 121.

[6] 'Hydrogen Safety Manual', Report TM-X-52454, 15. Washington, D.C.: National Aeronautics and Space Administration, 1968.

[7] D.P. Gregory, *et al. op. cit.*, p. VIII-9.
[8] Conference proceedings of the Hydrogen Economy Miami Energy (THEME) Conference, 18-20 March 1974, Miami Beach, Florida, ed. T.N. Veziroglu.
[9] B. Rosen, V.H. Dayan & R.L. Proffit, 'Hydrogen Leak and Fire Detection: A Survey', Rep. SP-5092, 6. Washington, D.C.: National Aeronautics and Space Administration, 1970.
[10] L.W. Jones, *Science*, **174**, October 1971, p. 367; priv. comm., 25 July 1974.
[11] D.P. Gregory *et al.*, *op. cit.*, p. VIII-37.
[12] F.A. Martin, 'The Safe Distribution and Handling of Hydrogen for Commercial Application', paper presented at the Intersociety Energy Conversion Engineering Conference, San Diego, September 1972.

CHAPTER 12

Material Aspects of a Hydrogen Economy

ON HYDROGEN-METAL INTERACTIONS

KNOWLEDGE of the relation between hydrogen and the material properties of metals has been gathered for a long time, and several thousand papers have been published on it.[1] It was realised at an early date[2] that one of the most sensitive metals to this embrittlement is low alloy steel. The all-pervadingness of the interaction of hydrogen with metals, with respect to their modes of failure, was gradually realised during the 1960s. Thus, hydrogen attacks iron not only when it is in contact with it in electrochemical evolution; but also when hydrogen is in the molecular form at high temperatures.[3] More subtly, the hydrogen evolution reaction occurs as a counterpart of some electrochemical metal dissolution reactions in corrosion. The condition of hydrogen discharge and recombination of hydrogen evolution[4] determines the condition with which the hydrogen is adsorbed on the metal, and this in turn greatly influences the rate of diffusion of hydrogen into the metal; and the effective final solubility, which may be increased many orders of magnitude above the thermodynamic level. The diffused hydrogen weakens the metal, and at critical potential conditions (in the range of which the corrosion potential may come), the hydrogen in the metal embrittles it. Moreover, the conditions which give embrittlement are more subtle than known until recent times, for they are greatly dependent upon stress, and the stress may vary complexly at special points on the surface of the metal.[5] Lastly, the corrosion and breakdown, which damages, does not appear on a superficial examination of a surface, but takes place inside the metal, at the moving tips of small propagating cracks and pits.[5] The conditions, both with respect to the pH of the solution inside them; and that of the stress at their tips, are complex.[5]

Stress corrosion cracking may, under certain circumstances, arise from hydrogen in metals.[6,7]

All these aspects of the damage of metals caused by hydrogen bring caution to a concept which bases the transfer of this medium of energy over long distances in pipes, the most likely constituent of which is iron or an alloy thereof. Were it not possible to use pipes made largely of steel, there would be a great increase in the cost of the transmission lines; and perhaps an absence of sufficient material.

THE MECHANISM IN EMBRITTLEMENT

Two theories have been put forward for the mechanism of hydrogen embrittlement – and some kinds of stress corrosion cracking. In the older one[8], hydrogen diffuses from the surface (where it has been deposited during hydrogen evolution as a result of corrosion; or from gaseous hydrogen) and accedes to voids in the structure. The origin of these voids is not clear, but it is agreed that they exist. Before the atomic hydrogen gets to them during its diffusion in the metal, they are of the order of 1000 Å in one dimension, and some tens of Å in others. They are associated with a pile-up of dislocations in the metal. When atomic hydrogen diffuses from the bulk to the interface between the bulk and a void, it adsorbs on the internal surface of the metal and is then evidently in contact with a vacuum. It sets up its equilibrium pressure of molecular H_2 in the cavity.

This equilibrium pressure may be very large. These large pressures arise as a result of the electrochemical concept of overpotential.[4 9 10] Essentially, the deviation of the surface from that of the equilibrium potential, which is forced upon it by the electrochemical kinetics of corrosion, increases the fugacity inside the void to an amount which corresponds to equilibrium between hydrogen in the void, and that on the metal surface in contact with solution. Due to the overpotential, this may be much higher than that of equilibrium.

An easy understanding of the situation can be given on a thermodynamic plane. Overpotential has to be present to make the reaction depart from equilibrium. Thus, it is present in respect to both partial reactions, at the corrosion potential. One could regard the overpotential as a shift of the equilibrium potential. (It is actually a shift away from the equilibrium potential.) In terms of pressure, the equilibrium potential of the hydrogen electrode depends on $(RT/2F)\ln p_{H_2}$. Equating this to the overpotential would appear to give the pressure of hydrogen in the void which would keep the H_2 in it in equilibrium with the H on the metal-solution interface. Thus, if the hydrogen overpotential of the electrode is -0.1 volt, the p_{H_2} would be 10^4 atm.

The description given fits some of the mechanisms of hydrogen evolution which give rise to hydrogen embrittlement, though it does not fit others.[4] The calculated internal fugacity can be above 10^6 atm, and there is a significant difference between fugacity and pressure (pressure is less than fugacity).

At a certain internal pressure of the hydrogen in the voids, the metal will begin to yield. When the critical yield stress has been reached, the void will spread. One assumes that many voids exist all over the metal. The presence of high pressure hydrogen within them will give a general spread of voids, connection up of some of them and reduction in strength of the metal. If this is under stress, it may break down.

This picture is over-simplified, nevertheless it gives the essence of the cause of the breakdown of metals by hydrogen.

Another mechanism was originated by Petch and Stables.[11] Here, the accent is less upon the pressure exerted by hydrogen within a void, and more upon the effect which adsorbed hydrogen has upon the surface of the individual crystals within the polycrystal. The surface tension is lowered, and crystal structure weakens.

This latter view has been developed less than the former model. It is likely that both models have importance. For example[5, 11], the surface tension of the metal is an important component of the first model and its value certainly will be altered if hydrogen adsorbs on the metal surface.

Within a Hydrogen Economy, the potential attack of hydrogen on metals may be from molecular hydrogen but because of water vapour coming in contact with pipes, thin layers of water on the pipe surface associated with moisture, are extensively present, conducting, because of the dissolution of traces of CO_2 (and etc.) in them. They may form corrosional situations, which produce adsorbed, atomic hydrogen on the surface and lead thereby to permeation and embrittlement.

The principal concern in the Hydrogen Economy situation would be from gaseous hydrogen. If this is at sufficiently high pressures, it may give rise to deleterious effects, as explored below.

THE EFFECT OF EXTERNAL PRESSURE

The permeation of hydrogen into a metal – and thus its aggregation at voids – depends upon the hydrogen pressure in a well-known way.[2] The solubility is proportional to the square root of the external hydrogen pressure. It may be that, in pipes at pressures of several hundred atm, the hydrogen pressures would be large enough to cause a degree of embrittlement.

However, it seems unlikely that this would be the case, for most steels have breakdown points at room temperature which demand local pressures greater than 10^3 atm, if the stress is one-time and not cyclical.[5] If direct effects of hydrogen gas at room temperatures, and in the range of pressures expected within pipes, seem unlikely, indirect effects may occur. Hydrogen may combine with elements of an alloy, thus beginning to dissociate and breakdown the alloy at the surface, and change its properties. Internally, hydrogen diffusion may cause a build-up of methane as a result of the reaction of hydrogen and carbon in the iron. However, the most likely effects come from those associated with local stresses.

Thus, the solubility of hydrogen as a function of pressure stated above[5], refers to the zero stress situation. There are usually local stresses. The effect of stress on solubility is important. Thus[12], the equation which represents solubility as a function of stress under thermodynamic conditions is:

$$c_\sigma = c_0 e^{\overline{V}_H \sigma / RT} \tag{12.1}$$

where c_σ is the local solubility at an area at which the local stress is σ, and $\overline{V}_H$ is the partial molar volume of H in the metal.

Calculation shows that, in practical ranges of σ, the solubility may become very large.

For a given external stress, there is a larger local stress near dislocations.[13] For certain parameters, Bockris *et al.*[13] calculated that the hydrogen solubility near a dislocation would be of the order of one hundred times more than the equilibrium solubility without stress. Herein may originate some of the unexpected difficulties met in the embrittlement of steels with gaseous hydrogen (see below).

These remarks refer to a surface of iron or steel which is not oxide covered. In so far as some coverage occurs, the surface may be protected, although the attack of hydrogen upon the oxide and its gradual, sometimes rapid, breakdown is a possibility, A thermodynamic calculation indicates whether surface oxides of iron are stable in the presence of hydrogen at various pressures, taking the surface and bulk oxides to have the same free energies, as a first approximation.

THE EFFECTS OF CRACKS AND PITS UPON DAMAGE AND EMBRITTLEMENT OF STEELS BY HYDROGEN

Pits in cracks occur in metals and these may induce special conditions which favour the spread of cracks. The subject is well-known in stress-corrosion cracking theory.[5]

The equation which relates solubility to stress is (12.1). However, the stress at the bottom of a pit must be distinguished from the macro stress put upon a sample; and from the stress which exists in the vicinity of dislocations.[13] Thus[14], the stress at the bottom of a pit, the radius of curvature of which is r, and the depth of which is l, is given by[14]:

$$\sigma_{\mathrm{tip}} = \sigma_{\mathrm{macro}}\, 2 \left[\frac{1}{r}\right]^{1/2} \qquad (12.2)$$

The solubility of hydrogen in iron has the order of magnitude of 10^{6} moles hydrogen per mole of iron, so that, if the exponential in equation (12.1) exceeds 10^{6}, about one hydrogen atom per iron atom can exist at the bottom of the pit and the iron-iron bond would then be greatly weakened. Utilising equation (12.2) and taking, as an example, $l = 0.1$ cm, $r = 10^{6}$ cm, $\overline{V}_{H} = 2.67$,[5] one finds that, if the stress is above 1.2×10^{8} dynes per sq cm ($\simeq 1$ Kg mm^{-2}), the H/Fe ratio in the surface layers of Fe at the bottom of the pit contains 1H to 1Fe – and hence breakdown and flow in the pit is likely, *even though the hydrogen in contact with the surface were present at room pressures.*

THE LIMITS OF THE USE OF CARBON STEELS IN THE PRESENCE OF HYDROGEN

This has been discussed principally by Nelson[15], and he gives the curves of Fig. 12.1.[15]

Small amounts of carbide-stabilising limits, chromium and molybdenum, provide increased resistance to attack. A chart illustrating this has been published.[16] 2.25% chromium, 1% molybdenum steel has considerable strength in resisting hydrogen.

Difficulties arise in respect to temperature rises of sensitive parts. For example, some of the cooling facilities in a plant may go out of action. Temperatures may then rise locally in carbon steel equipment to a level where hydrogen attack will begin to occur (probably at the base of small invisible cracks).

APPEARANCE OF SPECIMENS DURING EMBRITTLEMENT

Hofmann and Rauls[17] have made direct experiments upon the attack by high pressure hydrogen and one of their pictures is shown in Fig. 12.2.[17] They observe that the cracks forming on the surface of specimens occur largely along grinding grooves or at notches. They quote a specimen

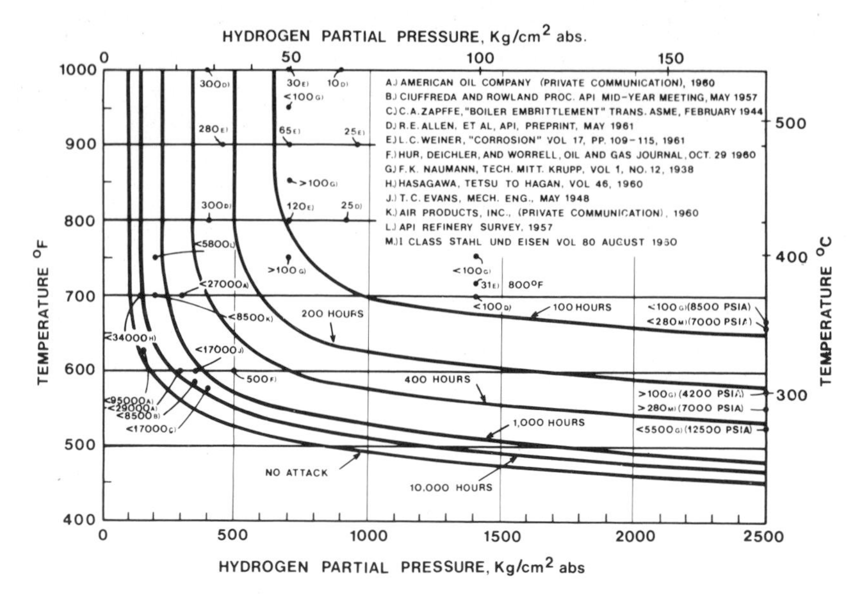

Fig. 12.1 Time for incipient attack of carbon steel in hydrogen service.[15]

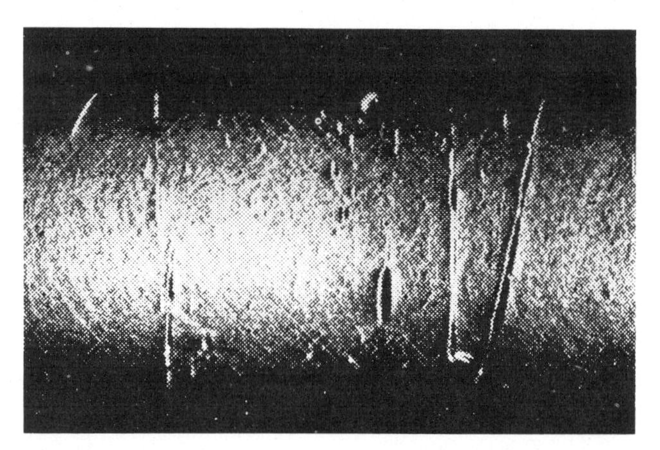

Fig. 12.2 Typical appearance of surface of a specimen suffering damage from plastic deformation during exposure to high pressure hydrogen.[17]

of ARMCO iron, normalised CK22 and normalised C45. Specimens were 8 mm in diameter with a circumferential notch of 60° notch angle, 1 mm deep, and root radius of 0.1 mm, undergoing breakdown in 150 atm of hydrogen when the stress was lower than that in air.

The effect of surface finish on embrittlement is important. Cracks in tensile specimens follow grooves formed in surface finishing. The notch effect of the grooves raises the local stress level. Cold working is an important influence, but is difficult to differentiate from geometric factors.[18]

EFFECTS OF HIGH PRESSURE STORAGE UPON VESSELS CONTAINING WELDS

This important subject has been examined particularly by Fairhurst.[19] The effect of the ductility of unnotched specimens is shown in Fig. 12.3.[19]

Fairhurst comments upon the greater susceptability to embrittlement of martensite rather than perlite; and the good resistance of high-alloy maraging steels. Austenitic steels are less sensitive, but they still embrittle.

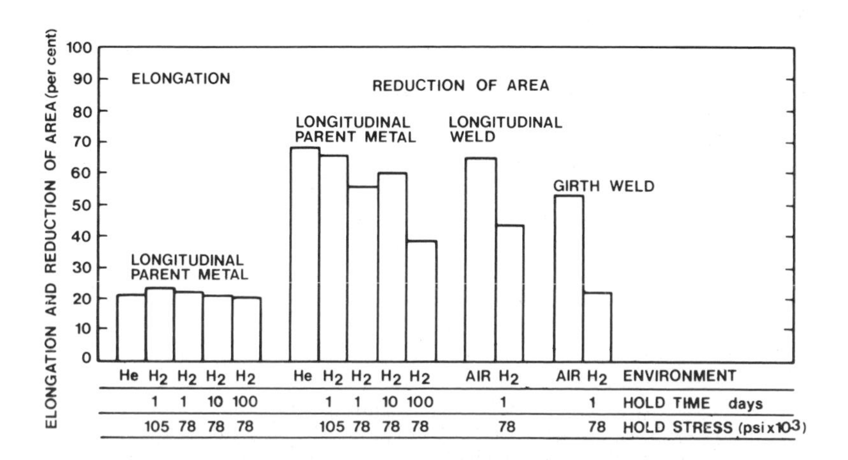

Fig. 12.3 Effect on the ductility of unnotched specimens of ASTM A-533 Grade steel (quenched and tempered 1.3% Mn, 0.5% Mo, 0.5% Ni) when held at room temperature in high-pressure hydrogen (10,000 psi, or 700 kg_f/cm^2).[19]

There is definitely danger in situations in which the hydrogen pressure rises, as with some storage devices, above about 5,000 psi. Hydrogen gives a reduction of strength and the main problem areas are near welds and notches.

Thus, Walter and Chandler[20] suggest that existing vessels should be used at lower pressures for hydrogen than for nitrogen or argon, and stress the importance of quality control to spot beginning surface flows. However, many of these will be invisible at the beginning and damage may be spreading inside the metal.

The possibility of hydrogen damage to pressure vessels during operation has been examined by Whitman *et al.*.[21] Vessels fabricated from low alloy steels are less likely to undergo embrittlement and carbon steels are more likely to fail.

A good practice is to use cold-wall pressure vessels. Refractory linings are used to lower the temperature of the vessel walls. Lower alloy steels can be used to combat hydrogen attack. Multi-wall vessels may be used in the wall in contact with the hydrogen has a lower solubility for it than the second wall.

Special designs of welds and multi-layer hot wall vessels have been evolved and diagrammed. One such design is shown in Fig. 12.4.[22]

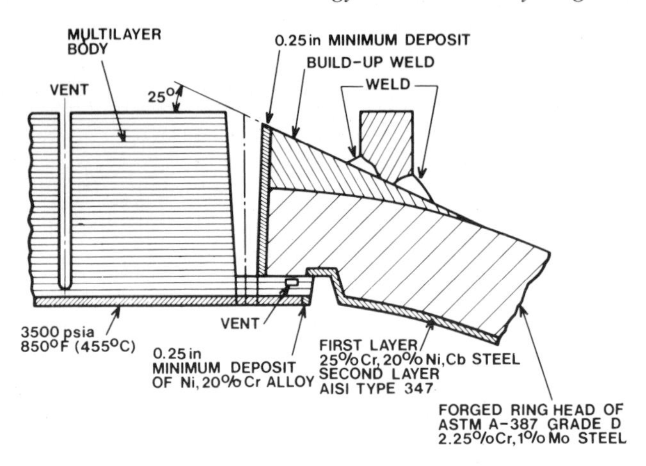

Fig. 12.4 Weld design of multilayer hot-wall vessel where hydrogen attack on the low-alloy weld metal is avoided by staggering the welds and by proper venting.[22]

HYDROGEN ENVIRONMENT EMBRITTLEMENT

The susceptability of metals to hydrogen gas has been emphasised during recent times. Types of embrittlement are shown in Table 12.1[23], and hydrogen environment embrittlement is the most relevant type to the Hydrogen Economy. Thus[24 25], storage tank failures which have occurred are interpreted in terms of hydrogen attack. The phenomenon is characterised by the formation of growth of surface cracks and the metal becomes deformed in a high hydrogen purity atmosphere. The disturbing feature is that the hydrogen may attack at even 1 atm.

The general solution is to use alloy steels, nickel, chromium and molybdenum being the important metals. There is no doubt that the difficulty can be avoided – the question is the cost of the steels.

TYPES OF METALS AND THEIR RESISTANCE TO ATTACK

The following indicates the groups of metals and their resistance to attack:

Very susceptible	High strength steels
Susceptible	Nickel and nickel alloys; titanium alloys; low strength steels
Fairly susceptible	Stainless steels
Non-susceptible	Monel; berylium-copper alloys; aluminum and copper

Research is needed in developing an appropriate 100 year-lasting material for the hydrogen pipes. Perhaps aluminum (almost inexhaustible in quantity because of its existence in clay) will become sufficiently cheap to be used. Perhaps a suitable hydrogen-resistant aluminum cladding of steel can be developed.

TABLE 12.1[23]

CHARACTERISTICS OF THE TYPES OF HYDROGEN EMBRITTLEMENT

Characteristics	Types of embrittlement		
	Hydrogen environment embrittlement	Internal reversible hydrogen embrittlement	Hydrogen Reaction Damage
Usual source of hydrogen	Gaseous (H_2)	Processing, Electrolysis, Corrosion (H)	Gaseous or atomic hydrogen from any source
Typical conditions	10^{-6} to 10^{8} N/m^2 gas pressure Most severe near room temperature Observed −100° to 700°C Gas purity is important Strain rate is important	0.1 to 10 ppm average H content Most severe near room temperature Observed −100° to 100°C Strain rate is important	Heat treatment or service in hydrogen, usually at elevated temperatures
Crack initiation	(Surface or internal initiation)*	Internal crack initiation	Usually internal initiation from bubbles or flakes
Rate controlling step	[Adsorption or * lattice diffusion = Embrittling step]	Lattice diffusion to internal stress raisers	Chemical reaction to form hydrides (e.g. in titanium) or gas bubbles

*Unresolved

IMPORTANCE OF OXYGEN IMPURITY

Johnson[26] has shown that hydrogen brittleness will often propagate at the bottom of flaws or cracks, even at very low stresses, when the apparent stress (local stress?) is one-tenth of the yield stress. High strength steels are the worst. The most important positive aspect of such embrittlement is that it can be taken care of, not only by the use of carbon steels containing certain expensive additives, but by adding a small quantity of oxygen to the gaseous hydrogen.

It is likely that the oxygen absorbs at the crack tip, displacing hydrogen from the surface. Absorbed oxygen is a barrier which prevents access of potentially embrittling hydrogen.

The field of protection against hydrogen environment embrittlement by oxygen is new and requires research. Thus, e.g. 200 parts per million of oxygen is the suggested content.[27] However, other impurities such as carbon dioxide, water vapour, ammonium and sulphur dioxide also work, although the limits of their effectiveness are not known.

Particularly dangerous in causing hydrogen environment embrittlement is hydrogen from liquid hydrogen storage, because this is hydrogen which is 'distilled off', and is particularly pure. With this hydrogen, an injection of impurities would have to be made before it is introduced into the pipeline.

EFFECT OF CYCLICAL STRESS

Walter and Chandler[28] have examined the effect of cycling with pressure on the hydrogen embrittlement of iron, steel and some of their alloys. The pressures were from 10 $MNm^{-2}\sqrt{m}$ (about 1400 psi) to 120 $MNm^{-2}\sqrt{m}$ (about 1700 psi). Results on inconel are shown in Fig. 12.5.[28] At

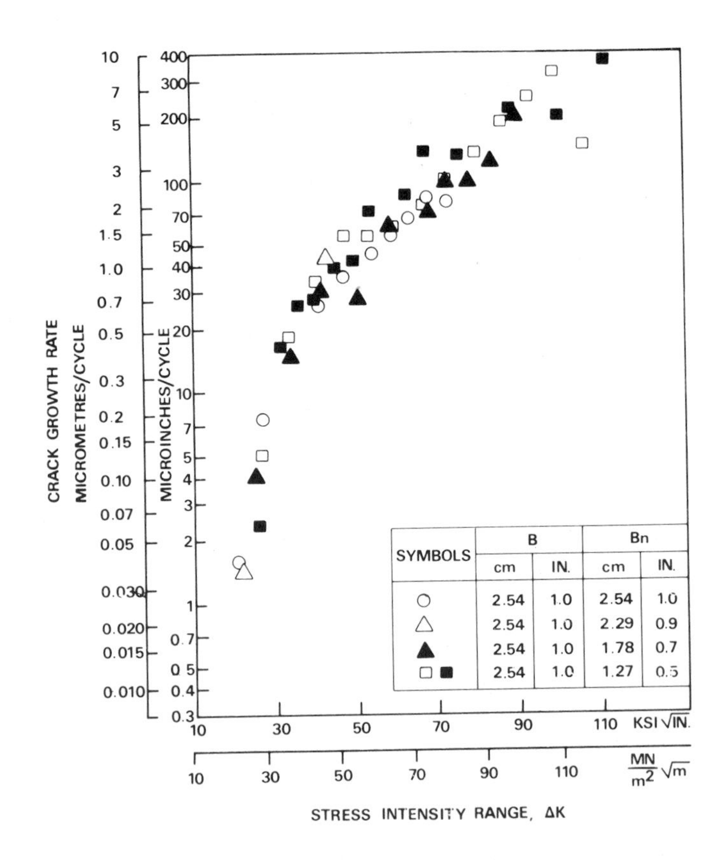

Fig. 12.5 Cyclic load crack growth rates as a function of stress intensity range for inconel 718 TDCB specimens with various side groove depths exposed to 34.5 MN/m^2 (5000 psi) hydrogen at ambient temperature $R = 0.1$, 1.0 cps.

4 cycles per sec., no environmental effects were noted on crack growth rate. The time per cycle must be at least 5 times greater for environmental effects. No influence on side grooves was shown.

If the results of Fig. 12.5[28] are extrapolated linearly down to 100 atm, 10^5 cycles would be needed to cause a 1 mm crack to grow. At 2 cycles of pressure change per day, this is 130 years. Low alloy steel gives about 50 years.

The major result here is that 100 atm is perhaps the highest practical pressure for economically useful life.

REFERENCES

1. E.C.W. Perryman, *J. Inst. Metals*, 78, 621 (1950/51).
2. E. Heyn, Stahl & Eisen, **20**, 837 (1900); **21**, 913 (1901); *Metallographist*, **6**, 39 (1903).
3. R. Staehle, paper presented at the International Conference on Stress Corrosion Cracking and Hydrogen Embrittlement of Iron-base Alloys, Unieux-Firminy, France, 12-16 June 1973.
4. J.O'M. Bockris & P.K. Subramanyan, *Electrochim. Acta*, **16**, 2169 (1971).
5. P.K. Subramanyan, 'Hydrogen in Metals', in M.T.P. International Review of Science, Electrochemistry, Series 1, Volume 6, ed. J.O'M. Bockris, Butterworths, London, 1973, p. 181.
6. J.O'M. Bockris & A.K.N. Reddy, *Modern Electrochemistry*, Rosetta Edition, Plenum, New York, 1973, p. 1338.
7. B.F. Brown, C.T. Fujii & E.B. Dahlberg, *J. Electrochem. Soc.*, **116**, 218 (1969).
8. C.A. Zappfe & C.E. Sims, Trans. A.I.M.E., **145**, 255 (1941).
9. J.O'M. Bockris & D. Drazic, *Electrochemical Science*, Taylor & Francis, London, 1972.
10. J.O'M. Bockris, paper presented at the International Conference on Stress Corrosion Cracking and Hydrogen Embrittlement of Iron-base Alloys, Unieux-Firminy, France, 12-16 June 1973.
11. N.J. Petch & P. Stables, *Nature*, **169**, 842 (1952).
12. W. Beck, J.O'M. Bockris, J. McBreen & L. Nanis, *Proc. Roy. Soc.*, **A290**, 220 (1966).
13. J.O'M. Bockris, W. Beck, M.A. Genshaw, P.K. Subramanyan & F.S. Williams, *Acta Met.*, **19**, 1209 (1971).
14. R. Oriani, priv. comm.
15. G.A. Nelson, *Hydrocarbon Processing*, **44**, No. 5, 185 (1965).
16. G.A. Nelson, *Werkstoff Korrosion,* **14**, 65 (1963).
17. W. Hofmann & W. Rauls, *Welding Research Supplement*, p. 225 (May 1965).
18. H.J. Wiester, W. Dahl & H. Hengstenberg, *Arch. Eisenhuttenwes.*, **34**, 915 (1963).
19. W. Fairhurst, *Chemical & Process Engineering*, p. 66 (October 1970).
20. R.J. Walter & W.T. Chandler, 'Effects of High-Pressure Hydrogen on Storage Vessel Materials', paper W8-2.4, Westec (March 1968).
21. G.D. Whitman *et al.*, 'Technology of Steel Pressure Vessels for Water-Cooled Nuclear Reactors', ORNL-NSIC-21 (December 1967).
22. D.W. McDowell & J.D. Milligan, *Hydroc. Proc.*, **44**, 119 (1965).
23. R.L. Savage, L. Blank, T. Cady, K. Cox, R. Murray & R.D. Williams, edd., 'A Hydrogen Energy Carrier', Vol. II – Systems Analysis, NASA-ASEE, September 1973, p. 110.
24. W.B. McPherson & C.E. Cataldo, 'Recent Experience in High Pressure Gaseous Hydrogen Equipment at Room Temperature', Technical Report No. 8-14.1, American Society for Metals, Metals Park, Ohio, October 1968.
25. J.S. Laws, V. Frick & J. McConnel, 'Hydrogen Gas Pressure Vessel Problems in the M-1 Facilities', NASA CR-1305, Washington, D.C., March 1969.
26. H.H. Johnson, paper presented at The Hydrogen Economy Miami Energy (THEME) Conference, Miami Beach, Florida, 18-20 March 1974.
27. R.L. Savage *et al.*, *op. cit.*, p. 67.
28. R.J. Walter & W.T. Chandler, 'The Influence of Hydrogen on the Facture Mechanics Properties of Metals', Pub. 572-K-22, Rockwell International, Canoga Park, California 91304, 1974.

CHAPTER 13

Modes of Transduction and Usage of Hydrogen

INTRODUCTION

IN A HYDROGEN ECONOMY, the hydrogen is likely to come from a distant highly insolated source, in pipes as a gas, or perhaps in tankers in a liquid form, and be transduced locally to mechanical or electrical energy.

There are three ways in which this could be achieved:

(1) In an extension of the present system, hydrogen may be used to fuel internal combustion engines. Only minor changes are needed in petroleum-fueled engines to make them run on hydrogen (Chapter 15). An evolution within the automotive industry *via* internal combustion to hydrogen as a fuel seems a likely path for the development of a Hydrogen Economy. The disturbance of the industry would be less, and therefore more economically attractive than if the other non-polluting option, that cars run on electrochemical power sources (which would mean largely rebuilding the means of production) were adopted.

For household electricity, one could continue the present system, distributing electricity (produced from hydrogen introduced to a central point in a town) to points of use in houses by means of cables. However, one would lose the total energy concept made possible by a Hydrogen Economy, accompanied by the in-house production of heat and electricity. In our present system, electricity-producing stations are cooled by water which is then discarded, that is, the system gratuitously heats rivers and the sea instead of houses and people. Rationality must not be sought in our present energy system, because this grew up in an era in which fossil fuels were cheap and seemed inexhaustible.

(2) Secondly, one could develop advanced modifications of combustion engines. These include the Wankel and Sarich rotary engines which have the advantage of high power to weight ratios (their tendency to pollute more than classical chemical transducers would no longer be important if they were fueled by hydrogen). One of the disadvantages of the internal combustion engine, its tendency to have an efficiency below 30%, is unchanged with new power plants, with the exception of the rocket engine, which needs hydrogen and pure oxygen.

(3) Alternatively, hydrogen could be used to produce electricity in fuel cells at local sites. How local the site would be is a matter for systems analysis, for instance, for household energy, it might be better to have substations for every few hundred houses, rather than in each

house. The advantage would be that of ease of maintenance, and the total energy concept could be preserved by piping heat to the surrounding buildings. Mechanical energy would arise from electric motors. Running on hydrogen, fuel cells show efficiencies up to 75%, easily greater than 60%. A catalyst may not be necessary in hydrogen-air fuel cells.

THE CHEMICAL CONVERTER

The internal combustion engine running on hydrogen is examined in Chapter 15. The principal advantage of continuing to use it is that it allows continuation of the amortisation of the capital invested in the means of production in the present transportation industry. Further, internal combustion engines have higher power to weight ratios than electrochemical engines at present developed, and this balances their poorer performance in respect to energy density and efficiency. The power per unit weight with internal combustion engines is about 0.5 hp per lb, whereas for the best hydrogen-air fuel cells it is about 0.1 hp per lb. The capital cost of an i.c. engine is said to be smaller than that of fuel cells, but this is because of the present mass-production facilities available for the former transducer, and the absence of any for the fuel cell. In fact, the fuel cell is simpler in mechanism than the combustion engine, particularly if the latter is driving a generator; it has few parts, none moving, and seems likely to be producible more cheaply if the same numbers per plant are considered. Correspondingly, the electric motor which would transduce the fuel cell energy to mechanical power is a far more rugged mechanism than the combustion engine. On electric railway engines, a million miles between major overhauls is commonplace.

Continuation of the use of i.c. engines in a Hydrogen Economy depends less on technological factors, and more on the socio-economic one of existing technology and the necessity of protecting capital. On the scientific side, the absence of a substitute for noble-metal electrocatalysts is important. Catalysts are desirable for the air electrode in fuel cells. Noble metals are unacceptable, not only because of price considerations, but owing to the lack of a sufficient supply.[1] The lifetime problem still poses difficulties.

The principal disadvantage associated with internal combustion engines, the fact that they are the primary sources of air pollution, is removed by the use of hydrogen. (Though NO remains a pollutant, it is produced in smaller quantities from hydrogen combustion compared with those of the gasoline-driven engines.) Noise pollution, however, would remain and also the vibratory properties of internal combustion engines leading to a more rapid erosion-corrosion factor among vehicles driven by them as opposed to the vibration-free functioning of electrically powered vehicles, the latter resulting in useful lives of 20 to 30 years (another powerful disadvantage from a corporate standpoint).

The balances which will determine the optimal mode of transduction depend upon the application, with the following factors:

(1) What is the importance of high power density (chemical engines best) *versus* high energy density (electrochemical engines best)?
(2) Is fuel cheap enough so that high efficiency of conversion (electrochemical best) is not important?
(3) For the application in question, is the need for electricity (many

metallurgical applications) or mechanical power (transportation)?

(4) To what degree are economic factors (electrochemical best) important? Thus, for space flight, capital costs are of minor importance, although the weight of fuel is of supreme importance.

Some minor factors, for example, novelty, availability of experience among engineers for chemical engines, and their lack of availability for electrochemical ones, will be factors which will diminish in importance with time, but lack of electrochemical engineers will be of decisive importance at first.

OTHER CHEMICAL ENGINES

Several heat engines exist, apart from the internal combustion engine, in particular; modifications of reciprocating engines, such as the rotary engine, the Sterling engine in which an external combustion heats a gaseous working substance, the jet, and the rocket motor.

Rotary engines such as the Wankel have a higher power density than have piston engines. The diesel and Sterling engines are less bad polluters than petrol i.c. engines, and hence provide a lesser advantage upon conversion to hydrogen fuel. The jet engine works well on hydrogen (see Chapter 15). It does not have a competitor in respect to aircraft in electrochemical engines, because here the power to weight ratio is of dominating importance and is much higher (20 kW kg^{-1}) in a jet engine than in an electrochemical engine (which does not at present have the prospect of $>$ 1 kW per kg in power).

THE APHODID STEAM ENGINE

The use of pure oxygen with hydrogen greatly improves the prospects for steam engines. The Carnot efficiency will be greater (perhaps over 50%) because of the lack of heat removal by nitrogen.[2] One would use the steam onto turbine blades. The problem of the temperature which these could withstand whilst remaining economical would be the main one.

Another advantage would be the reduction in size. There is no boiler in this steam engine (Fig. 13.13). The reduction in size could be *c*. 100 times.[3] Pollution would be only NO.

ROCKET ENGINES

Spacecraft fly on rocket engines fueled by H_2 and O_2. Their application on the ground is, in prospect, particularly for electricity production. Liquid hydrogen is pumped to 1000 psi and brought into a thrust chamber, where its temperature is increased sharply. It is then brought into a turbine which produces the pressure both for the H_2 and the O_2 (i.e. the liquid hydrogen expansion provides the pump energy). The hydrogen oxygen combustion provides the thrust which can generate the power.

Pollution would be only NO and the efficiency would be far above classical air operated engines.[4]

These engines are in their early stages of development. They need pure oxygen. Their use over more than 50 miles from a central power plant would be impractical for this reason. Their use at the central plant would be pointless, because here electricity will be available.

MHD CONVERSION

Seikel *et al.*[5] have made estimates of the use of MHD plants using H_2 sent from coal at remote locations. The MHD station would be in the city and the local distribution would be by wire to plants and houses. Gaseous oxygen would be produced locally from air. The major advantage is the absence of pollution. Of course, CO_2 and SO_2 removal are still problems. As a step on the way to a final Solar-(and nuclear) Hydrogen Economy, the scheme is worthy of further consideration. Costs (1974) show 5-11 mils $(kWh)^{-1}$.

THE FUNCTIONING OF AN ELECTROCHEMICAL ENERGY CONVERTER

Although 'fuel cells' have become a newspaper topic, there are many engineers who do not appreciate the mode of operation of an electrochemical converter, so that some statements will be made here with the aim of changing that situation. A full account of the theoretical basis of the converters will be found in Bockris and Srinivasan[1], and an elementary account in Bockris and Nagy.[6]

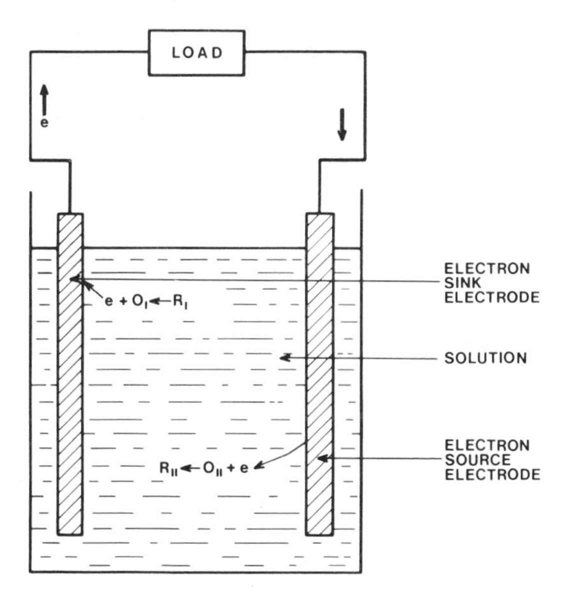

Fig. 13.1 Schematic diagram of an electrochemical energy producer.[7]

A very simple diagram which illustrates the basic steps in the work of a converter is shown in Fig. 13.1.[7] The overall reaction (say, the combination of hydrogen and oxygen to form water) must go spontaneously at the temperature and pressure concerned. Hydrogen is brought into contact with one electrode, and oxygen (it may be air) in contact with the other. There are numerous ways of doing this, they will not concern us at present. The hydrogen – in general 'the fuel' – will undergo some preliminary reactions on the electrode, e.g. hydrogen molecules will dissociate and give rise to adsorbed hydrogen atoms, after which there will be ionization on the surface (the electrochemical step) and the electrons therefrom will contribute themselves to the circuit, flow through the load (e.g. an electric motor) and end up at a lower potential, in the oxygen electrode, where they will donate themselves to oxygen in the solution. This is a simple description of an electrochemical converter. In the case of hydrogen and oyxgen, the electrode reactions are:

$$2H_2 \rightarrow 4H^+ + 4e \qquad (13.1)$$

and

$$O_2 + 4H^+ + 4e \rightarrow 2H_2O. \qquad (13.2)$$

Water is thereby formed in the overall reaction, and, with appropriate temperature control, this evaporates, condenses, and may be used as potable water (see Chapter 14).

Many other fuel cells may be envisaged, e.g. natural gas may be a fuel, with oxygen at the counter electrode. If the hydrocarbon is cracked beforehand, the reactions at the electrodes will be as above.

POLLUTION AND ELECTROCHEMICAL CONVERSION

Electrochemical converters can work at relatively low temperatures. There is no heat pollution. In a fuel cell which oxidises hydrocarbons, CO_2 and H_2O are the only products, there are no unsaturated hydrocarbons to cause pollution, and no NO.

Thus, the fuel cell represents a completely ecologically acceptable energy source and can be considered, when run on hydrogen fuel, as zero polluting. In contrast with the hydrogen combustion engine, no NO is produced.

HEAT GENERATION IN FUEL CELLS: QUALITATIVE

In a chemical converter, the essential way in which energy is produced is that the chemical reaction gives rise to products which have a lesser potential energy than the reactants. The maximum amount of energy which may be obtained ideally (neglecting the practical Carnot factor) from a chemical reaction is the free energy.

In an ideal electrochemical converter the major part of the heat, which would be produced in a chemical converter, is eliminated. There is no collision between reactants to give heat as the difference between the potential energy of the product and reactants. The corresponding energy (diminished by the entropy loss) appears directly as electricity.

Thus, one could refer to electrochemical conversion as 'cold combustion'.[8] The heating or cooling which may accompany the action of an electrochemical converter is connected, on the thermodynamic side, with the $T\Delta S$ heat of the reaction; and with the existence of certain kinetic hold-ups, see below. The unavoidable $T\Delta S$ component usually amounts to less than 10% of the ΔH component in a reaction, so that there is very little heat given out in fuel cell action from the thermodynamic aspects of its activity.

That there is in practice some heat produced in fuel cells – so that, in the 'total energy concept', they may also be used as heat sources

– is because kinetic inefficiencies (associated with the electrochemical concept of overpotential[9], and the ohmic loss in the solution) enter into the total practical heat production of the cell.

Equations concerning the heat production in practical electrochemical converters will be developed below.

THE BASIC RELATION BETWEEN THE RATE OF FUNCTIONING OF AN ELECTROCHEMICAL CONVERTER AND ITS POTENTIAL

The current, I, which flows from an electrochemical converter, output of which is across a resistance R_e:

$$I = \frac{(V_c - V_a) - \eta_{ohm}}{R_e} \tag{13.3}$$

Here V is the electrode potential of the cathode or anode respectively. The symbol η_{ohm} represents the potential drop which occurs in the electrolyte due to the applicability to it of Ohm's law, when a current passes between cathode and anode.

We may write the potentials as consisting of two parts, a thermodynamic one and a 'kinetic' one. The thermodynamic part is the Nernst, or thermodynamic, potential and is directly related to the standard free energy of the electrode reaction. The kinetic part is the overpotential[9], and is a function of the rate at which the reaction occurs.

Thus, for a cathode:

$$V_c = V_{r\,c} - \Sigma\, \eta_c \tag{13.4}$$

$$= V_{r\,c} - \eta_{c\ act} - \eta_{c\ conc}. \tag{13.5}$$

For an anode, the corresponding equations would be:

$$V_a = V_{r\,a} + \Sigma\, \eta \tag{13.6}$$

$$= V_{r\,a} + \eta_{a\ act} + \eta_{a\ conc}. \tag{13.7}$$

The suffix 'act' and 'conc' denote types of overpotential. The 'activation' type is one which is associated with intrinsic aspects of the situation, i.e. the carrying out of a surface electrode reaction, or passage of electrons across the interface. The concentration aspect refers to the fact that, as the electrode reaction works, the concentration of ions at the interface will be above or below that in the bulk and, if the converter is run at sufficiently high rates, will begin to rate-determine the electrode reaction (and will determine the maximum power per unit area of electrode which it can produce).

One now has to introduce an important equation, known as the Tafel equation, which, under certain conditions, relates the overpotentials to the current produced in the cell. This Tafel equation involves the η_{act} terms. There is another equation, of lesser importance, except under conditions of extremely high current, which refers to η_{conc}. The relevant four equations are given as follows:

$$\eta_{c\ act} = -\frac{RT}{\alpha_c F} \ln i_{oc} + \frac{RT}{\alpha_c F} \ln \frac{I}{A_c} \tag{13.8}$$

$$\eta_{c\ conc} = -\frac{RT}{nF} \ln \left[1 - \frac{I}{A_c i_{L\ c}}\right] \tag{13.9}$$

$$\eta_{a\ act} = -\frac{RT}{\alpha_a F} \ln i_{oa} + \frac{RT}{\alpha_a F} \ln \frac{I}{A_a} \tag{13.10}$$

$$\eta_{a\ conc} = -\frac{RT}{nF} \ln \left[1 - \frac{I}{A_a i_{L\ a}}\right] \tag{13.11}$$

In these equations, the coefficient n refers to the total number of electrons which take part in the overall reaction, and this would be two for the reaction of hydrogen plus oxygen to water. The symbol i_o is the exchange current density, the physical meaning of which is the rate of reaction, in terms of current per unit area, which occurs equally in both directions at equilibrium in an electrode reaction. R is the current over an electrode whose area is A.

Correspondingly, the η_{ohm} is given by R_i, where the R_i is the internal resistance of the cell.

Putting these equations together with equations (13.6) and (13.7), one obtains:

$$V_a - V_c = IR_e = E = \left[E_r - \frac{RT}{\alpha_c F} \ln \frac{I}{A_c i_{o\ a}} + \frac{RT}{nF} \ln \left[1 - \frac{I}{A_o i_L}\right] \right.$$

$$\left. - \frac{RT}{\alpha_a F} \ln \frac{I}{A_a i_{o\ a}} + \frac{RT}{nF} \ln \left[1 - \frac{I}{A_c i_{o\ c}}\right] - IR_i \right]. \tag{13.12}$$

Thus, the quantity IR_e is the potential difference which the electrochemical converter can apply to an outside load, R_e. It is the effective cell potential for a given load, and it is noteworthy that it is the total thermodynamic potential diminished by internal potential losses, mainly the overpotentials.

The make-up of the net potential produced by an electrochemical converter can be seen in Fig. 13.2.[10] Here, it will be understood that, if the current from the cell (proportional, thus, to the rate at which the cell is being used), is kept extremely low, less than I_1 in Fig. 13.2, then one can obtain an applied potential near to $V_{R\ c} - V_{R\ a}$, where those terms represent the thermodynamic cell potentials. Thus, appropriately, at limitingly small rates, the fuel cell output is near to that thermodynamically indicated in an ideal situation.

The loss of potential which occurs as the rate of working is accelerated is shown in the figute for each electrode. It is due to the overpotential terms. With increase of current density, these increase. It is obvious that the central quantity of electrochemical converters is the exchange current density, i_0, and, for an ideal situation, with high i_0, the slope of the potential-current density lines in Fig. 13.2[10] would be very small, so that, at even very high current densities, the potential which could be applied to an outside load would be near to the thermodynamic. The fall-off of power which would occur at sufficiently high rates would be that connected with concentration overpotential.

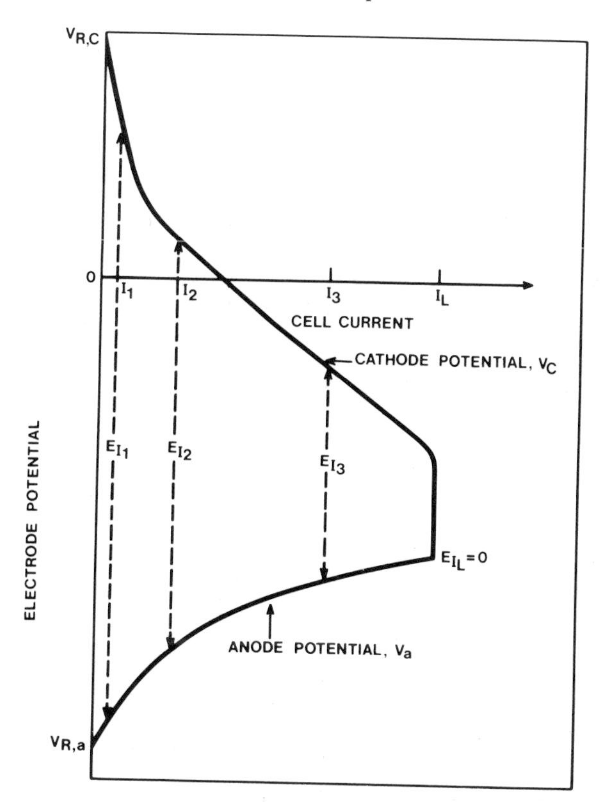

Fig. 13.2 Typical relation of individual electrode potentials V to cell potential E and to current I in an electrochemical energy converter.[10]

In Fig. 13.3[11], an analysis of the various regions of action of fuel cells is given, and one can see the effects of the thermodynamic potential, the activation overpotential, the ohmic overpotential and the transport limitation which causes the converter to go to zero potential (and therefore power) at sufficiently high rates.

EFFICIENCY

The efficiency of energy conversion of a fuel cell is defined, qualitatively, by the amount of electrical energy which can be obtained from one

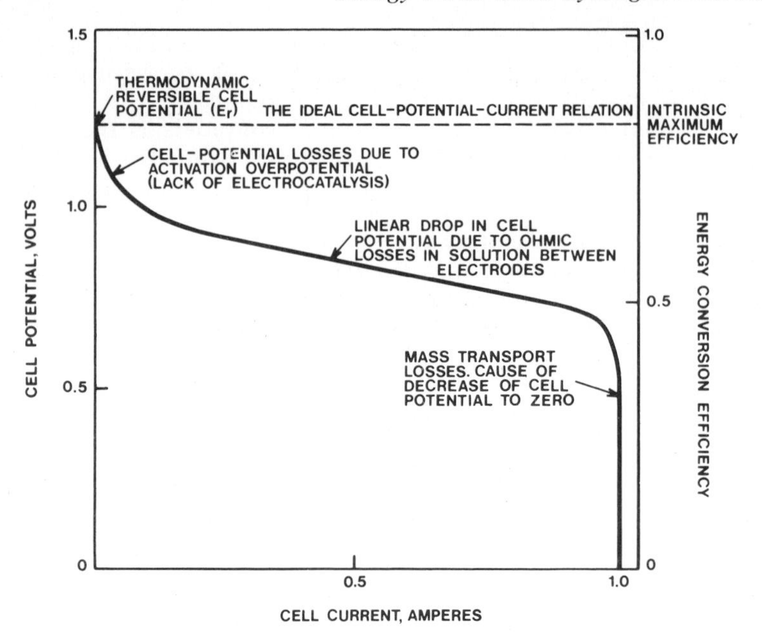

Fig. 13.3 Typical cell potential and efficiency vs. current relation of an electrochemical electricity producer showing regions of major influence of various types of overpotential losses.[11]

mole of the overall reaction, divided by the heat content change for the reaction concerned.

Thus, the basic, ideal expression for efficiency of a fuel cell, would be:

$$\epsilon_{\text{ideal}} = -\frac{nFV_e}{\Delta H} \tag{13.13}$$

This figure is liable to be near to 1. Indeed, if the efficiency is defined – as it sometimes is – as:

$$\epsilon_{\text{ideal}} = -\frac{nFV_e}{\Delta G} \tag{13.14}$$

the efficiency is in fact 1, because of the thermodynamic relation between the potential of a cell and the free energy.

Of course, this efficiency neglects the real overpotentials, which are a fact of life in electrochemical energy conversion and which reduce the efficiency of the electrochemical converter. Thus, the actual efficiency could be written as:

$$\epsilon_{\text{real}} = -\frac{nFV_e}{\Delta H}\frac{V}{V_e} \tag{13.15}$$

where V is the actual potential of the converter, and is given by the equation:

$$V = V_{rev} - \Sigma \eta \tag{13.16}$$

as explained in the kinetics of the cell deirved in the last section.

The ratio of:

$$\epsilon_p = \frac{V}{V_e} \tag{13.17}$$

is called the voltage efficiency.

There is also another efficiency factor called the Faradaic efficiency, which takes into account the possibility that fuel and reactant may pass through the electrodes without reacting. Thus, the efficiency of a fuel cell, complete, is:

$$\epsilon = (\epsilon_{ideal}\, \epsilon_p)\, \epsilon_f. \tag{13.18}$$

One can often take ϵ_f as virtually 1.

A typical plot of the efficiency of a fuel cell with varying parameters is given in Fig. 13.4.[12] The figure given is calculated for the still partly idealised circumstances of planar electrodes (see also Table 13.1).

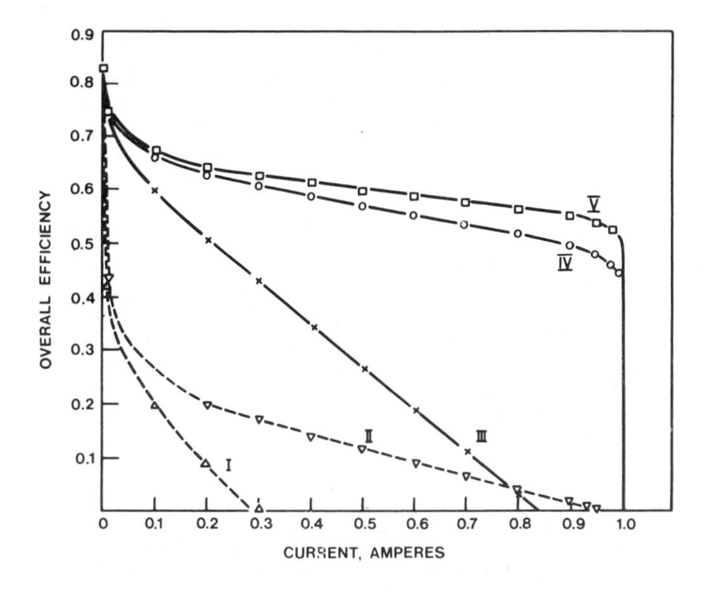

Fig. 13.4 Plots of overall efficiency ϵ *vs.* current *I*.[12] (The parameters which have led to this figure are given in Table 13.1[13]).

We have shown that the efficiency of an electrochemical reactor is very high (in practice, > 75%) at low rates, but is reduced at higher rates (50-60%). However, what is the most important aspect is the energy produced in an electrochemical reactor for unit weight, compared with that of other transducers.

This is one of the most important aspects of electrochemical energy conversion: it is unrivalled as far as efficiency is concerned. Thus, the energy per unit weight of a fuel cell compared with batteries, and two kinds of internal combustion converters, is as shown in Fig. 13.5.[14]

Even when the fuel has to be processed to form hydrogen before it is introduced into the fuel cell, the efficiency is still higher than that of any other known converter by at least 15%. When the fuel is hydrogen, the working efficiency is likely to be > 60%, about twice times the efficiency offered normally by internal combustion converters, and some 50% higher than that offered by internal-combustion engines.

TABLE 13.1[13]

ASSUMED ELECTRODE KINETIC PARAMETERS IN THE THEORETICAL ANALYSIS OF THEIR EFFORTS ON THE PERFORMANCE OF ELECTROCHEMICAL REACTOR

Calculation Number	$A_a i_{o\,a}$ amp	$A_c i_{o\,c}$ amp	$A_a i_{L\,a}$ amp	$A_c i_{L\,c}$ amp	α_a	α_c	R_i ohms
I	10^{-3}	10^{-6}	1	1	$\frac{1}{2}$	$\frac{1}{2}$	1
II	10^{-3}	10^{-6}	1	1	$\frac{1}{2}$	$\frac{1}{2}$	0.1
III	>1	10^{-3}	1	1	∞	$\frac{1}{2}$	1
IV	>1	10^{-3}		1	∞	$\frac{1}{2}$	0.1
V	>1	10^{-3}	1	1	∞	$\frac{1}{2}$	0.01

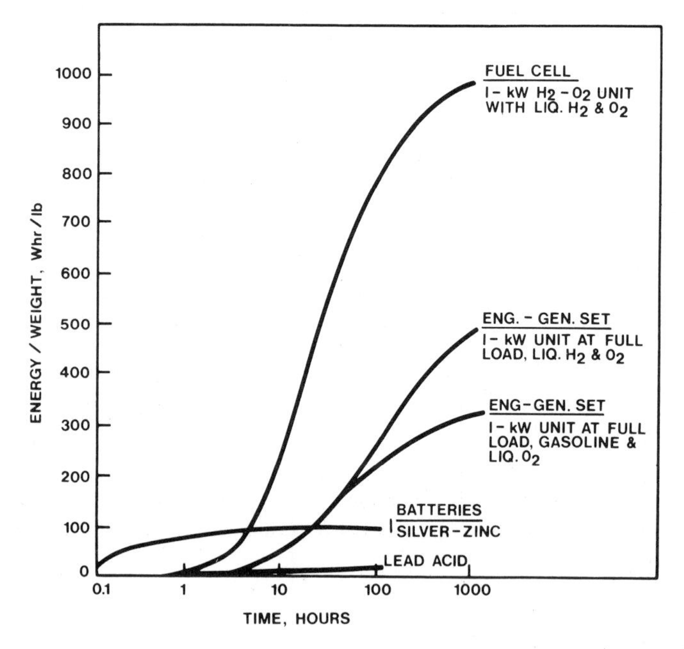

Fig. 13.5 Plot of energy/weight ratio as a function of time for some energy converters.[14]

POWER

The power of an electrochemical reactor is expressed as:

$$P = IE, \tag{13.19}$$

where I is the total current and E is the effective potential.

With the equations derived earlier, it is easy to show that the power is:

$$P = I\left[E_r - \frac{RT}{\alpha_c F}\ln\frac{I}{A_c i_{o\,c}} + \frac{RT}{nF}\ln\left[1 - \frac{I}{A_c i_{L\,c}}\right]\right.$$

$$\left. - \frac{RT}{\alpha_a F}\ln\frac{I}{A_a i_{o\,a}} + \frac{RT}{nF}\ln\left[1 - \frac{I}{A_a i_{L\,a}}\right] - IR_i\right] \tag{13.20}$$

The corresponding function of this power and how it varies with current is shown in Fig. 13.6.[15]

Mathematical conditions which relate to the power in an electrochemical converter can be deduced, but, if they are to be deduced analytically, and in a simple way, there must be severe simplifications. The most obvious one is to assume that the relation between the over-voltage and current is that stated above, namely that the activation overvoltage

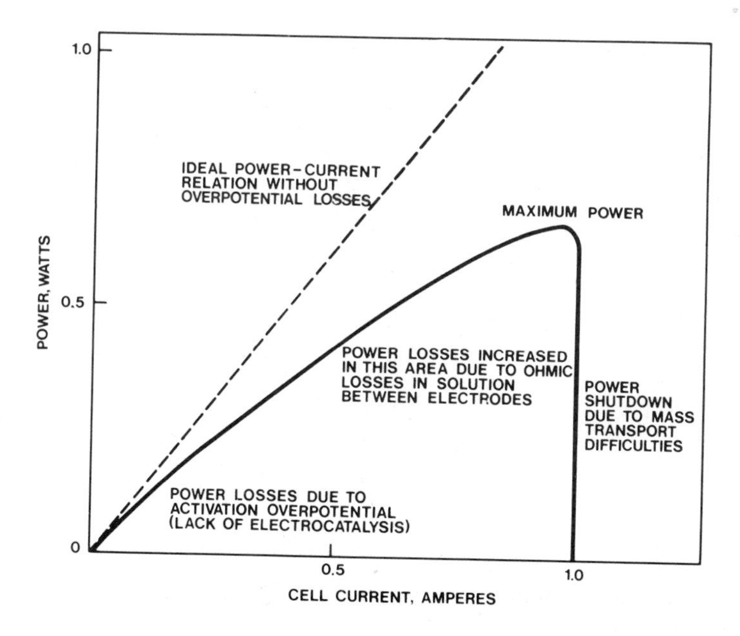

Fig. 13.6 Typical power density *vs.* current density relation for an electrochemical electricity producer showing reasons for power losses in various regions.[15]

is linear with the logarithm of the current density, and to neglect other kinds of polarisation. Under these simplifying conditions, one can write the power as:

$$P = I[E_r + (a_1 + a_2) - (b_1 + b_2) \log I], \tag{13.21}$$

where the a and b terms come from the Tafel equation, written in the form

$$\eta_{act} + a \pm b \log I. \tag{13.22}$$

Thus, referring to the equations in the efficiency section, the term a has the significance:

$$\alpha = \frac{RT}{\alpha F} \ln Ai_o. \tag{13.23}$$

Correspondingly,

$$b = \frac{RT}{\alpha F}. \tag{13.24}$$

Thus, one obtains the conditions for maximum power by setting:

$$\frac{dP}{dI} = O. \tag{13.25}$$

Thus,

$$\frac{dP}{dI} = E_r + (a_1 + a_2) - (b_1 + b_2) \log(I_m) - (b_1 + b_2) = O. \tag{13.26}$$

It follows that:

$$\log I_m = \frac{E_r + (a_1 + a_2) - (b_1 + b_2)}{b_1 + b_2} \tag{13.27}$$

$$E_m = (b_1 + b_2) \tag{13.28}$$

$$P_m = (b_1 + b_2) \exp \frac{E_r + (a_1 + a_2) - (b_1 + b_2)}{b_1 + b_2} \tag{13.29}$$

Re-establishing the equation for the power in terms of the appropriate real parameters, one obtains:

$$P_m = \frac{RT}{\alpha_o F} +$$

$$\frac{RF}{\alpha_a F} \exp\left[\frac{E_r + (RT/\alpha_c F) \ln A_c i_{Oc}) + RT/\alpha_a F \ln(A_a i_{O\ a}) - (RT/\alpha_c F) - (RT/\alpha_a F)}{(RT/\alpha_c F) + (RT/\alpha_a F)}\right] \quad (13$$

Making the even further simplification that the activation overpotential is predominant at only one electrode (e.g. the other electrode may have a reaction with a very high exchange current density), the most simple expression for the maximum power is given by:

$$P_m = \frac{RT}{\alpha_c F} A_c i_{o\,c} \exp\left[\frac{E_r \alpha F}{RT} - 1\right] \tag{13.31}$$

Such an equation differs from one which would apply to any real reactor in several ways, not only because of the approximations made in the deduction, but also because the electrodes have been taken to be planar, whereas in all practical fuel cells they are porous (see below). However, the equation has the following usefulness: it allows us to sense the way in which the various factors influence the power. Thus, this is proportional to the exchange current density and exponentially proportional to the reversible potential. These conclusions will not apply in a simple way in a real fuel cell, but they indicate trends.

DIFFUSE POWER CONCEPTS POSSIBLE WITH ELECTROCHEMICAL CONVERTERS

Electrochemical converters lend themselves more easily to the diffuse power concept: the hydrogen would flow through pipes to housing and industrial sites and the needed electric power, and heat, would be produced at the site. There would be no loss of energy by rejecting the heat in cooling water to the sea or to rivers. Thus, in the fuel cell, the production of electric power would be silent, and at high efficiency, whereas small local internal combustion plants in buildings would not be practical because of noise, vibration and air pollution.

TOTAL ENERGY CONCEPT

This ties up well with the preference for an electrochemical converter in the Hydrogen Economy. The heat production in an electrochemical converter will be treated below, and it will be shown that, although the electrochemical converter is twice as efficient as most heat engines, it still has perhaps 40% of its converted energy to give out as heat (overpotential losses). This heat, produced in the building along with the electricity, can be used as the building's heat source.

TARGET PROGRAMME

The American gas industry has for some years had an ambitious undertaking entitled 'The TARGET Programme'. This aims at achieving the diffuse power concept, and the total energy concept. The aim is to distribute natural gas – this is, then, converted to electricity at the site of use in a fuel cell. The pricing prospects for natural gas, however, mean that the TARGET programme upon a big scale could be a beginning of a Hydrogen Economy.

In 1975, the TARGET programme was in an advanced testing stage, with fuel cells in apartment buildings on test.

THE ELECTROCHEMICAL ENGINE

This term, first used by Henderson[16], is that given to the fuel cell-electric motor combination.

In the electrochemical engine, the stress would be upon power per unit weight, rather than efficiency per unit weight. At present, the electrochemical converter can do as well as a diesel heat engine, and the lower power per unit weight which it represents compared with other heat engines is compensated by its freedom from pollution, and greater energy efficiency.

POROUS ELECTRODES

As has been shown above, the efficiency and power of an electrochemical converter depends not only upon the exchange current density, i_o, but upon the limiting current, i_L. This is the maximum rate of transfer of reactants to an electrode surface and depends firstly upon the solubility of the reactant in solution. However, the elementary equation which relates i_L to solubility does so through the boundary layer thickness, δ. This term depends upon geometry and the configuration of the reactor. At a planar reactor in the absence of stirring, it has a numerical value of about 0.05 cm, and can be reduced to *c.* 0.001 cm for a high degree of agitation, but such stirring will (though invaluable in increasing the limiting current by decrease of δ) be difficult to carry out in practice, because it will consume much mechanical power.

If one carries out elementary calculations of the power per unit apparent area of fuel cell electrodes, one obtains a poor prognostication for electrochemical converters. The progress which has been made in fuel cell technology has been as much in the realisation of how to engineer porosities of greater effectiveness (in terms of the equations, to increase i_L) as has been put into the other important fuel cell aim: the increase of i_o, i.e., electrocatalysis.

Two main names are relevant here, those of Bacon and Justi. Bacon was the first – in 1955 – to engineer a fuel cell of practical power (5 kW), and one of the major things he did was to increase the porosity of the electrodes which he had been using, and to make them more conducting by utilising, e.g., doped nickel oxide. Justi, correspondingly, concentrated upon the pore within such oxides, and pointed out the importance there of the three phase boundary between gas, solution and electrode material. Here, the reaction rate was at a maximum.

Earlier, the efficacy of having porous rather than planar electrodes was thought to be a surface area question, and it was not until Wagner[17] pointed out that little of the pore could be active, that a different view was taken. The essence of this view can be seen from an examination of Figs. 13.7[18] and 13.8.[19] In Fig. 13.7, one sees a pore without structure in it, i.e. a perfect cylindrical pore with no anodes of any particular structure at the boundary between the gaseous reactant and the liquid. An *IR* drop develops down the pore, and this *IR* drop diminishes the potential obtained from deeper within the pore. Therefore, on this ground, only a small amount of the pore is effective in producing power. This already indicates, therefore, that the efficacy of a multi-pored system lies more in the direction of the three-phase boundary between the gaseous reactant, the solid electrode and the liquid phase, than in a simplistic

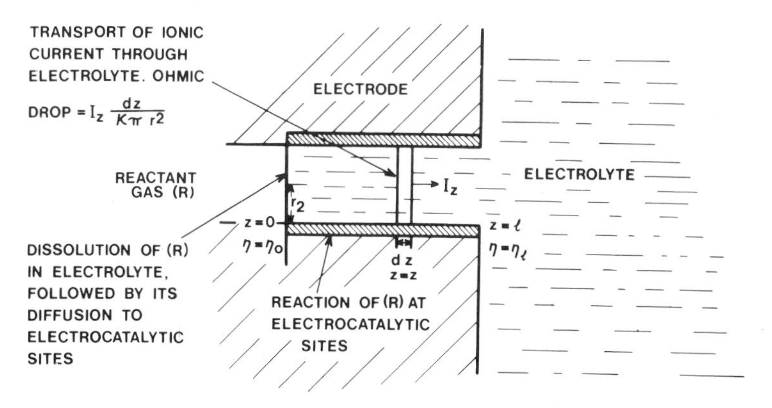

Fig. 13.7 Mode of operation in a single pore of a porous gas-diffusion electrode using the simple-pore model.[18]

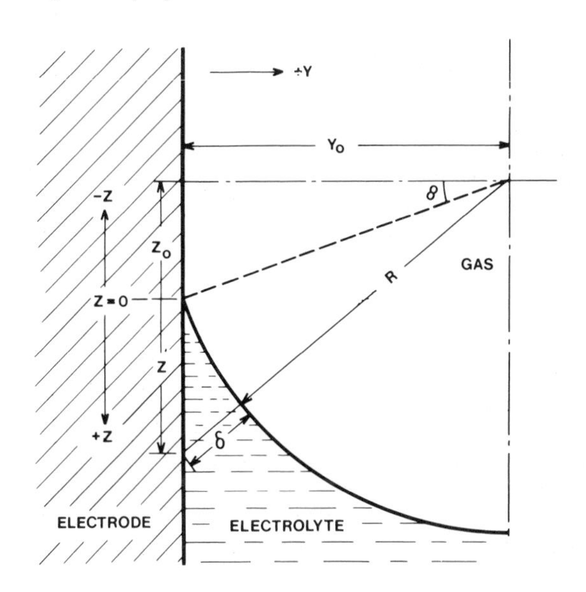

Fig. 13.8 Finite-contact-angle model for a porous gas-diffusion electrode. Co-ordinate system for theoretical calculations is shown.[19]

view of 'extra area' per apparent unit area. Thus, regard the limiting current aspect of electrochemical energy converters, and refer to the important term δ, which relates the limiting current, i_L, to the diffusion current $DnFc/\delta$. Suppose that one could make very thin layers of liquid on the surface of the electrode, then the δ would not be 0.05cm – that for a planar surface in unstirred electrolyte but much lower. The situation which has been established[19] at a meniscus is shown in Fig. 13.8. Much of the meniscus is a thin film and it is clear that the δ is small, perhaps a few hundred Å. Along these layers, the maximum reaction rate is high, and the fact that such an active area occurs at every pore, is the essential reason for the efficacy of porous electrodes. Another way of showing this is illustrated in Fig. 13.9[20], where a cone type of pore is shown. Here, too, there are very thin layers of solution at the 'end' of the contact between liquid and solid.

In practice, the considerations are more complex because of wetting. Thus, the capillary attraction of the small pore sucks in the solution. Unless the gas pressure is exactly equal to the attractive capillary force, the solution will pass through the pores to the outside of the cell and 'flood' the electrode, eventually leaking the solution out of the cell. To avoid this, hydrophobic materials are introduced into the cell and the most important of these is the fluorinated polyethylene called tephlon.

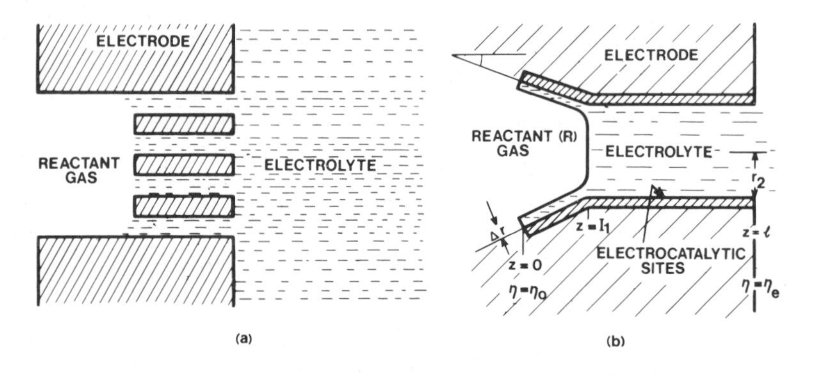

Fig. 13.9 (a) Single pore of a double-porous structure of a porous gas-diffusion electrode; (b) cone model for a porous gas-diffusion electrode. Only a single pore is shown.[20]

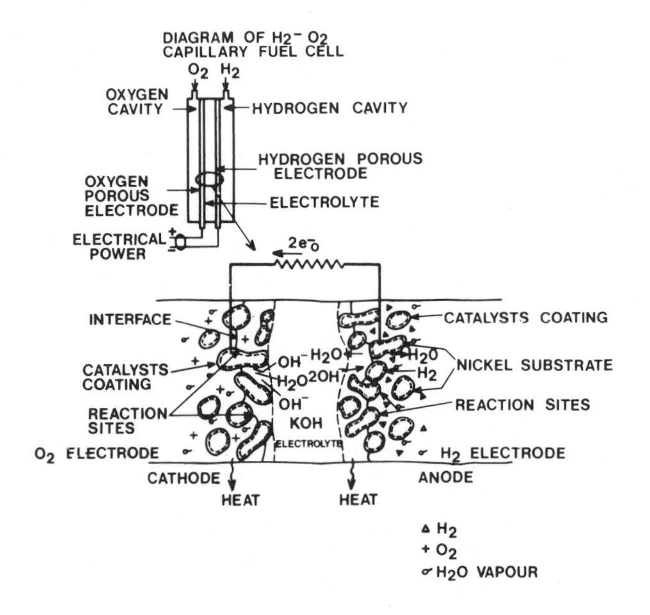

Fig. 13.10 An Allis-Chalmers hydrogen-oxygen fuel cell and a schematic representation of its mode of operation.[21]

An approximate diagram of what may be the state of affairs in a tephlon-containing electrode is shown in Fig. 13.10.[21] Here, the individual pieces of tephlon form boundaries between the gas and the solution and on these tephlon pieces – which are non-conducting – there are pieces of catalyst which allow conductivity to take place throughout the electrode. The reaction can occur at the very large number of three-phase boundaries thus formed within the porous mass.

FORMS OF REACTORS

Although the porous electrode is the main form for electrochemical reactors, there are other possibilities. The principle is to maximise the contact between fuel and electrode, under the situation which allows minimal ohmic loss in the cell, and maximal impelling of the reactant to the interface. One such idea is a jet electrode, in which one electrode, say the anode, rotates, and in front of it are a series of jets which contain, e.g. oxygen in solution, with an electrode in the mouth of the jet.

Another possibility is the spaghetti tube model in which the electrodes are in effect tubes of porous material, so that the gas can come from the entire length of the cylinder through the porous material to contact catalyst which is in the outside of it. Such an electrode[22] has the advantage that there is no ohmic drop down the pore and little diffusion hold-up.

Other ideas exist, including the development of electrodes carried out by the French company, Alsthom, in which a parallel layer of solution is separated by a membrane, on either side of which is the catalyst relevant to cathode and anode. Cathode and anode are impelled in contact with the common permeable membrane, eliminating many concentration and ohmic difficulties.

HEAT GENERATION AND FUEL CELLS: QUANTITATIVE

It has been pointed out above, that the thermodynamic heat generated by a fuel cell is a very small fraction of that generated in a heat converter, because the energy difference between the reactants and products of a reaction is directly converted to electricity.

In the ideal sense, were there no overpotential, and no i^2R losses, there would be a rate of heat evolved (for negative ΔS) or absorbed (for positive ΔS) equal to:

$$Q = \left[\frac{4.18\ T\Delta S}{nF}\right] i. \qquad (13.32)$$

A term which is as important, often more so, is that with the overpotential and $i^2 R$ loss. The total equation is:

$$Q = \pm\left[\frac{4.18\ T\Delta S}{nF}\right] i + i\Sigma\eta + i^2 R \text{ watt cm}^{-2}. \qquad (13.33)$$

In the hydrogen-oxygen fuel cell, ΔS is negative, i.e. the cell evolves heat. Even under ideal conditions of zero potential loss, there would be a small heat emitted from a hydrogen-oxygen cell. In reality as much as 40 to 50% of the energy of the reaction may be emitted as heat in a fuel cell.

ADVANTAGES AND DISADVANTAGES OF ELECTROCHEMICAL TRANSDUCTION

(1) No moving parts, and hence potentially a longer life than that of combustion engines. The actual life of fuel cell electrodes at the present time is short. The cause is partly due to poisoning of the electrode and the resultant decrease of catalytic properties on its surface. Another, more important, cause is recrystallisation. Overcoming the former difficulty can be expected, but the reduction of recrystallisation rates by an order of magnitude may be more difficult.

(2) High efficiency.

(3) No noise pollution.

(4) Electrochemical transducers can be made in any size and modules assembled to give any power. There is only a small scale-up factor. Internal combustion engines are not satisfactory at < 1 hp and cannot be scaled up by module addition.

(5) Chemical engines need starting by electrochemical sources, and waste fuel on idling. Electrochemical engines switch on and off without the need of auxiliary power sources; they have no warm-up period and do not need to idle.

The negative aspects of an electrochemical transducer centres on its lower power density (and, at present, its shorter lifetime). It will not be easy to obtain electrochemical transducers with power densities greater than 1 kW per kg, because the factor is limited by diffusion *in solution*, which must be less in velocity than diffusion in the gas phase. Even with excellent electrocatalysis, there will always be a greater diffusional barrier to high rates in a solution-phase reactor.

Secondly, negligible capital is as yet invested in fuel cell plants, so that they do not have this considerable influence encouraging their use: the capital is in the older, competing devices. That is a principal reason why conversion to the non-polluting electrochemical devices will be difficult.

Thirdly, the only period of continued and significant support fuel cells have had (with the exception of the strong effort of Pratt & Whitney Aircraft) is that between 1962 and 1968, when some $20 million per year of U.S. Government funds was spent on research and development in fuel cells in the U.S.* From about 1969, this (space-oriented) programme was curtailed: government support of electrochemical energy conversion and storage was removed, just as chemical reactors as a main cause of air pollution came into public consciousness.

Fourthly, few engineers understand the mechanism of electrochemical engines, whereas all engineers understand chemical engines in principle and

---

* There are fuel cell programmes in Western European countries and a very large programme in the Soviet Union. However, in the writer's experience, the direction of the huge Russian technological programmes is greatly influenced by what is seen to be successful in the U.S.A. A U.S. cut-back in an area of technology, has strong influence on what Russian scientists advise their government.

hundreds of thousands of engineers and mechanics have real experience of them. What seems right to nearly all engineers are systems which contain pistons, produce heat and noise and have moving parts. Silent, motionless, electrochemical systems have a strangeness factor.

The last three difficulties are worth overcoming. The efficiency, potential life, and absence of noise and moving parts, give such great advantage to fuel cells that the transducer eventually used for hydrogen is likely to be the electrochemical one. Most of the non-space oriented researches in fuel cells in the American programmes carried out during the '60s were attempts in laboratories of oil corporations to transduce hydrocarbons by means of fuel cells to electricity. However, hydrocarbon fuels give low rate constants, and, therefore, high overpotentials. Efficiencies were hence not much better than those of internal combustion engines. With hydrogen as a fuel, the advantages of the electrochemical transducer is realisable, because hydrogen reacts rapidly at an anode, causes little overpotential and hence low efficiency loss.

Situations where electrochemical transducers would have the advantage, compared to chemical transducers, would depend upon the net indications among the following:

(1) What is the relative importance to the application of high energy density (electrochemical engines best) compared with high power density (chemical engines best)?
(2) Is the cost of fuel sufficiently high so that a doubling of the efficiency of its conversion to useful energy would be sufficiently attractive to make it economically better to buy the new technology?
(3) Is the application one in which *electricity* is needed (electrochemical conversion is direct) or *mechanical energy* (chemical conversion direct)?
(4) Economic factors. These are more difficult to estimate, because electrochemical engines are not yet in mass production. A figure for 1974 in the region of $400 per kW is heard, not much more than that of internal combustion generators. The lifetime of the converters is the key point of the reality of this figure. The gain in efficiency, absence of noise pollution, and vibration, is to be laid against possible increases of initial cost.

APPLICATIONS OF FUEL CELLS

(1) *Household heat and electricity*
This may be supplied by roof-top collected solar energy in latitudes between 30N and 30S. In other latitudes, hydrogen would be piped to the house, and the fuel cell would be advantageous because the total energy concept (heat and electricity from the same amount of fuel) is then possible with zero noise or vibration.

(2) *Central electricity production*
A coal-fired fuel cell is an excellent prospect for the intermediate range future (1980 to 2020 say) in which coal may have to be used as a primary source of energy. It would suit the development of a Hydrogen Economy to go the path of hydrogen production from such electricity.

According to the Office of Coal Research[23–26], a 1000 megawatt coal plant could produce 0.3 kWh^{-1} (1966) cents electricity.

Table 13.2[23] shows an estimate of comparative costs.

TABLE 13.2[23]

COMPARISON OF FUEL CELL COSTS WITH OTHER MEANS OF ELECTRIC POWER PRODUCTION (in 1971 cents per kilowatt-hour)

Types of costs	Fuel cell plants		Fossil fuel plants		Water		Fast breeder	High temperature gas turbine	Magneto-hydrodynamic power plan
	1980*	1990†	1980	1990	1980	1990	1990	1990	1990
Annual capital	0.67	0.34	0.54	0.67	0.84	1.05	1.05	0.62	0.62
Fuel	0.36	0.36	0.57	0.71	0.23	0.29	0.10	0.48	0.48
Maintenance and operation; Insurance	0.04	0.04	0.03	0.04	0.06	0.07	0.07	0.04	0.04
Thermal effects	0.02	0.02	0.03	0.04	0.04	0.06	0.04	0.03	0.03
Total	1.09	0.76	1.17	1.46	1.17	1.47	1.26	1.17	1.17

* These figures consider that the initial cost equals that of fossil fuel plants.

† These figures consider that the high efficiency and ease of manufacturing of fuel-cell plants brings their initial cost to one-half that of fossil fuel plants.

(3) *Transportation*

In the air, conversion will be made towards hydrogen fueled jets because of the high power to unit weight ratio compared with those of other transducers. Electrochemically driven helicopters might be produced, because of the importance of reducing noise pollution if these vehicles are to come into general use as passenger carriers in towns. Helicopter noise is partly chopper noise from the rotors, but near the ground much comes from the engine and this would be eliminated.

Ground and sea transportation: The time of amortisation argument favours chemical engines. However, the rising price of fossil fuels will make the efficiency of conversion increasingly important. Fuel cells will tend to be the final energy source for both land and sea transports.

(4) *Manufacture*

This offers a mixed picture. An aluminum plant runs on dc power, at low volts, and would be an ideal application for fuel cell power from hydrogen if the plant is not being run from a nearby hydro-electric source. Hydrogen could be a fuel for many chemical processes (see Chapter 14) and near silent fuel cell-driven electric motors would lack the pollution (NO) associated with chemical converters. Accompanied by the economic advantage which would arise from the higher efficiency, there may be many cases of advantage.

USES OF ELECTROCHEMICAL CONVERTERS IN THE PRE-HYDROGEN ECONOMY TIME

Although the first uses of hydrogen as a fuel may well be in the internal combustion engines, with fuel cells developing finally as the main transducers because of their higher efficiency, it is possible to see contemporary use of fuel cells, in special purpose applications. There are some areas where their application has no rival, e.g. auxiliary power in lunar missions.

Another aspect of fuel cells which seems right for immediate exploitation is in portable generating sets, where the lesser weight of the fuel cells for the unit of energy transported would be of great advantage.

A third example would be for power sources for isolated centres. Here, the main cost of the power is the cost of transportation of the fuel. Directly fuel becomes expensive, the fuel cell with its greater efficiency, becomes a more desirable transducer.

Indoor vehicles, too, would be an immediate application for fuel cells, for the reason of the lack of pollution, and they would then give a much longer range than the present used battery vehicles. In shunting engines, where a great deal of idling wastes fuel if they are powered by diesel engines, fuel cells would show an advantage.

There are two immediate applications in naval affairs. Smaller submarines, where atomic power may be uneconomic, would be possible subjects for fuel cells. Absence of noise would decrease detectability. Deep submergence vessels at present use batteries and would have a longer range on fuel cells.

In the Army infrared detection of tanks and other transport would be avoided by fuel cell power. Lastly, recreational applications in campsites, road carriers, scuba-diving, etc., would prove a fertile area for the development of fuel cells before they become of general use in a Hydrogen Economy.

DIFFICULTIES OF FUEL CELLS IN THE 1970s

With only one company having a significant team researching fuel cells within the U.S., it is remarkable that the progress has remained so high. The principal difficulties of fuel cells at this time are three:

(1) Electrocatalysis: The hydrogen dissolution reaction has a high rate constant with non-noble metal catalysts. For this reason, it does not have to have a special investigation. Conversely, the equally important oxygen dissolution reaction is a very slow reaction and the origin of much power and efficiency loss. It does need investigation. When catalysed by platinum in acid solution, and silver in alkaline solution, the reduction of overpotential, i.e. increase in efficiency, is good. However, there is not enough of these metals to make a large scale application possible.[27]

What is wanted is a realisation of the dream of a 'substitute for platinum'. It seems unlikely that this will be found. The reason for the great efficacy of platinum as a catalyst, qualitatively stated, is easy to understand. One has to have, in an electrocatalyst, a *certain degree* of adsorption of the radicals. If this is too low, it will not give sufficient reactivity, because insufficient sites are covered per unit area. If the coverage is too high, there will be an insufficient arrival of new material on the surface from the solution per unit time, and a low reaction rate per unit area will result. The coverage will depend largely, compared with other catalyst materials for the same reaction mechanism, upon the strength of the bonding between radicals of the solute and the catalyst. Obviously, the maximum rate will be reached at some intermediate value of bonding, i.e. of M-M bonding. Platinum has a bonding intermediate, among the noble metals, between the small bond strength offered by gold, and the high bond strength offered by osmium. A substance with the same, or nearly the same lattice characteristics as Pt is needed: from general chemical considerations that is not likely outside the noble metals.

The second consideration is stability. Most of the reactions for which catalysts are required occur in an anodic or positive situation within the electrochemical source (expressed on the European scale). The danger with most materials is that the bonding to the substrate will be inadequate to withstand the anodic electric field and this will dissolve instead of remaining to act as catalyst. The result of this requirement is that practically only the noble metals have been used for the catalysis of oxygen reduction and hydrocarbon oxidation.

However, there may be a solution in the use of small traces of noble metals – less than mono-layers – adhering to a substrate of cheaper materials, e.g. conducting tungsten bronzes. It seems[28, 29] that there is an enhancement of reactivity of the platinum per atom under these situations and this may come from the synergistic effect between the platinum and the substrate. The first partial reaction in the reduction of oxygen, say, occurs with the platinum. The *subsequent* reaction steps in the overall reaction for the reduction of oxygen to water occur more quickly upon the substrate, which is known to be a poor catalyst of the first step. Considerations along these lines may produce relatively cheap catalysts which give the same order of magnitude reaction rates as does platinum. There is no doubt that catalysis of oxygen reduction is the area most urgently demanding funding in fuel cell research.

(2) Another difficulty (and perhaps a more serious one) which urgently needs research is the aging effect in electrodes. For all electrode reactions, there is a decrease with time of the reaction rate. The speed of the decrease is, of course, the economics-determining factor. A fuel cell, the electrodes of which can remain stable for 2 to 4 years, may be a cost-effective device. A six months life would not be acceptable.

The decay with time of reactivity at electrodes has been associated with impurities in solution.[30] They are gradually adsorbed on the surface. They are organic in origin, perhaps oxidation fragments of hydrocarbons. A gradual spread over the surface of an inactivating organic polymer is likely.

Recently however, it has been found[31] that a more common cause of the decay of the efficacy of fuel cells with time lies in a *recrystallisation* process on the surface of the electrode. There is a thermodynamic tendency for small crystals to aggregate with time into larger ones, and these present a lesser surface area to the reactants. Correspondingly, there are active areas upon electrodes which depend upon the precise configuration of the electrode surface, in particular the concentration of certain types of dislocations and the rotation of the crystallographic spirals on the catalyst surface which may give changes in the topography of the surface with time.[32, 33]

The decrease of electrode activity with time is the most pressing problem in fuel cell technology.

(3) Effective geometry of transducer devices. Basic electrochemical engineering of porous electrodes, particularly the theory thereof, rotating wheel transducers, and other devices, have received only sporadic research. Equations which describe the design criteria for porous electrodes[18,19] in terms of the shape, form and dimensions of this electrode, as related to the parameters of the reaction, are too idealised. The making of porous electrodes is still very largely empirical. Thus, e.g. the occupancy of pores, and the placing of catalysts in positions in pores where they may be 100% useful, is a subject which requires more study.[34]

ASSEMBLY OF ACTUAL FUEL CELLS

The considerations given above have been for electrodes and two electrode *cells*. In practice, fuel cells must be multi-cell to give the potential needed. They must have separators, supply tubes, pumps for gases and liquids, devices for the control of vapour pressure, heat exchangers, etc. The assembly of a single cell in the General Electric ion-exchange cell is shown in Fig. 13.11.[35] Correspondingly, the Gemini 1 kW cell is shown in Fig. 13.12.[35]

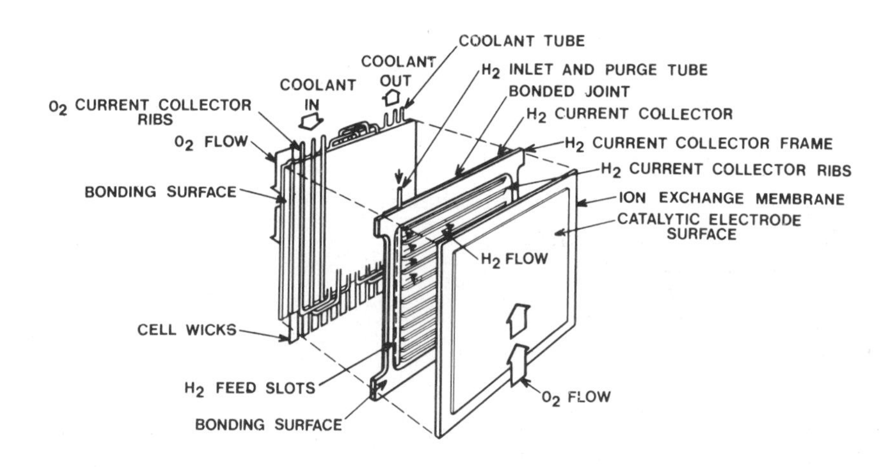

Fig. 13.11 Assembly of a single cell in the General Electric ion-exchange-membrane hydrogen-oxygen fuel cell.[35]

OTHER DEVICES NEEDED DURING THE USE OF HYDROGEN

Apart from the actual transduction of hydrogen to energy by chemical or electrochemical means, many new devices will be necessary. One of these will be the burners to use hydrogen. Thus, if hydrogen burns in oxygen, the reaction is a straightforward formation of water. If hydrogen combines with oxygen in air, 3.76 molecules of nitrogen are mixed with 2 molecules of water, 1 of oxygen and 2 of hydrogen. The nitrogen

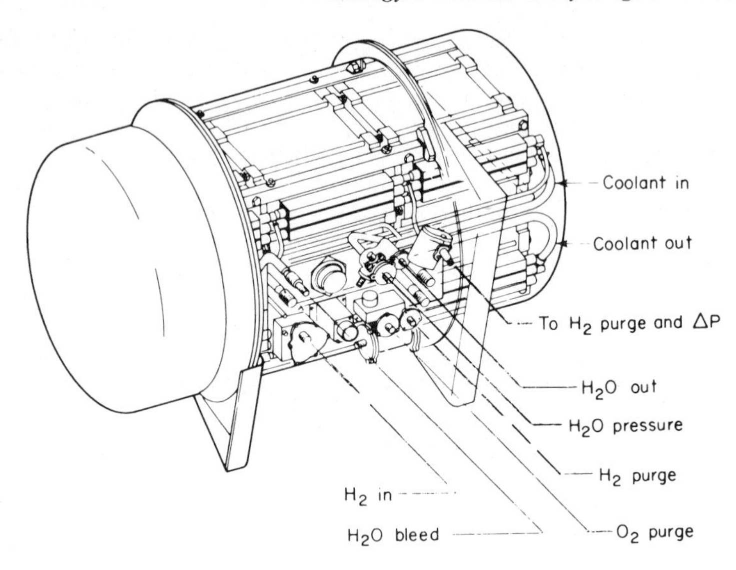

Fig. 13.12 The General Electric ion-exchange-membrane hydrogen-oxygen fuel cell battery. Layout of a three-stack fuel cell.[35]

absorbs a portion of the combustion energy and this results in a lower flame temperature. Thus, in pure oxygen, the burning temperature is about 3080° K; whilst for the burning in air it is about 2400° K.[36]

The formation of NO_x has to be taken into account. The lower combustion temperature in the presence of nitrogen will decrease the Carnot cycle efficiency.

One of the ways in which hydrogen and oxygen can be burnt together effectively is to use the aphodid burner, Fig. 13.13.[2] The burner produces high temperature steam that can later be used by any one of the steam-energy converters. The Carnot cycle efficiency is high – 60% has been mentioned – and this would give competition to the fuel cell.

Catalytic burners may be of interest. Hydrogen reacts with oxygen in air, whilst in contact with the catalyst and heat can be produced without flame.[37] Very thin catalytic layers, perhaps palladium, would be suspended on a ceramic substrate. The operating temperature is under 750° C. No nitric oxide is formed. The burner is inherently safe for use in houses without flues. It could form the basis of a non-polluting space-heating system.

Flame burners will have to be different for hydrogen. There are at present two types of combusters, the catalytic type and the flame combuster, the latter being, of course, the usual type. Its satisfactory performance depends upon the stability of the flame and its lack of flash-back. If the fuel velocity is too high, the flame advances away from the burner and is extinguished. If it is too low, it enters the burner and tends to flash back.

Thus, hydrogen burners must be produced which allow for these conditions. The flow rate of hydrogen through the air must be about three times that for methane. A larger flame stability range must be allowed for.[38] A thermal sink could be used to extinguish the flame.[39] Pilot ignition lights would be easy to arrange.

Natural gas burners will have to be replaced because their port size will be too big, their air mixing devices will be unsatisfactory, and because the flame temperature will be higher.

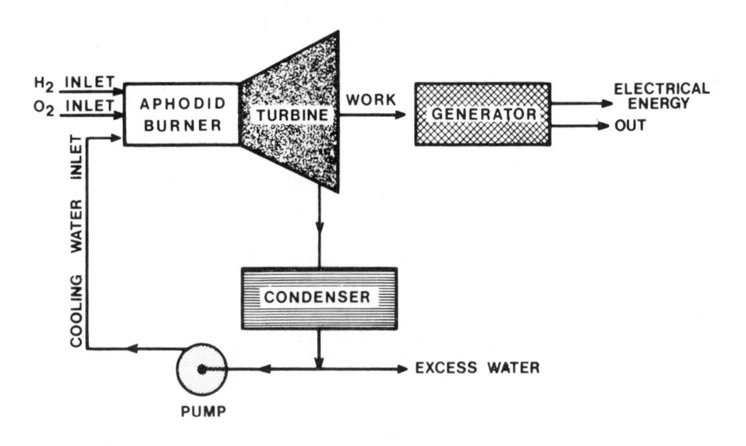

Fig. 13.13 Aphodid burner steam power plant.[2]

CONCLUSIONS

The mode of transduction adopted will be that which optimises the profit of the larger industrial groups.* The former factor will favour chemical engines at first, and a change to electrochemical engines will take place in so far as the price of fuel is high, so that the efficiency advances give gains to counteract the cost of rebuilding the means of production.

For certain corporate groups, however, particularly the gas industry, fuel cells will be, in the long term, a powerful help to their financial stability. Electrical utilities are developing giant fuel cells, now, because of the greater efficiency of conversion and this incentive will rise as the fossil fuel price rises continuously during the exhaustion phase of such fuels which has been entered.

In the long term extrapolation, one should see electrochemical trans-

---

* This does not, of course, imply that socialist countries will develop a technology which optimises the benefits of the population. There, too, a vested capital argument is effective. It may be, theoretically, the people's capital, but vast, automated plants for making internal combustion engines will not be taken down and rebuilt until there has been amortisation. Therein lies one of the many advantages of inflation.

ducers in general use with hydrogen fuels, except jet driven aircraft: the economics of the greater efficiency of conversion and the simplicity are likely to become irresistible.

MAJOR RESEARCH RECOMMENDATIONS

The major one is to increase the rate of development of the fuel cell. Its funding has been very sporadic, and, when compared with internal combustion engines, so tiny, that 20 to 30 years of strong development with good funding in companies which would profit by success of the endeavour should bring remarkable improvements. It is at once made worthwhile if hydrogen fuel is available. As fossil fuels rise in price, the development of hydrogen transducing fuel cells will become unavoidable. One of the subjects within the fuel cell area which urgently needs fundamental and mechanistic work is oxygen electrocatalysis. Another is the effect of recrystallisation on the useful lifetime (hence, economics) of electrodes. Lastly, the limitations of mass transfer, though well understood at the fundamental level, have been little researched with applications of the technique available from modern heat transfer engineering (heat transfer depends on diffusion and convection as does mass transfer).

There are some other areas which need research. The double layer at solids and the role of water thereat are vital topics and here the research has been tiny in extent and utterly fundamental, although the area is so very practically important.

The economics of efficiency in conversion, and the need for long-term developments which are non-pollutive, will find that a Hydrogen Economy becomes a Hydrogen-Electrochemical Economy (see Chapter 17).

REFERENCES

[1] J.O'M. Bockris & S. Srinivasan, *Fuel Cells: Their Electrochemistry*, McGraw-Hill, New York, 1969, Appendix A, p. 633.
[2] W.L. Hughes & S.O. Brauser, 'Energy Storage Systems', U.S. Patent No. 3459953, 5 August, 1969.
[3] 'Rocket Power to Light the Cities', *Skyline Magazine*, **30**, 4 (1972).
[4] R.B. Scott *et al.*, *Technology and Uses of Liquid Hydrogen*, Pergamon Press, 1964.
[5] G.R. Seikel, J.M. Smith & L.D. Nichols, 'H_2-O_2 Combustion Powered Steam-MHD Central Power Systems', paper presented at The Hydrogen Economy Miami Energy (THEME) Conference, 18-20 March 1974.
[6] J.O'M. Bockris & Z. Nagy, *Electrochemistry for Ecologists*, Plenum Press, New York, 1974.
[7] J.O'M. Bockris & A.K.N. Reddy, *Modern Electrochemistry*, Rosetta Edition, Plenum Press, New York, 1973, p. 1361.
[8] E. Justi & A. Winsel, *Kalte Verbrennung-Fuel Cells*, Steiner, 1962.
[9] J.O'M. Bockris, *J. Chem. Ed.*, **48**, 352 (1971).
[10] J.O'M. Bockris & S. Srinivasan, *op. cit.*, p. 181.
[11] *ibid.*, p. 182.
[12] J.O'M. Bockris & S. Srinivasan, *J. Electroanal. Chem.*, **11**, 350 (1966).
[13] J.O'M. Bockris & S. Srinivasan, *op. cit.*, p. 183.
[14] W. Mitchell, jnr., *Fuel Cells*, Academic Press, New York, 1963.
[15] J.O'M. Bockris & S. Srinivasan, *op. cit.*, p. 190.
[16] R. Henderson, priv. comm., 1966.
[17] C. Wagner, priv. comm., 1963.
[18] S. Srinivasan, H. Hurwitz & J.O'M. Bockris, *J. Chem. Phys.*, **46**, 3108 (1967).
[19] J.O'M. Bockris & B. Cahan, *J. Chem. Phys.*, **50**, 1307 (1969).
[20] J.O'M. Bockris & S. Srinivasan, *op. cit.*, p. 252.
[21] J.L. Planter & P.D. Hess, Allis-Chalmers Research Publication, September 1963.
[22] J.O'M. Bockris and S. Srinivasan, *op. cit.*, p. 219.
[23] S. Baron, 'Options in Power Generation and Transmission', presented at Brookhaven National Laboratory Conference, 'Energy, Environment and Planning, The Long Island Sound Region', October 1971.
[24] News release, Department of the Interior, Office of Coal Research, Washington, D.C., 12 December 1966.
[25] T. Aaronson, *Environment*, 13, No. 10, 10 (December 1971).

[26] W.J.D. Escher, 'Prospects for Hydrogen as a Fuel for Transportation Systems and for Electrical Power Generation', ORNL-TM-4305, Oak Ridge National Laboratory, September 1972.
[27] J.O'M. Bockris & S. Srinivasan, *op. cit.*, p. 633.
[28] J. McHardy & J. O'M. Bockris, *J. Electrochem. Soc.*, **120,** 53 (1973); K. Kinoshita, J. Lundquist & P. Stonehart, *J. Catalysis,* **11,** 325, 1973.
[29] *ibid.*, 61 (1973).
[30] J.O'M. Bockris & S. Srinivasan, *op. cit.*, p. 421.
[31] P. Stonehart, priv. comm., May 1974, Fuel cell electrocatalyst considerations, ERPI Catalyst Workshop, Pao Alto,California, 1975.
[32] J.O'M. Bockris & S. Srinivasan, *op. cit.*, p. 289.
[33] H. Kita, M. Enyo & J.O'M. Bockris, *Can. J. Chem.*, **39,** 1670 (1961).
[34] B. Cahan, Thesis, University of Pennsylvania, 1968.
[35] J.L. Schanz & E.K. Bullock, paper presented at the American Rocket Society, Santa Monica, California, 25-28 September 1962.
[36] R.L. Savage, L. Blank, T. Cady, K. Cox, R. Murray & R.D. Williams, edd., 'A Hydrogen Energy Carrier', Vol. II – Systems Analysis, NASA-ASEE, 1973.
[37] D.P. Gregory, assisted by P.J. Anderson, R.J. Dufour, R.H. Elkins, W.J.D. Escher, R.B. Foster, G.M. Long, J. Wurm & G.G. Yie, 'A Hydrogen-Energy System', prepared for the American Gas Association by the Institute of Gas Technology, Chicago, August 1972, p. VII-17.
[38] H.F. Coward & G.W. Jenks, 'Limits of Flammability of Gases and Vapors', U.S. Bureau of Mines Bulletin, 503, 1952.
[39] P. Stonehart, priv. comm., July 1974; *Catalysis Review* (with P.N. Ross), **11** (1975); J.A. Bett, K. Kinoshita & P. Stonehart, *Journal of Catalysis,* **35,** 307 (1974).

CHAPTER 14

Some Consequences of the Availability of Massive Amounts of Hydrogen

INTRODUCTION TO USES OF HYDROGEN

IT WILL BE JUSTIFIED to refer to a 'Hydrogen Economy' if there is a widespread use of hydrogen, in addition to its use as the medium of energy in transmission. There would be substantial household, industrial and transportation applications.

In this chapter, some of the probable applications are exemplified. These include household uses, the powering of transports, uses in chemical technology and metallurgy, and the availability of significantly increased amounts of potable water as a by-product (see Fig. 14.1[1]).

A principal gain would be a reduction in pollution.[2]

METALLURGY[3]

(1) *Ferrous metallurgy*[4]

There is interest in non-polluting iron production, i.e. ferrous metallurgy which avoids CO and CO_2 injection into the atmosphere. Reduction of Fe_2O_3 with H_2,

$$Fe_2O_3 + 3H_2 \rightarrow 2Fe + 3H_2O, \quad (14.1)$$

is well known and there are various processes by which it can be carried out.[1] Among these is the H iron process, the Hygas process and the Purofer process. 1 Kg of iron needs about 1 cubic foot of hydrogen for its reduction. The temperature of the fluidised bed in which the reduction is carried out ranges between 500 and 800°C. Water in significant quantities would be the condensable product.

The difficulty of changing from the normal reduction to hydrogen is that the capital in present blast plants could not be recovered once the swing to hydrogen reduction set in.

The stimulus for change to a hydrogen reduction process for iron is the decrease in pollution, but this is not a factor from which an increase in profit can easily be made. It is important not to use the process:

$$Fe_2O_3 + CO \rightarrow 2Fe + CO_2. \quad (14.2)$$

Marchetti[1] quotes 2,500 megawatts as the necessary energy to give

3.5 million tons per year of iron in sponge form from a hydrogen reduction plant.

The set-up for hydrogen in steel making can be seen from Fig. 14.2[4], which shows a diagram for one process which supplies 90% reduced iron to electric steel making furnaces. Fig. 14.3[4] shows the reaction kinetics for the reduction of hematite by hydrogen, and in Fig. 14.4[4], the flow of hydrogen and water in another process for hydrogen reduced iron is shown.

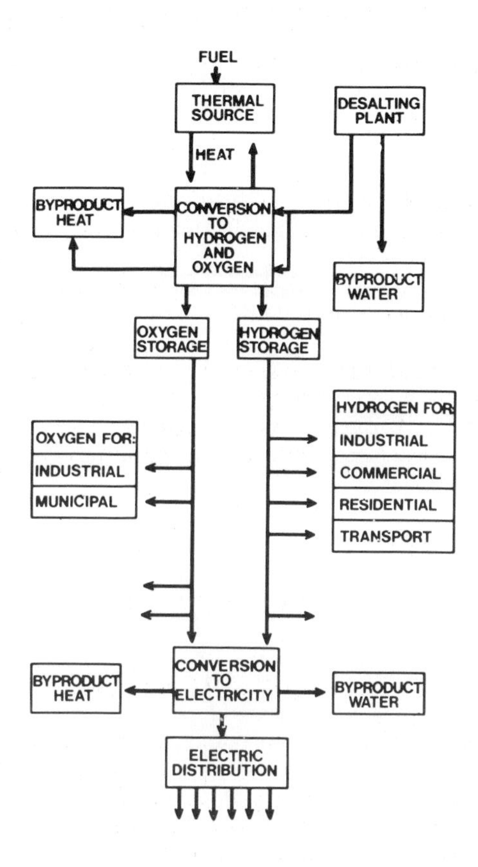

Fig. 14.1 A Hydrogen Economy.[1]

The achievement of widespread hydrogen reduction in ferrous metallurgy will affect the geography of the ferrous industry: it will avoid the necessity of having a plant at a point which takes account of the availability of both the ore and coke. Hydrogen will arrive in pipes from the distant nuclear or solar stations. Alternatively, a nuclear reactor may be associated with large steel production plants and produce hydrogen directly at the plant, by electrolysis. Cost savings should arise from the elimination of the costs of the transportation of coke.

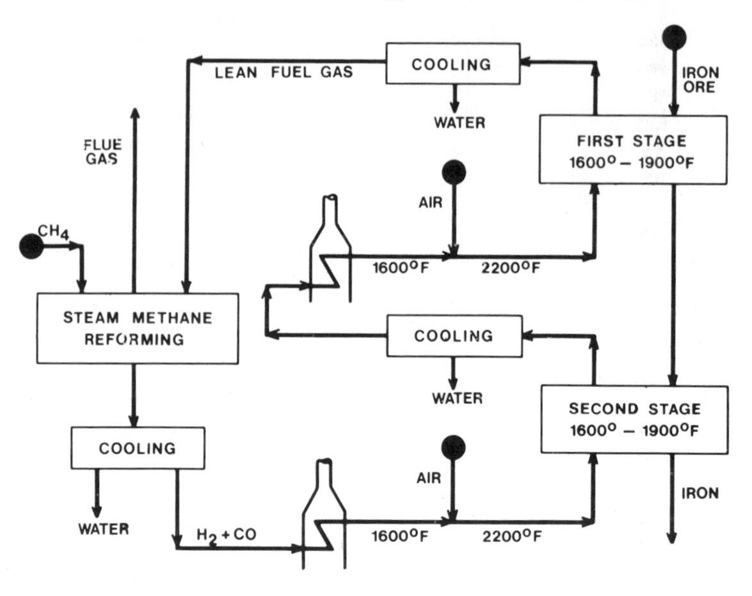

Fig. 14.2 Schematic diagram of the HyL process for supplying 90% reduced iron to electric steel making furnaces.[4]

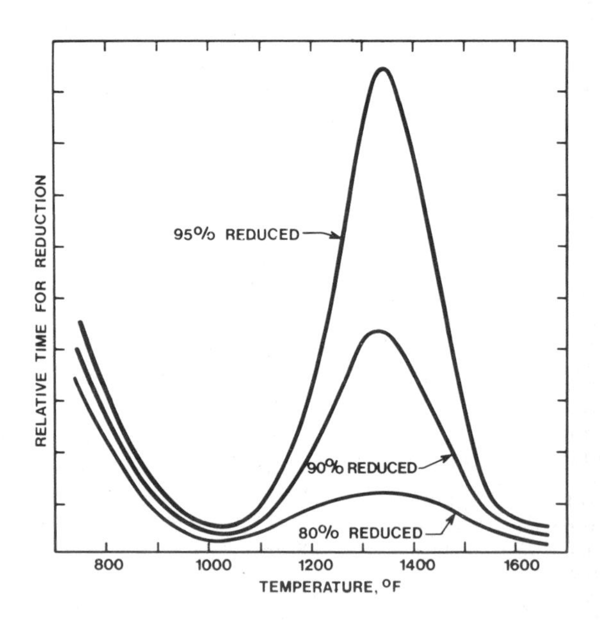

Fig. 14.3 Representative reaction kinetics for reduction of a high-grade Australian hematite by hydrogen.[4]

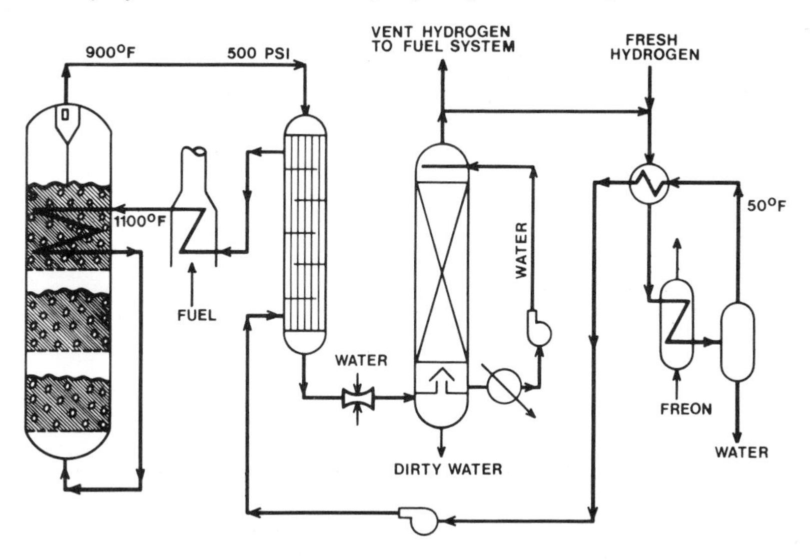

Fig. 14.4 Flows of hydrogen and water in the H-iron process.[4]

(2) *Other metallurgical processes*[5]
The possibilities of a future hydrogen technology for iron production infer similar possibilitles for other technologies which at present pollute the atmosphere. Copper, zinc, nickel and lead are metals where the technology of production could be achieved by a hydrogen reduction to give H_2S and then a reaction of this substance to give sulphur and water. (Other possibilities, e.g. molten salt electrolysis, exist for the pollution-free production of these materials.)

CHEMICAL TECHNOLOGY[6]

Hydrogen is extensively used in chemical technology now, e.g. in the production of oil and fats (Table 14.1)[6]. The use for these is about 48%; 38% is used in fertilizer production and 14% in methanol production. An expansion in the use of hydrogen in chemical technology would follow a reduced price of hydrogen, in constant dollars, which can be expected either from research on new methods or as a result of reasonably declining energy costs or as a result of breeder or solar technology.

With the rising price of natural gas supplies, resulting from the approaching exhaustion of that resource, there will be an expansion of the need for hydrogen as the replacement. Organic chemical technology will face a future without oil or natural gas as base products. For some 50 years*

---

* Such estimates must be bracketed with quite wide limits. They depend on the degree to which solar and atomic sources are developed and the comparative cost situation. If the energy supply were largely non-fossil fuel by 2000, the remaining fossil fuels would supply us with organics and plastics for hundreds of years. At present, the prospect is that we shall be hard pressed not to be retarded in our economic development by high and rising prices for energy which began in 1973.

it may be possible to use organic chemicals from coal. After this, the production of organic chemicals will involve hydrogen because much of it must start with carbon dioxide (from carbonate rocks and the atmosphere) by the process of reduction:

$$CO_2 + H_2 \rightarrow HCHO. \tag{14.3}$$

Alternatively, such reduction of carbon dioxide could be carried out using electrochemical reduction.[7 8]

The use of H_2 from electrolysis to produce CH_3OH at \$2.00 to \$2.50 per MBtu has been proposed and analysed by Steinberg, Beller and Powell.[9]

TABLE 14.1[6]

CONTINGENCY FORECASTS OF DEMAND FOR HYDROGEN BY END USE, YEAR 2000
(billion standard cubic feet)

End use	Estimated demand 1968	U.S. forecast base 2000	Demand in year 2000			
			United States		Rest of world*	
			Low	High	Low	High
Anhydrous ammonia	872	3,060	2,460	4,490	7,200	12,700
Petroleum refining	775	4,580	2,340	32,640	6,000	36,000
Other uses**	413	1,450	1,450	24,660	2,000	25,000
TOTAL	2,060		6,250	61,790	15,200	73,700
Adjusted range			15,500 (Median	52,530 34,015)	24,950 (Median	63,950 44,450)

* Estimated 1968 hydrogen demand in the rest of the world was 2,995 billion cubic feet.

** Includes hydrogen used in chemicals and allied products, for hydrogasification of coal and oil shale, in iron ore reduction, and for miscellaneous purposes except plant fuel.

OTHER INDUSTRIAL APPLICATIONS

The largest single area of energy consumption (30% of the whole) is in the processing of materials: heating, mechanical drive; primary metal industry; chemicals; petroleum refining; food, stone, clay and glass products; and paper and allied products. The 1970 distribution of energy to these suppliers is shown in Fig. 14.5.[10] Natural gas supplies about half of these energy needs[10] and, thus, many products of industry will be directly affected by natural gas price rises.[11]

The growth pattern here is 3% per year and it is estimated[11] that conservation efforts will reduce the demand by only 5 to 8%.

Hydrogen could be used in many of these processes, where natural gas is used at present. Hydrogen could be used in internal combustion engines (Chapter 15), so that it can be used for the working of these in industry. Gas turbine power plants, at present fueled by natural gas, could be fueled by hydrogen without significant problems. The hydrogen could be used for blade cooling. Much depends upon the fuel cell: improvement in this would lead to a takeover of many of the tasks carried out in industry by natural gas and internal combustion.

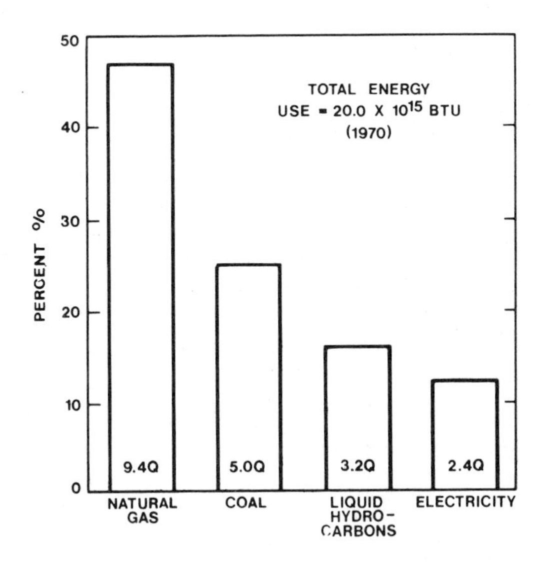

Fig. 14.5 Comparison of source fuels for industrial-fuel, 1970.[10]

TRANSPORTATION ON THE GROUND (Chapter 15)

Hydrogen as a fuel for transportation is the subject of chapter 15. Will it be possible to avoid the use of hydrogen as a fuel for cars? It seems unlikely that this will be methanol on grounds of cost. But an attempt could be made to develop electric cars. Effective electric cars need a massive research effort on the development of new high energy density batteries begun in the brief, flush period when NASA interest in electrochemical energy conversion and storage gave rise to research funding of these areas. Whether such research will ever be sufficiently funded in the U.S.A. at institutions and universities where the results of research could not be controlled, seems unlikely. Another attitude might be in effect in Japan: less resistance to hydrogen cars would be offered by the corporations, because production could take place with present jigs and tools.

If combustively-driven, hydrogen-fueled cars come into widespread use, the fact that electrochemical drive would reduce the amount of fuel needed (greater efficiency) would lead to the gradual introduction of fuel-cell driven cars.

TRANSPORTATION IN THE AIR (Chapter 15)

Above mach 3.5 hydrogen is the only possible fuel because of weight problems with other fuels, and because of the light weight materials which could be used if hydrogen cooling were available.

HOUSEHOLD USE OF HYDROGEN[12]

A Hydrogen Economy could evolve out of the U.S. Gas Industry's plans* to reform natural gas within a household and use the derived hydrogen directly for the following purposes:

(1) *Light*
Phosphorus could be placed inside tubes, hydrogen combines with the phosphorus and gives out a bright light in the cold.[13]

(2) *Heat*
Hydrogen is diffused through a ceramic containing a catalyst and increases in temperature, thus acting as a heating source.[13]

(3) *Cooking*
The idea is similar to that for heating. Fine regulation is possible. The materials through which hydrogen will be passed here would be ceramics. No *flame* would be present.

The catalysts to be used in the above devices will be, e.g. copper, copper oxide and MnO_2.

(4) *Electricity*
A minimal amount of household electricity may be used, e.g. for the powering of household appliances, and this will be produced from fuel cells in the house. What degree of household operation will be carried out electrically will be a matter of individual choice. Alternators can be introduced for functions which need ac.

Household use of energy amounts to some 25% of energy needs. The present inefficiency of household space heating – resistance-heating using electricity produced after efficiency has been lost by the application of a fuel to a heat engine – could be increased by the use of fuel cells operating heat pumps.[14] Supposing that it is doubled, the saving of energy would be *c.* 12% of the total energy budget. *There is no other single step which could bring about so much energy conservation.*

(5) *Water*
Astronauts drink fuel cell-produced water and a hydrogen supplied house would have a significant, potable water production. Suppose that the individual household uses 35 kWh per day, then the water to which this is equivalent will be (using 1 volt as the cell potential) about

$$\frac{3{,}500 \times 1 \times 60 \times 60}{2 \times 10^5} \qquad (14.4)$$

moles of water per day. This is about two gallons of distilled water per household day.

According to Hammond[15], the use of electricity in the U.S. by 2000 will be about five times the 1972 use. Thus, if Hammond's estimate

---

* 'Project TARGET', for *diffused* power, must clearly be developed into one in which hydrogen, rather than natural gas, is the fuel distributed.

holds, production within the house would amount to 10 gallons per day, a significant fraction of needs for pure water. However, in a full Hydrogen Economy, all functions (transportation and industrial, as well as household) will be consuming hydrogen and producing fresh water. For a 10 kW per person country, this would mean about 10 gallons per person per day of potable water, were it all collectable.

The per capita water supply for Adelaide, Australia, in 1974 is shown in Table 14.2.[16, 17]

Thus drinking water and water for hospitals could come from a full Hydrogen Economy. A secondary supply for some other supplies might be of slightly lesser quality. However, it would have to be bacteriologically safe for, e.g. the effect on garden and farm food products.

Introduction of a Hydrogen Economy could cause a significant increase in the water supply. By condensation, collection and circulation, new communities could be built up with significantly reduced needs for potable water from normal sources.

TABLE 14.2[16]

AVERAGE CONSUMPTION FIGURES PER CAPITA PER DAY FOR VARIOUS USES OF WATER IN ADELAIDE, SOUTH AUSTRALIA (litres)

Use	litres
Drinking and culinary	45
Laundry, bathing, sanitary	140
Industry and commerce	95
Garden watering and primary production	170
Public parks and gardens	15
Hospitals and institutions	20
	485 litres

FURTHER REMARKS ON A RESIDENTIAL ENERGY SUPPLY

Sufficient work has now been done to ensure the feasibility of the use of hydrogen to supply residences.[18]

One may use converted fossil fuel vented combusters, unvented hydrogen-air combusters, catalytic burners and fuel cells. The devices now used in furnaces, water heaters, etc., could be converted to hydrogen simply by changing burners, controls and pilot lights.

Leak-proofing of systems would have to be ensured and it seems likely that conversion to hydrogen would often imply replacement of the entire gas pipeline in the house.

Refrigeration could be carried out with an absorption type of unit.

The cost of the changeover here would be a considerable objection, for it will mean purchase of new appliances and replacement of piping throughout the residence. Difficulties would also be associated with the invisible flame of hydrogen, making it more difficult to detect leaks – and its lack of smell (Chapter 11).

ELECTRIC POWER GENERATION

The distribution of electric power generation for 1970 is shown in Fig. 14.6.[14] The growth is 7% per year. The residential energy percentage supplied by electricity is 16%, but this is expected to grow to 30 by 1985 and 35 by 2020.[20] Commercial electricity will grow at about 3.5% per year.

Conservation measures may decrease the use of electric power: 25% of the generation demand in 1985 could be reduced in this way.[19]

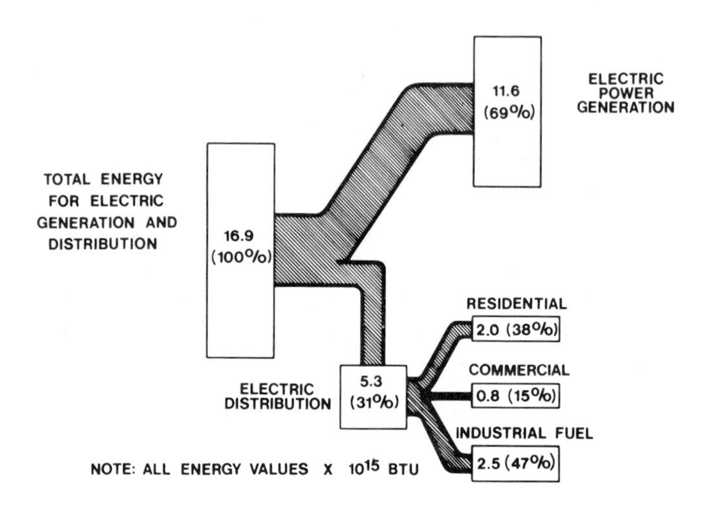

Fig. 14.6 Electric demand patterns, 1970.[19]

ELECTRICAL ENERGY STORAGE[20]

An advantage of the Hydrogen Economy is that the pipes in which the electricity, in the form of hydrogen, is brought from generator to site of use may also be storage vessels. There are characteristic variations in the need for electricity and this leads to an under-utilisation of the high capital cost generating facility, during the night hours, together with the need to have peaking devices for a short time in the early evening. The gas turbine is now used for this purpose is relatively efficient, but uses high grade or gaseous fuel.

Hydrogen would make a good medium for storage of energy to get over these peaks. Salzano *et al.*[21] have shown that, when the oil supply is constrained, hydrogen storage peaking plants are economic alternatives and minimise the cost of operating the energy system. Correspondingly, the use of hydrogen generated from off-peak electricity, stored and then used in fuel cells to meet peak demand is economic and attractive.[22] Significant benefits would accrue for hydrogen derived from low-use period electricity in other applications. The possibility of using this low-use period electricity lies at the basis of many of the earlier potential possible advantages of the Hydrogen Economy.[21–25]

Correspondingly, recent authors[26, 27] have considered the question of hydrogen storage of the low use night-time power. They find about

0.7 for the recovery efficiency, although the 1974 available machinery would probably not do better than 0.25. A hydrogen energy storage system would be a good start to a more widespread hydrogen energy system. Thus, some of the advantages of the hydrogen system, noted by Kippenhan and Corlett[26], include the absence of incomplete combustion products, the minimisation of nitric oxide formation and good heat sink capacity for such purposes as turbine blade cooling.[28–32]

HYDROGEN FOR FOOD PRODUCTION[33 34]

Hydrogen can be combined with carbon dioxide to produce formaldehyde and methanol, and combination with nitrogen in the presence of enzymes can produce proteins and other foods. Correspondingly, hydrogenomonas or clostridium aceticum react with hydrogen and live off inorganic substrates. From carbon dioxide, food production is possible.

In terms of animal food, this has been done. The synthesis of proteins and hydrocarbons is possible with an efficiency of 50%.

Schlegel[33 34] in particular, has developed the concept of electrolysis and the production of hydrogen thereof, giving rise to food. The basic reaction would be to use the aerobic hydrogen-utilising bacteria called 'Knallgasbacteria'. These are suited to the transformation of the energy of molecular hydrogen to organic material. The reactions can be summarised by:

$$6H_2 + 2O_2 + CO_2 \xrightarrow{\text{enzyme}} (CH_2O) + 5H_2O. \qquad (14.5)$$

Up to now, only a few chemolithotrophic hydrogen-oxidising bacteria have been isolated from enrichment of cultures. The main direction of the research is seeking new strains of the Knallgasbacteria.[33]

Recently Schlegel and Lafferty[34] have developed methods of production of 'Biomass' for the union of hydrogen with carbon dioxide and discuss the enrichment in growth, the basic metabolism and the way in which internal generation of a hydrogen-oxygen mixture can occur.

It appears that a reasonable research expectation is the large-scale synthetic production of food from carbon dioxide and hydrogen. The former would be recovered from the atmosphere and the latter produced electrochemically from water.

SEWAGE AND TRASH[35]

The Pittsburg Coal Research Centre of the Bureau of Mines has reported on the pyrolysis of trash in hydrogen to give methane and ethylene. These materials could be used as source materials for plastic manufacture. Glass and metals settle out of the reactor for further use.[35]

ECONOMICS OF HYDROGEN

It is obvious that the price of hydrogen compared with the price of fossil fuels is the key consideration (other than that of pollution) in all these considerations. Various estimates of the present and possible future prices of hydrogen have been given in this book (see Chapter 9). Difficulties of the date of the estimate, together with the changing value of the monetary unit, together with the speculative character of many estimates, makes a solid comparison difficult. In 1973, Winsche, Hoffmann and Salzano[36] published a table showing the estimated price comparison of hydrogen to gasoline when the gasoline price was 19 cents per gallon (Table 14.3[36]). The implications for the use of hydrogen are clear, for the price of gasoline from exhausting oil is bound to increase continuously, whereas hydrogen from coal (Chapter 4), nuclear or solar sources, will increase at a smaller rate during the intermediate period before nuclear power will become abundant. Table 14.3[36] implies that a doubling of the 1974 price of gasoline would bring gaseous hydrogen to about the same price as hydrogen produced by present technology with 10 mils electricity. Such an event seems likely to occur long before a Hydrogen Economy could be implemented.

TABLE 14.3[37]

THE INFLUENCE OF THE COST OF ELECTRICITY ON THE RELATIVE COST OF ELECTROLYTIC HYDROGEN AND GASOLINE OF EQUAL ENERGY EQUIVALENTS AT THE PUMP

The ratio of the cost of electrolytic hydrogen to gasoline is given by:

$$\frac{\text{Cost of electrolytic } H_2}{\text{Cost of gasoline}} = 0.65 + 0.23g$$

where g is the cost of electricity in mils per kilowatt hour. The value 0.65 represents the cost of transportation and distribution.

Cost of electricity (mils per kilowatt hr)	Electrolytic hydrogen/gasoline* (ratio of costs)
0.0	0.65
0.2	0.69
0.4	0.74
0.6	0.78
0.8	0.83
1.0	0.88
1.5**	0.99 ***
2.0	1.1
4.0	1.6
6.0	2.0
8.0	2.5
10.0	2.9
12.0	3.4

* No credit has been given for the by-product oxygen. The cost of the hydrogen fuel tank, that is, hydrogen stored in a Mg_2NiH_4 alloy, will be $70 to $100 per year. This assumes that the tank costs two to three times the price of the alloy and is amortised over 10 years at 10% interest.

** This is the break-even point where electrolytic hydrogen at the pump is competitive with gasoline which is assumed to cost (1972) $0.19 per gallon at the pump exclusive of state and local taxes. (The corresponding Sept. 1974 price would be about 1.5 times this figure, its doubling about three times, cf. 2.9 in the column in Table 3).

*** The calculations given here can be made consistent with those of Chapter 9 for the cost of 1 MBtu of hydrogen if fixed costs are 80c and $E \simeq 1.5$ volt.

INTRODUCTION TO USES OF OXYGEN

Large amounts of oxygen would be available as a consequence of a Hydrogen Economy: both chemical and electrochemical methods proposed to produce hydrogen give oxygen as a by-product.

Little consideration has been given to uses of this oxygen. The possibilities depend upon the cost of piping the oxygen to the site of use. In 1973 costs, the piping is economical for only quite short distances, e.g. 50 miles. At larger distances, it is cheaper to extract oxygen from air.

USES OF OXYGEN[37]

A massive use of oxygen takes place now in the steel industry. In the following a few newer ideas are suggested.

(1) *Aerobic sewage treatment*

At present, sewage treatment involves air oxygen. However[38], emphasis is increasingly placed upon pure oxygen. This would enable a plant for sewage purification to be 3 to 4 times smaller than that at present, with gains in the cost of plants.

(2) *The reversal of pollution*

A clean-up of many areas of the world is now necessary, for example, river areas such as the Rhine in Germany and the Hudson River in the U.S..

These waters decay because the biological organisms which give rise to normal ecological balance and lead to the destruction of pollutants have themselves decayed through lack of oxygen. This is because pollutant and organism compete for oxygen. If the pollutant wins, the organism dies.

The Union Carbide Corporation[37] has developed a high pressure oxygen treatment for lakes. Lake water is passed through a pipe and oxygen is injected at high pressures therein. The oxygenated water is re-introduced into the lake and its oxidifying properties revivify the latter. With surplus cheap oxygen available in a Hydrogen Economy, such a process might become economic.

Injection of ozone breaks up detergent materials.

(3) *Trash*[39]

If incinerators were fed with pure oxygen, instead of air, there would be a higher flame temperature, the plants would burn the trash more efficiently, CO would be eliminated, and smog rising from trash combustion reduced. Metals would be produced in oxide form as a residue from the combustion of the trash, e.g. zinc, tin and aluminum could be recovered as oxides.

(4) *Other uses*[40]

The widespread use of oxygen in industry in processes which now use air would reduce the size of plants, and thus reduce fixed costs. The elimination of NO as a pollutant from internal combustion engines would occur if they were run on hydrogen-oxygen mixtures.

Much *speculative* thinking needs to be done in respect to the application of pure oxygen when better estimates of its price as a function of distance are available.

REFERENCES

[1] C. Marchetti, *Chemical Engineering and Economic Review*, **5**, No. 1, 23 (1973).

[2] G. deBeni & C. Marchetti, *Euro Spectra*, **9**, 46 (June 1970).

[3] G.A. Mills & J.S. Tosh, Paper No. ASME-NAFTC-4, contributed by ASME-Fuels Division, for presentation at the North American Fuel Technology Conference, Ottawa, Canada, 31 May-3 June, 1970.

[4] A.M. Squires, 'Iron and Steel with Hydrogen', The City College of the City University of New York.

[5] L.W. Jones, paper presented at the Cornell International Symposium and Workshop on the Hydrogen Economy, August 1973.

[6] A.S. Mann & C. Marchetti, paper presented at The Hydrogen Economy Miami Energy (THEME) Conference, March 1974.

[7] 'An Electrochemical Carbon Dioxide Reduction – Oxygen Generation System having Only Liquid Waste Products', Phase I, prepared by F.B. Leitz & H.I. Viklund, Ionics Incorporated, February 1967.

[8] 'An Electrochemical Carbon Dioxide Reduction – Oxygen Generation System having Only Liquid Waste Products', Phase II, prepared by F.H. Meller, Ionics Incorporated, April 1968.

[9] M. Steinberg, M. Beller and J.R. Powell, 'A Survey of Applications of Fusion Power Technology to the Chemical and Material Processing Industry', Brookhaven National Laboratory, New York, May 1974.

[10] R.L. Savage, L. Blank, T. Cady, K. Cox, R. Murray & R.D. Williams, edd., 'A Hydrogen Energy Carrier', Vol. II – Systems Analysis, NASA-ASEE, 1973, p. 94.

[11] *ibid.*, p. 95.

[12] L. Lessing, *Fortune*, p. 138 (November 1972).

[13] R.L. Savage *et al.*, *op. cit.*

[14] 'Technology of Efficient Energy Utilisation', Report of a NATO Science Committee Conference, Les Arcs, France, 8-12 October 1973.

[15] R.P. Hammond, in *The Electrochemistry of Cleaner Environments*, ed. J.O'M. Bockris, Plenum Press, New York, 1972, p. 207.

[16] Engineering and Water Supply Department, Adelaide, Australia, priv. comm., 31 July 1974.

[17] J. Johnston, 'An Identification of Water Pollution', South Australian *Education Gazette*, August 1971.

[18] D.P. Gregory, assisted by P.J. Anderson, R.J. Dufour, R.H. Elkins, W.J.D. Escher, R.B. Foster, G.M. Long, J. Wurm & G.G. Yie, 'A Hydrogen Energy System', prepared for the American Gas Association by the Institute of Gas Technology, Chicago, August 1972.

[19] R.L. Savage *et al.*, *op. cit.*, p. 119.

[20] R.G. Murray, paper presented at The Hydrogen Economy Miami Energy (THEME) Conference, 18-20 March 1974.

[21] F.J. Salzano, E.A. Cherniavsky, R.J. Isler & K.C. Hoffman, paper presented at the Hydrogen Economy Miami Energy (THEME) Conference, 18-10 March 1974.

[22] Arthur D. Little, Inc., 'Study of Base-Load Alternatives for the Northeast Utilities System': a report to the board of the Trustees of Northeast Utilities, 5 July 1973.

[23] M. Lotker, E. Fein & F.J. Salzano, paper presented at the IEEE PES Winter Meeting, New York, 27 January-1 February 1974.

[24] J.A. Casazza, R.A. Huse, V.T. Sulzberger & F.J. Salzano, paper presented at the Conference Internationale de Grands Reseaux Electriques a Haute Tension, Paris, France, August 1974.

[25] P.A. Lewis & J. Zemkoski, paper presented at the IEEE Intercon, 1973.

[26] C.J. Kippenhan & R.C. Corlett, paper presented at The Hydrogen Economy Miami Energy (THEME) Conference, March 1974.

[27] W.R. Parrish, paper presented at The Hydrogen Economy Miami Energy (THEME) Conference, March 1974.

[28] R.L. Costa & P.G. Grimes, Chemical Engineering Progress Symposium Series, **63**, 45 (November 1971).

[29] M.L. Kyle, E.J. Cairns & D.S. Webster, 'Lithium/Sulfur Batteries for Off-Peak Energy Storage: A Preliminary Comparison of Energy Storage and Peak Power Generation Systems', Argonne National Laboratory Report No. ANL-7958 (March 1973).

[30] S. Baron, Proceedings American Power Conference, **35,** 451 (1973).
[31] L.A. Wenzel, 'LNG Peakshaving Plants – A Comparison of Cycles', Paper No. c-1, Cryogenic Engineering Conference, Atlanta (August 1973).
[32] W.C. Kincaid & C.F. Williams, 'Storage of Electrical Energy through electrolysis', paper presented at the 8th IECEC, Philadelphia, (August 1973).
[33] H.G. Schlegel, in *Fermentation Advances*, Academic Press, New York, 1969.
[34] H.G. Schlegel & R.M. Lafferty, in *Advances in Biochemical Engineering*, Vol. 1, Springer-Verlag, Germany, 1971.
[35] R.C. Corey, 'Pyrolysis, Hydrogenation and Incineration of Municipal Refuse – A Progress Report', Proceedings of the 2nd Mineral Waste Utilisation Symposium, ITT Research Institute, Chicago, Illinois, March 1970.
[36] W.E. Winsche, K.C. Hoffman & F.J. Salzano, *Science,* **180,** No. 4093, 1325, June 1973.
[37] L.O. Williams, Austronautics and Aeronautics, p. 42 (February 1972).
[38] M.D. Rickard & A.F. Gaudy, Jr., *Journal of Water Pollution Control,* May 1968.
[39] W.F. Schaffer, Jr., 'Cost Study of the Treatment of Sewage Sludge by the Wet-Air Oxidation Process, using Oyxgen Produced by Low-Cost Electricity from Large Nuclear Reactors', Oak Ridge National Laboratory, February 1968.
[40] J.E. Browning, *Chemical Engineering,* p. 88 (26 February 1968).

CHAPTER 15

Transportation

INTRODUCTION

TRANSPORTATION is beset with the difficulty of the *rising price* of fossil fuels, together with that of the atmospheric pollution, of which it is the largest cause. Air transportation suffers from the first two, but also from the need for a lighter fuel for the SST and its successors to increase their range, and hence commercial viability.

The use of hydrogen as a fuel for internal-combustion engines solves the difficulties of pollution and weight. It seems likely that hydrogen from coal will be cheaper than gasoline long before a Hydrogen Economy could be implemented. The use of hydrogen in transportation may form its first large-scale application.

RELEVANT PROPERTIES OF HYDROGEN AND OTHER FUELS

This is shown in Table 15.1.[1]

TABLE 15.1[1]

COMPARATIVE FUEL PROPERTIES

Property	Hydrogen	Other Fuel
Heating value, Btu/lb.	53,000	20,000 (gasoline)
Minimum ignition temperature, °F	1065	1000 (butane)
Theoretical flame temperature in air, °F	3887	3615 (butane)
Flammability limits, % by volume in air	4.0-74.2	1.9-8.6 (butane)
Maximum flame velocity ft/sec	9.3	1.03 (butane)
Specific volume liquid litres/kg	14.3	1.33 (gasoline)
Energy density: Btu per cu ft	$2.50\ 10^5$	$9.38\ 10^5$ (gasoline)

HYDROGEN-BASED TRANSPORTATION

The kWh/mile for various types of vehicles is shown in Fig. 15.1.[2] The results show the energy requirements as a function of vehicle range and speed with zero grade road load. With such data, Hietbrink *et*

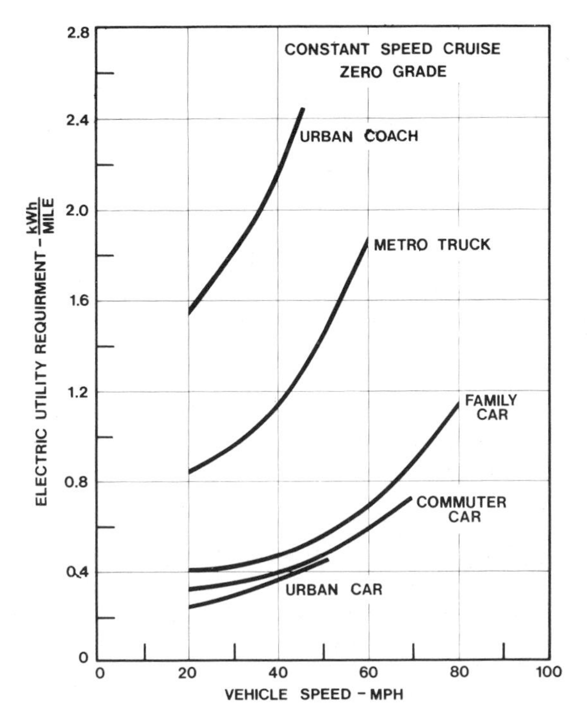

Fig. 15.1 Energy requirement from the electric utility source as a function of vehicle range and speed.[2]

al.[2] calculated the effect on the electric utilities of the introduction of a substantial degree of electrification of cars by 2000. The basis of the results are the projected car registrations in the U.S.A.[3], and estimates of future electric generation.[4] Other assumptions which Hietbrink *et al.*[2] made are that cars average 8,000 miles per year; 1 kWh/mile (Fig. 15.1); the diurnal load pattern will be the same at 2000 as in 1972; and battery charging (correspondingly, hydrogen production) at night. Table 9.3 (see Chapter 9) resulted. 52% of the generating capacity is in use all the time.[5] It is reasonable to expect at least 20% of the total power load could be available for the production of hydrogen. The table shows that *the complete automobile load could be obtained from low use-period power at 2000.* (For 1969, 28.5% of the automotive load could have been supported[6]). This battery-oriented calculation is conservative and would apply with little change to a situation in which cars were hydrogen driven, the low use-period electricity being used to generate hydrogen at, say, 75% efficiency (easily attainable, see Chapter 9). Schoeppel[7] calculates that about half the total automotive fuel needs could have been obtained from hydrogen from night-time power in 1972.

The hydrogen could be carried in cylinders, or liquefied. More space (3.75 times more) would be needed for the containment of liquid hydrogen than for the gasoline.

Four advantages would accrue from the introduction of hydrogen as a fuel for transportation.

(1) It could be developed now, on the basis of low-use period power and classical hydrogen generators. The price in 1975 at which hydrogen could be expected therefrom ($2.50-3.50) would not be substantially different from the price of gasoline ($2.44 per MBtu). The price of gasoline is likely to rise parallel to increases brought about by OPEC countries. Such rises need not be so severe for coal and therefore electricity.

(2) It would eliminate more than half of present air pollution.

(3) Conversion could be carried out gradually, gas stations would carry both gasoline and hydrogen until the changeover was complete.

(4) The energy sources would be at first coal, and later increasingly nuclear and solar. Continuity would be preserved.

HISTORY OF HYDROGEN-BASED TRANSPORTATION

The Rev. W. Cecil[8] proposed the application of hydrogen to produce moving power in machinery to the Cambridge Philosophical Society of 1820.

Ricardo[9] noted that, if the hydrogen-air mixture was rich, there was violent ignition and backfiring. A basic advance was made by Burstall[10] of Cambridge University. He pointed out that the chief difficulty was the rapidity with which hydrogen burns. *If a small ignition advance was used*, satisfactory burning could occur. Erren's is the most well-known name[11-14] associated with the advance of the hydrogen engine. He worked with dirigibles, and used hydrogen normally vented in flight to run their engines during descent. He used improved cylinder-wall

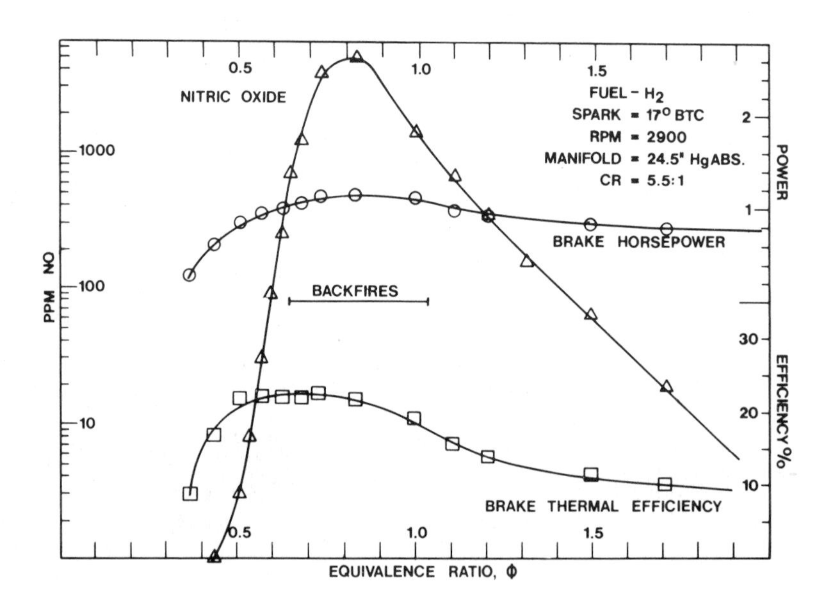

Fig. 15.2 Wide open throttle response of a 4-cycle hydrogen fueled engine to variation of ϕ.[17]

hydrogen injection techniques, which reduced the possibility of backfiring.[12] There was controversy between Erren and King.[14 15] King stated that, if weak mixtures are used, it is not necessary to use the special injection technique suggested by Erren. According to King[16], backfiring was due to the aggregation of carbon particles in the cylinder. King thought it might be acceptable to use carburetted hydrogen engines, so long as they were frequently cleaned. Fig. 15.2[17] shows that the backfire regions depend upon the equivalence ratio. A lean mixture is satisfactory, the displacement must be more than about two-thirds.

These works of the 1930s were the basis of the hydrogen engine and the work has come alive again with the contributions of Schoeppel and Murray.[1 6 7 18-20] They started where Erren had stopped. They have found that high pressure injection of hydrogen near the top centre of the cylinder causes ignition to occur immediately, injection being continued during a portion of the power stroke. The injection technique used by Schoeppel and co-workers gets away from the carbon nucleation problem observed by King. Ignition is caused to occur as soon as enough hydrogen is in the cylinder to support combustion. This is at the beginning of injection.

An attempt was made by the Perris Smogless Automobile Association to build a vehicle in which hydrogen and oxygen were both used. However, the oxygen dewars are too weighty.[21] Billings and Lynch[17] used metal hydrides for storage in vehicle propulsion by hydrogen in the 1970s.

Swain and Adt[22] developed, at the University of Miami, a method of port injection which utilises the manifold side of the intake valve heads to face the injectors. Their work resulted in a 1970 Toyota running 22 miles on one pound of hydrogen (equivalent to 51 mpg of gasoline).

Storage options are shown in Fig. 15.3.[7]

Hydrogen engines may certainly be run with carburettors[23], but the carburettor design does have to be different from that for internal gas or gasoline, because the volumetric mixture ratio is unsuitable for use with hydrogen. The redesign needs a change in the area ratio of the gas and air passages. Apart from the ignition systems, other changes in engines are zero. Thus, it was demonstrated in 1948[16] that a carburettor engine free from carbon deposits will run on hydrogen with a compression ratio of 10:1.

The relative cost of hydrogen *versus* gasoline will increasingly favour hydrogen. Were emergent electrolyser technology to be used, and the

Fig. 15.3 Hydrogen storage options.[7]

price for electricity given by Philip Sporn[24] for 1974 (9.4 mils), hydrogen could be some 60% above the price of gasoline now: using the low use-period power which should be rationally available, it could be cheaper at this time as a gas, or some 25% more expensive with conventional electrolytes. As a liquid it could be marginally more expensive.

The remaining problems with hydrogen automobiles are with the handling and distribution system. The development of hardware to store and utilise hydrogen has still to be done, as well as the overcoming of public resistance to the use of hydrogen, because of lack of understanding that the passage and transport of hydrogen on highway and rail has long been a common practice.

THE U.S. FEDERAL POLLUTION STANDARDS AND THE HYDROGEN-FUELED AUTOMOBILES[23]

A strong encouragement towards hydrogen-fueled cars is the difficulty of maintaining U.S. Federal Standards if gasoline continues to be used as a main fuel. The standards are shown in Table 15.2.[7] The automobile industry will have difficulty in attaining the Federal Emission Standards in 1976 *at a cost acceptable to the consumer.* The situation could be met if:

(a) The pollution level is relaxed. According to Murray and Schoeppel[19], lung cancer in urban areas is some 25% above that in country areas.

(b) Tax monies are used to subsidise the price of a purer gasoline, and the price of clean air equipment to keep the old system in operation.

TABLE 15.2[7]

FEDERAL STANDARDS FOR COMPOSITE EMISSIONS FROM LIGHT-DUTY VEHICLES ON A DRIVING CYCLE, GRAMS/MILE

Model year	Hydrocarbons	Carbon monoxide	Oxides of nitrogen
1973	3.4 (3.0)*	39.0 (28.0)*	3.0 (3.1)*
1974	3.4 (3.0)*	39.0 (28.0)*	3.0 (3.1)*
1975	0.41	3.4	3.0
1976 and later	0.41	3.4	0.4

* Numbers in parentheses represent standards being considered for adoption.

The key is in the economics. Thus, the technology exists for a reliable engine with 1976 emission standards, but not at an acceptable cost. Even were the Federal Standards attained, the growth of the car population (if not cut-back) will cause an increase in the amount of emission per year after 2000 (Fig. 15.4[25]). The increasing price of fossil fuels will make a change away from gasoline-powered cars inevitable. The long-term alternative to hydrogen-driven cars is electrochemically-powered cars.

However, U.S. Government funded research on high energy-density batteries has been sporadic and a dynamic national research attack on them has never been made. Research by automotive companies on any new system of propulsion which would imply the devaluation of the capital invested in the present plants is likely to be defensive. The large corporations which manufacture cars could change to hydrogen-fueled cars with no more re-tooling than that for a model-year change. To commence the manufacture of electric cars would be much less acceptable.

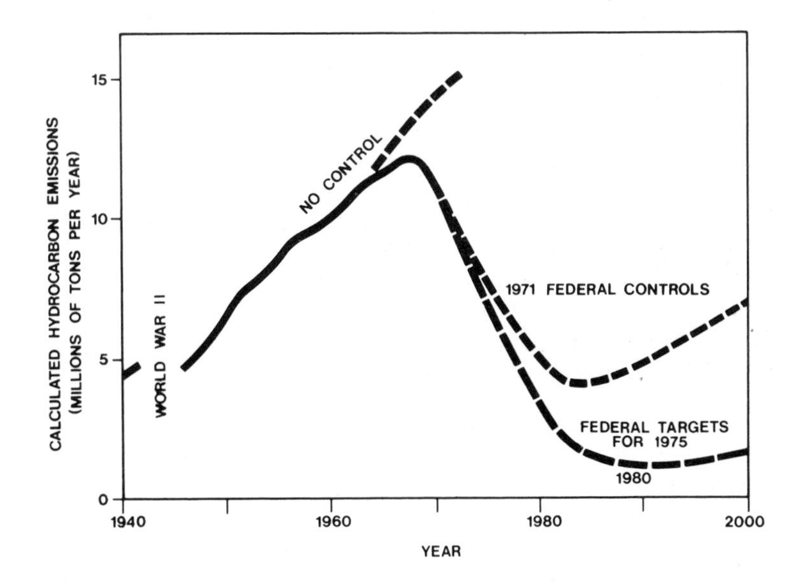

Fig. 15.4 Estimated effect of Federal controls in the U.S.A. on hydrocarbon emissions from passenger vehicles. A rising trend is expected after 1985 because of the increasing number of cars.[25]

THE PRESENT HYDROGEN ENGINE FOR CARS

The Oklahoma State University engine (Fig. 15.5[19]) uses direct cylinder gaseous hydrogen injection. The engine is a modified 4-stroke cycle, air cooled, single cylinder, 3.5 hp engine. Air alone is compressed which limits hydrogen infiltration past the piston rings and prevents pre-ignition. Use of air as the working fluid eliminates the possibility of an accumulation of an explosive mixture in the intake manifold or crank case. When further co-ordinated with an injection means, as used in the design by Murray and Schoeppel[19], combustion commences at once. It continues smoothly without detonation, excessive pressure or temperature.

The brake horse power-torque and brake specific fuel consumption for wide open throttle conditions, is shown in Fig. 15.6.[19] The engine requires about one-third the fuel mass flow rate of its fossil fuel counterpart.

Difficulties recently met have concerned heat transfer characteristics and cast iron cylinders in the neighbourhood of the exhaust valve.[12 19] Localised hot spots in this area tend to produce cylinder distortion. A cooling system is necessary.[15 16 19] These difficulties have been overcome.

Fig. 15.5 Clinton engine converted to hydrogen operation.[19]

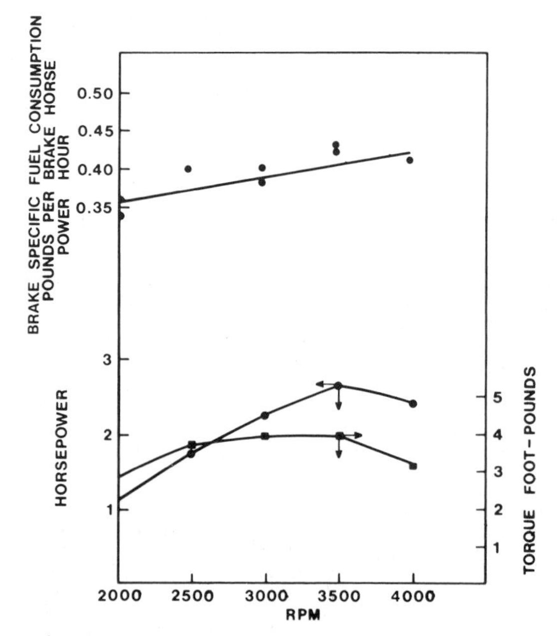

Fig. 15.6 Operational characteristics of Murray and Schoeppel engine.

SAFETY

A number of safety devices will be required for hydrogen-driven cars to avoid obvious hazards. Most vehicles are left to stand for several days from time to time and (if the fuel is to be liquid) a safe venting system will have to be devised. The buoyancy of hydrogen is great and the diffusion rapid. It will be safe simply to vent it slowly into the air, so long as the space is not enclosed. Venting into enclosed spaces with ceiling ventilation will need testing. Other methods of dealing with the hydrogen which has to be vented during storage, would be by various means of combustion to water, e.g. catalytic or electrochemical. The catalytic ignition will require some research, for instance into catalysts which are cheaper than platinum, but have the same effect.[26]

Leaks will have to be guarded against by regular checks: colouring or addition of an odorant has been suggested (Chapter 11).

One difficulty is the heavy collision. The major consideration would be to locate the tank within the car in such a position as to minimise its involvement.[27] Fire resulting from piercing of the tank is possible, as with gasoline. *Explosion,* however, could not be expected, except in an enclosed space.

CHARACTERISTICS OF PRESENT HYDROGEN ENGINES

An internal combustion engine with fuel injection develops a higher output with hydrogen than is possible with gasoline.[6 19 20 28] Carburetted engines produce about half the power.[29] Its emission characteristics are in Fig. 15.7.[1] They average 2.0 g of NOX per brake-hp per hr, 10%

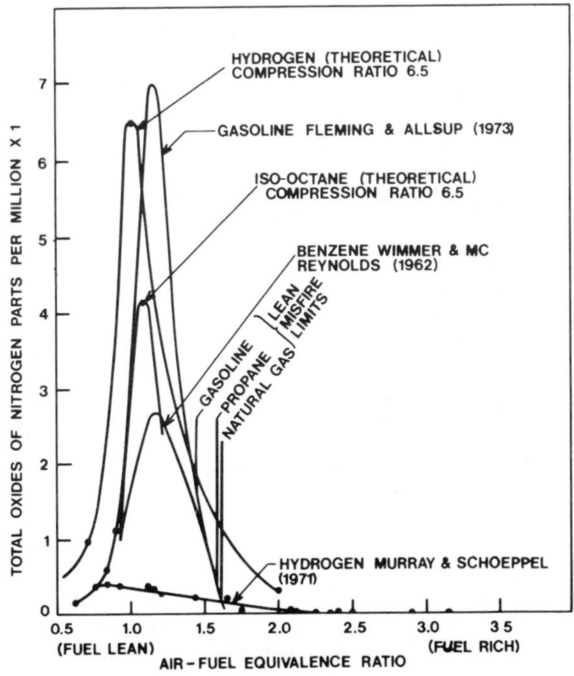

Fig. 15.7 Emission characteristics for various fuels as a function of air-fuel mixture.[1] (See also Watson, Milkins and Deslandres.[47])

of that from gasoline. Ignition can be achieved using compression, glow or spark. Hydrogen engines are easy to start, respond rapidly to different rates of fuel injection and run cooler than their gasoline equivalent.

Hydrogen engines could easily meet the 1975-76 Federal controls.

COST OF LIQUID HYDROGEN

(1) The present cost of hydrogen, using *old* electrolyser technology, paying 9.4 mils per kWh (cf. the breakdown of costs in Chapter 9), and the full maintenance costs for the plant, is about $5.53 per MBtu (2.2 volts for the electrolyser).

(2) The cost of liquefaction. The theoretical energy cost is about 5,000 Btu per lb. The real one is 15,000 Btu per lb. 1 lb of hydrogen is equivalent to 53,500 Btu, so that the Btu necessary to liquefy an amount of hydrogen which would produce a million Btu is:

$$\frac{10^6}{5.35\ 10^4}\ 1.5\ 10^4\ \text{Btu},$$

i.e. 0.28 MBtu. But one MBtu $\simeq$ 300 kWh: therefore, 0.28 MBtu $\simeq$ 84 kWh or 78 cents per million Btu of liquid hydrogen. This is the cost of the energy alone. We take the total cost as about 85 cents per MBtu of hydrogen, an upper figure.

(3) Hence, the cost of liquid hydrogen from classical electrolysis should be about $6.40 per MBtu.

(4) The 1974 cost of U.S. gasoline is about $2.44 per MBtu (mean, ex-refinery cost, U.S.A., September 1974, no tax, distribution cost or profit).

(5) Using equation 9.32, the price of emergent technology hydrogen, using $E = 1.5$ and 9.4 mils electricity, is $3.62 per MBtu gaseous, or about $4.47 per MBtu of liquid.

(6) Using classical electrolysers ($E = 2.0$), but power from low-use time electrical energy at 5 mils kWh^{-1}, the price of gaseous hydrogen would be $3.09 per MBtu, and liquefied about $3.94.

(7) Using emergent technology ($E = 1.5$, fixed costs 40 cents per MBtu), low use-time power at 5 mils, the cost of hydrogen would be: $2.12 per MBtu and about $2.97 per MBtu, liquefied.

Thus, if low-use rate electricity were to be sold at 5 mils per kWh (see reasoning behind price, Chapter 14), classical present electrolysers could produce liquid hydrogen at this time, about 50% more expensive than that of gasoline. Electrolysers which use technology already demonstrated (and low use-time power) could produce liquid hydrogen at a price about 20% higher than that of present U.S. gasoline.

The values discussed above may be compared with those in a NASA study done by Hallett.[30] The values which Hallett adduced in 1968 are shown in Table 15.3.[30]

However, the hydrogen prices obtained by Hallett were obtained assuming 2.72 mils per kWh and must be appropriately corrected for much higher electricity prices in the region of 5 (for night power) to 7 to 10 mils per kWh (cf. Chapter 9 where 9.4 mils has been rationalized for the 1974 price). Using 5 and 9 mils, respectively, the electrolysis figures would move to between $3.50 and $5.20 per MBtu. Correspondingly, the Hallet study[30] assumed a $2.40 per ton for the price of coal, whereas the price is now many times higher.

Witcofski[31] concludes (1974) that a liquid hydrogen price between $2.75 and $3.18 per MBtu is reasonable.

A recent estimate has been made by Jones[32], based upon reports by Johnson[33] and Michel.[34] They are given in Table 15.4.

Other estimates and projections of the prices of liquid hydrogen are given in refs. 32, 35-37. The average estimated price of liquid hydrogen in 1974 dollars is $3.88 per MBtu.

TABLE 15.3[30]

COST OF LIQUID HYDROGEN, adduced by Hallett, 1968

Steam reforming of methane	8.1¢/lb LH_2 ($1.57 per MBtu)
Partial oxidation of heavy oils	8.1¢/lb LH_2 ($1.57 per MBtu)
Electrolysis	13.8¢/lb LH_2 ($2.69 per MBtu)
Coal gasification	11.4¢/lb LH_2 ($2.21 per MBtu)

TABLE 15.4[32]

JONES' ESTIMATE OF LIQUID HYDROGEN COSTS, DOLLARS PER MBtu (1974)

Gasoline	$2.74*
LH_2 from coal	$2.48 to 3.17
LH_2 from nuclear	$5.51 to 5.37

CONCLUSIONS CONCERNING HYDROGEN-FUELED LAND-BASED TRANSPORTATION

There are few technical difficulties in producing vehicles which are powered by internal combustion engines fueled by hydrogen. Present engines could be used and the changes in their construction would have to be only a change in the ingition system. Possible new ignition systems have been engineered by Schoeppel[1 6 7 18-20], Adt[22] and others.

Technical difficulties may be seen in the mode by which the automobile carries its hydrogen. A liquid hydrogen tank would be attractive, and the projected cost is about $400. Gaseous storage may be practical. Solid storage would be too expensive (Chapter 11).

A difficulty would be in obtaining the acceptance of the public. There would have to be an educational programme in which the scientific facts concerning safety and the safety record of working with liquid hydrogen were disseminated.

The most important aspect is the cost of liquid hydrogen. Estimates of this vary from $2.90 to $5.50 per MBtu. The 1974 ex-refinery price of U.S. gasoline was about $2.44 per MBtu. When, and if, the AEC's latest prediction of cheap nuclear power of 3 mils per kWh is attained, the cost of liquid hydrogen would be about $2.00 per MBtu.

---

* This is 12% higher than the $2.44 per MBtu quoted elsewhere for September 1974 U.S. gasoline, ex refinery. The latter figure is a U.S. mean for the date.

Gasoline will continue to increase in price extra-inflationally, because it reflects crude oil exhaustion. Hydrogen is electricity-based and the cost of coal-based electricity should be relatively constant (in constant dollars), because it will not represent an exhaustion situation till after 2000. Precisely when hydrogen will become cheaper than gasoline is a function of the funding of development work in means of producing hydrogen; and the extra-inflational rate of climb of the gasoline price. Liquid hydrogen would seem likely to be manufacturable at a price cheaper than gasoline in the U.S. sometime during the 1980s, i.e. before the other requisite changes in technology (building and distribution system) have time to be completed, even if these began in the mid '70s.

All these considerations are based on the use of electricity. If photo-oriented methods succeeded, then there could be hydrogen well below the cost of gasoline; and indeed below the cost of crude oil in 1974 (*c.* $1.80 per MBtu).

ADVANTAGES OF HYDROGEN AS AN AIRCRAFT FUEL

The most important advantage of hydrogen as an aircraft fuel is that it gives a greater payload. Thus, in Fig. 15.8[38], the subsonic situation is shown. The corresponding payload comparison is shown in Fig. 15.9[38], and in Fig. 15.10[38], the range of Mach 3 transport performance of hydrogen and JP fuel is shown. Hydrogen powered supersonic aircraft would have a range of $1\frac{1}{2}$ to 3 times that of the JP fueled plane, with corresponding implication for the economic viability of supersonic and hypersonic aircraft. For example, the doubtful economic viability of the Concorde SST would be greatly improved if the vehicle were redesigned to run on liquid hydrogen. *Hypersonic* aircraft, on the other hand, could not be operated unless they ran on hydrogen. Hydrogen

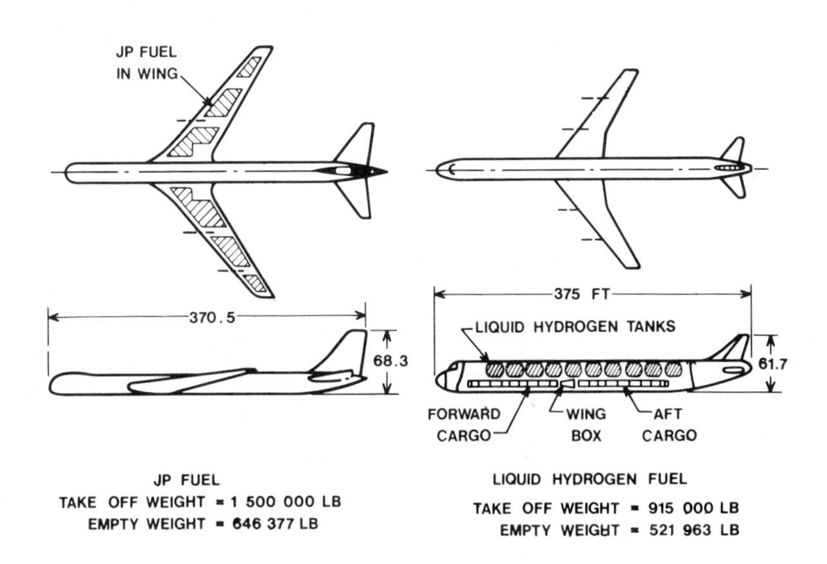

Fig. 15.8 Subsonic cargo aircraft, payload = 265,000 lb., range = 5070 n.m..[38]

fueled hypersonics would make the reaching of any part of the globe possible from any other point within about 6 hours.

A 10 yr lead time is necessary in aircraft design. Liquid hydrogen is likely to become cheaper than gasoline within a decade. Post-Concorde supersonic aircraft should, therefore, be designed for hydrogen. Hyper and supersonic vehicles would then lose their difficulties of range, payload and pollution.

Another advantage of using liquid hydrogen to fuel aircraft is airframe cooling which would become feasible. In the presence of such cooling, it would be possible to use aluminum and boron alloys, and not the titanium and other heavier alloys planned for at present, because only they have the strength needed at the high temperatures which the outside of supersonic aircraft will reach. The use of the light-weight alloys would further increase range and payload.

The diminution of the sonic boom would be another advantage in hydrogen fueled hypersonics. Thus, the hypersonic would fly at a greater altitude than the SST or subsonics. The sonic boom is 3 lb per sq ft for SST, but would be only 1 lb per sq ft for the HST at higher altitudes.

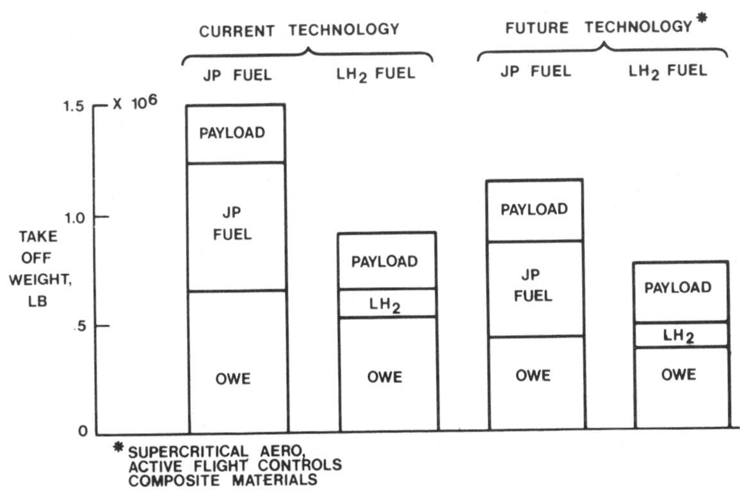

Fig. 15.9 Projected advance cargo transports, Mach-Cruise = 0.85, range = 5070 n.m., payload = 250,000 lb.[38]

SUPERSONIC AIRCRAFT

The biggest difficulty of the present supersonic aircraft is range limitation. Hydrogen would overcome this, as it would the high sonic boom. With hydrogen, Mach 3 transport would give 5,000 miles range at a gross weight comparable with large current transports.

Estimates concerning a Mach 3 transport with 300 passengers are compared in Fig. 15.11.[39] The payload increases by 50%. However, the vehicle would be bulkier than JP fueled vehicles, because of the lower density of hydrogen.

Take-off noise would be reduced because of lower gross weight and wing loading.

COOLING OF THE AIRFRAME

The concept of hot and cold structures in hypersonic aircraft must be taken into account in thinking of the future of air transportation. Hot structures rest upon the fact that the equilibrium radiation temperature at high cruising speeds of the order of Mach 9, are in the range where super alloys are still structurally stable. Heat shields may be used. But it is not known whether the airframe lifetimes corresponding to commercial aircraft operation will be able to be attained with these structures.[36]

It could therefore be of great advantage to consider the possibility

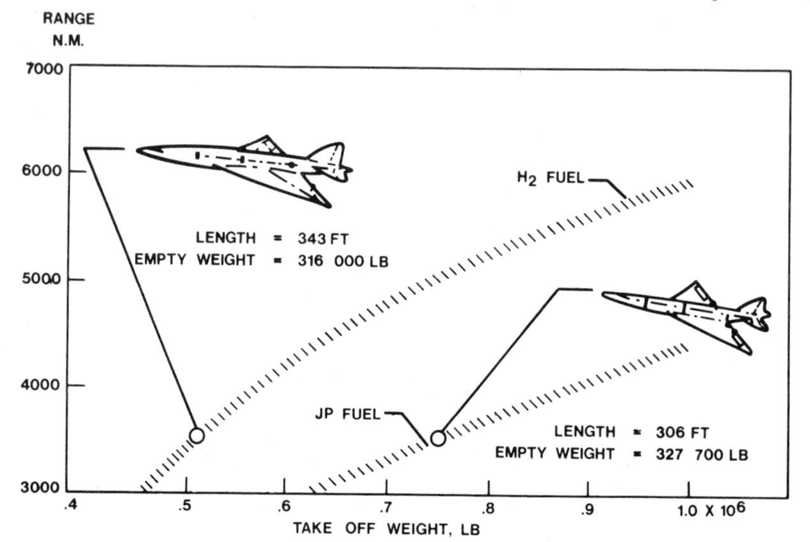

Fig. 15.10 Comparative Mach 3 transport performance, 300 passengers, titanium structure.[38]

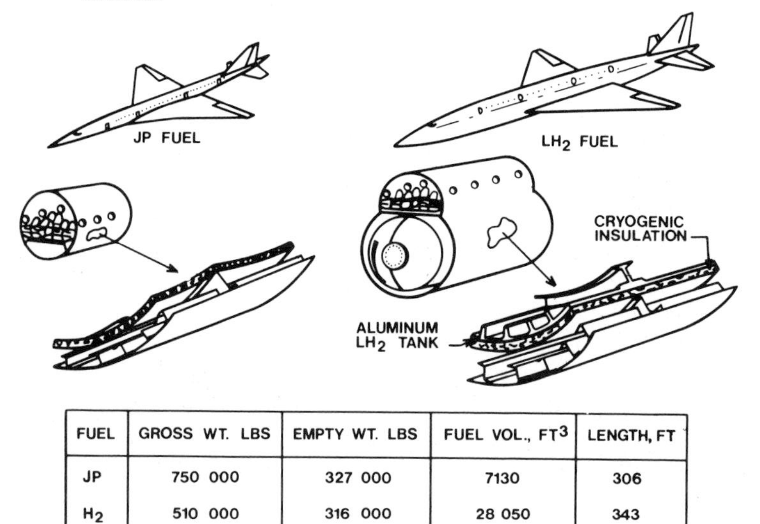

FUEL	GROSS WT. LBS	EMPTY WT. LBS	FUEL VOL., FT^3	LENGTH, FT
JP	750 000	327 000	7130	306
H_2	510 000	316 000	28 050	343

Fig. 15.11 Comparative JP/LH_2 aircraft characteristics, Mach = 3, 3500 n.m. range, 300 passengers.[39]

of cooled structures where the cooling would mean that the airframe would have an outside temperature of some 500° F. Then, titanium could be used. However, if it were possible to cool to 200° F, aluminum could be used and even the still lighter weight boron-aluminum alloys. Studies[40 41] have indicated that an airframe cooling system using a secondary coolant circulating internally and referring back to a hydrogen fueled heat exchanger could reduce the airframe of a Mach 6 air vehicle to the titanium level and, with limited heat shielding, to aluminum. A cooled wing panel is illustrated in a diagram from Witcofski's work (Fig. 15.12[36]). The weight of the plumbing and heat exchanger is more than offset by the reduced weight of the airframe. The arrangement is shown in detail in Fig. 15.13.[36]

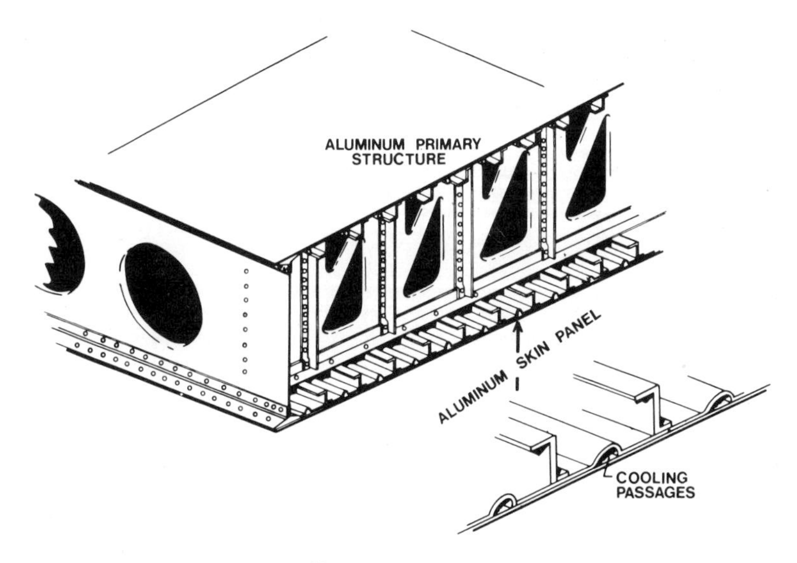

Fig. 15.12 Cooled wing structure.[36]

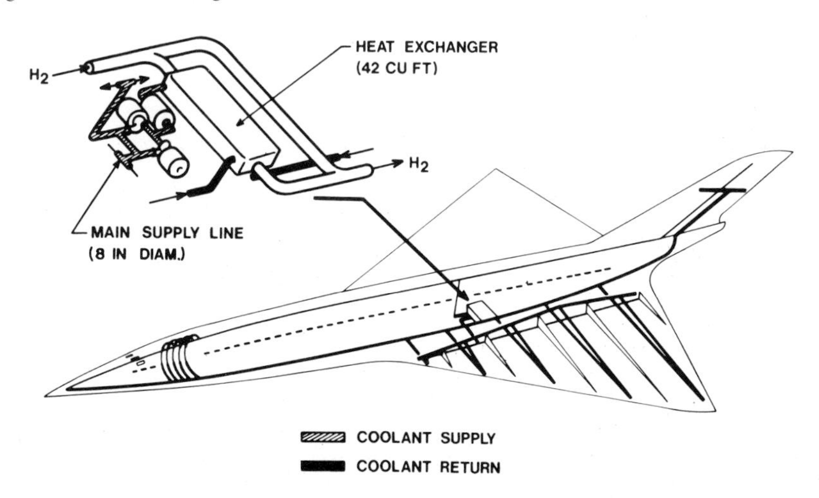

Fig. 15.13 Aircraft cooling system.[36]

COST PROJECTIONS

These have now been extensively studied for subsonic, supersonic and hypersonic aircraft with various kinds of fueling[42] (Fig. 15.14[36]). The economic competitiveness of the supersonic would be a matter of the liquid hydrogen price. The estimated price of hydrogen was taken in 1972 as about $2.50 per MBtu which is the lower limit of the 1975 projected cost.

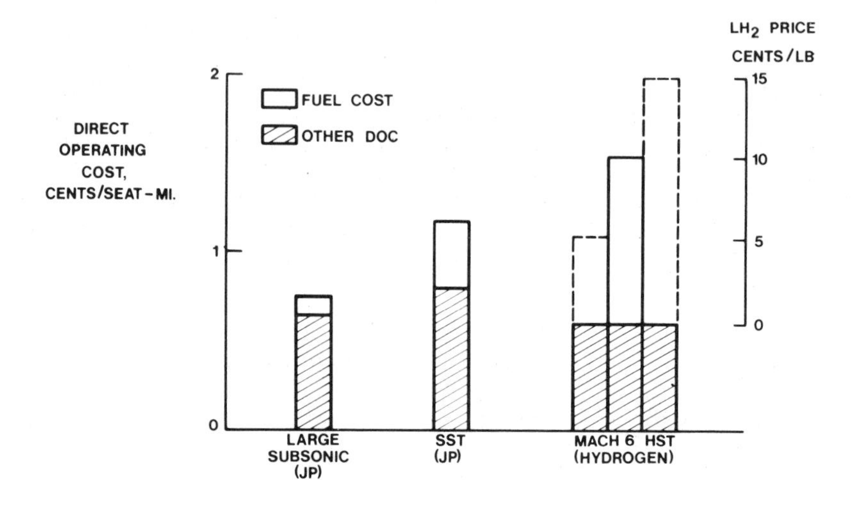

Fig. 15.14 Comparison of direct operating cost. Range, 4600 st miles.[36]

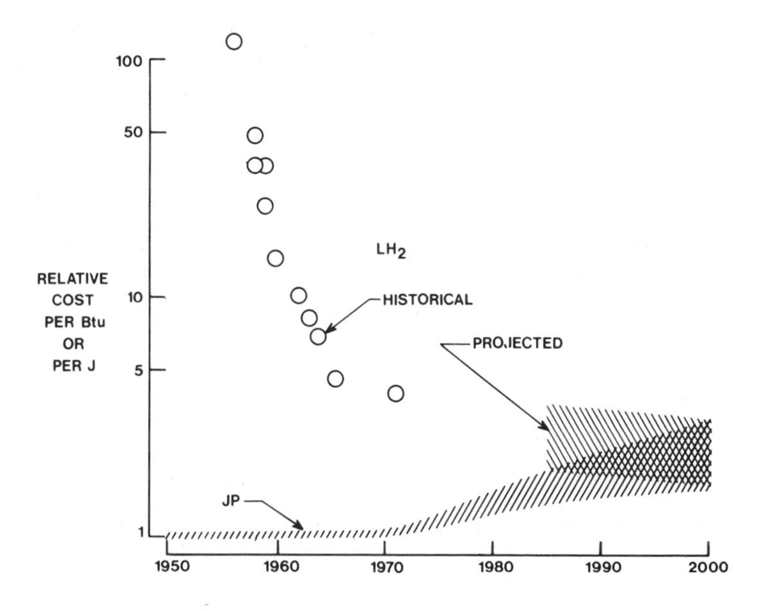

Fig. 15.15 Future fuel cost.[36]

The economics is likely to shift in favour of hydrogen compared with gasoline because of the rising price of fossil fuels. A 1972 projection is shown in Fig. 15.15.[36] The abnormal price rises which started in 1973 for fossil fuels make the attainment of price advantage for hydrogen likely earlier than formerly anticipated. Some parallel considerations are shown in Fig. 15.16.[43]

EXPERIENCE ALREADY GAINED IN HYDROGEN-POWERED FLIGHT

Apart from the early Zeppelin flights, where hydrogen was used to fuel engines during descent, a B57 aircraft was successfully operated on liquid hydrogen, using a J65 turbo-jet and a flight speed of Mach = 0.72[44] (Figs. 15.17 and 15.18[45]). A RAM heat air exchanger was used to gasify the liquid hydrogen, and helium was used to obtain appropriate pressure in the fuel tank and to purge the system. Ground tests showed that the fuel consumption of the engine using JP-4 fuel was about 2.7 times more in weight than the fuel consumption using hydrogen, in fair agreement with the ratio of the heats of combustion of 2.6 (Table 15.1).

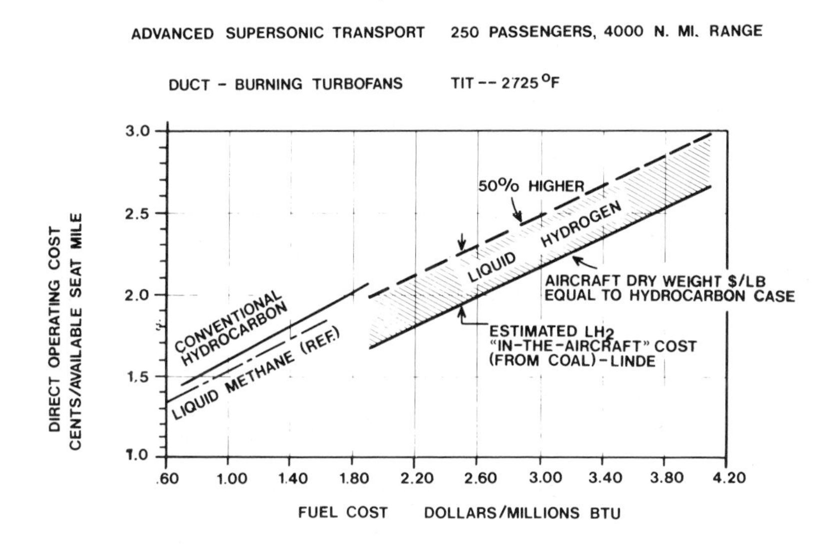

Fig. 15.16 Effect of fuel cost on direct operating cost.[43]

THE USE OF SLUSH-HYDROGEN

There would be an advantage in this form (e.g. a 50% mixture of liquid and solid hydrogen) in fueling aircraft. The fuel volume would decrease 14% and the additional heat capacity would eliminate fuel boil-off. However, there would be extra cost to produce the slush; ground facilities would have to be increased.

Fig. 15.17 Experimental hydrogen – convertible aircraft. In testing at NASA Lewis, Cleveland 1956-57.[45]

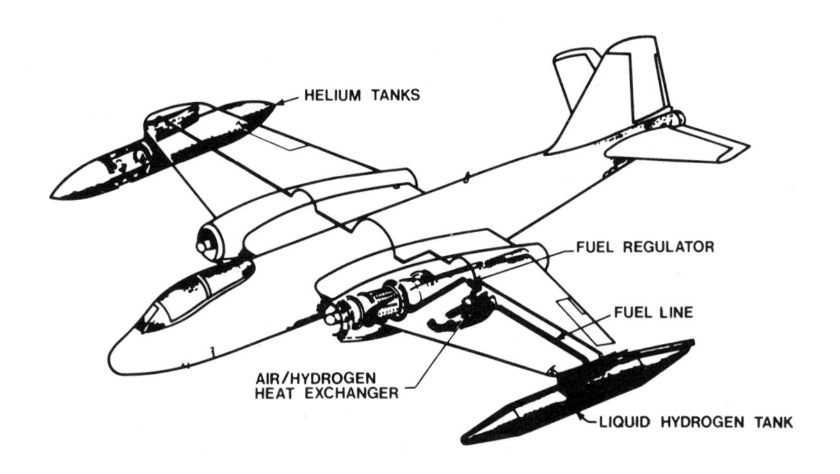

Fig. 15.18 Hydrogen system for experimental aircraft – tested as early as 1956. B-57 'Canberra' engine convertible to hydrogen.[45]

WITCOFSKI'S CONCEPT OF THE USE OF O-P CONVERSION

Witcofski[37] suggests that there are prospects for on-board power from the ortho-para conversion in hydrogen, which would take advantage of the endothermic reaction to increase cooling capacity. Realization would need catalysts to promote the conversion.

NEGATIVE EFFECTS WHICH WOULD ARISE FROM THE USE OF HYDROGEN IN AIRCRAFT[46]

(1) The principal difficulty is the density of hydrogen which is only about 90% of that of the JP fuels, so that about ten times the volume will be required for the same weight of fuel. This would be cut down greatly for equal ranges: the requirement for volume would then be about 3.5 times. Thus, there would have to be some re-design of aircraft to accommodate such volumes of fuel.

(2) The weight of the fuel tanks causes concern: these would be heavier than those for storage of JP fuel containing the same energy.

(3) The various dangers of hydrogen (Chapter 11) would necessitate extra training for ground personnel, special apparatus for detection of leaks, purging techniques to eliminate air and a change in fueling techniques.[46]

POSSIBLE SUPERSONIC TRANSPORT DESIGNS

There have now been many discussions of designs of supersonic and hypersonic aircraft run on hydrogen.[36, 37, 45, 46] Some possibilities are shown in Fig. 15.19.[43] Table 15.5[43] shows aircraft characteristics.

CONCLUSIONS REGARDING THE USE OF HYDROGEN AS AN AIRCRAFT FUEL

(1) The decisive reasons are economic. At some time in the '80s the price of liquid hydrogen is likely to become less than that of gasoline.

(2) The second most important reason is the improvement in aircraft performance which the use of liquid hydrogen would allow. It would mean a longer range and better payload. In particular, it would revolutionise the prospect for hypersonic transports by using airframe cooling to make light weight alloy use possible.

(3) Pollutional aspects of aircraft, both near airports and at high altitudes, would be reduced towards zero.

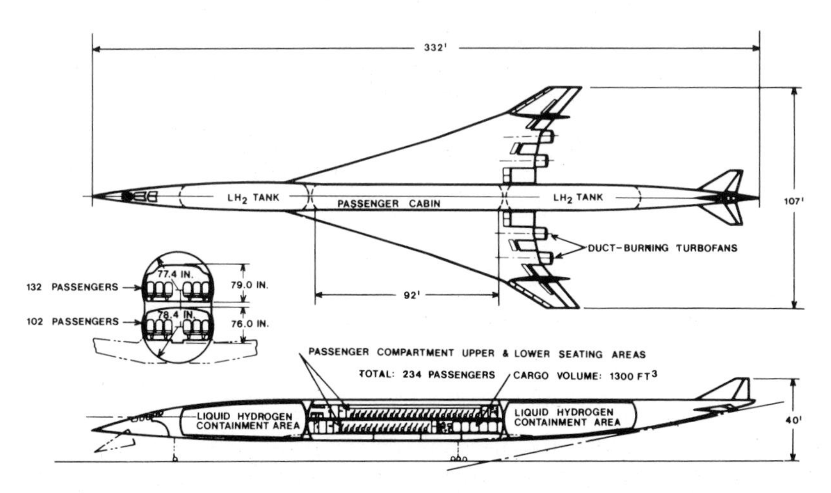

Fig. 15.19 Hydrogen-fueled supersonic transport design (NAS2-7732).[43]

TABLE 15.5[43]

AIRCRAFT CHARACTERISTICS (PRELIMINARY DATA)
Basis: 1981 Technology.

Aerodynamic configuration	Arrow wing
Mach No. (cruise)	2.7
Payload	234 passengers
Range	4200 n mile
Engines	Turbofan, duct-burning
Fuel	Liquid hydrogen
Wing-loading (takeoff)	50 lb/ft^2
Thrust/weight (sea level, static)	0.5
Gross takeoff weight	379,000 lb
Fuel weight (total)	98,000 lb
Aircraft zero-fuel weight	280,000 lb
Operator's empty weight	232,000 lb
Payload weight	49,000 lb

REFERENCES

1. R.G. Murray, R.J. Schoeppel & C.L. Gray, paper presented at the 7th Intersociety Energy Conversion Engineering Conference, San Diego, September 1972.
2. E.H. Hietbrink, J. McBreen, S.M. Selis, S.B. Tricklebank & R.R. Witherspoon, in *The Electrochemistry of Cleaner Environments*, ed. J.O'M. Bockris, Plenum Press, New York, 1972, p. 47.
3. 'Automobile and Air Pollution: A Programme for Progress' – Part II, Subpanel Reports to the Panel on Electrically Powered Vehicles (R.S. Morse, Chairman), U.S. Department of Commerce, December 1967, p. 103.
4. *ibid.*, p. 101.
5. E. Hines, M.S. Mashikian & L.J. van Tuyl, SAE Publication – 690441, Chicago, May 1969.
6. R.J. Schoeppel & S. Sadiq, paper presented at the Frontiers of Power Technology Conference, Stillwater, Oklahoma, 15-16 October 1970.
7. R.J. Schoeppel, *Chemtech*, p. 476 (August 1972).
8. Recorded by R.A. Erren & W.H. Campbell, *J. Inst. Fuel,* **6,** 277 (1933); W. Cecil, Trans. Cambridge Phil. Soc., **1,** part 2, ch. 14, p. 217-239.
9. H.F. Ricardo, Report of the Empire Motor Fuels Committee, *Proc. Inst. Automobile Engineers*, **18,** (1923-24).
10. A.F. Burstall, *Proc. Inst. Automobile Engineers*, **22,** 358 (1927).
11. R.A. Erren, 'Method for Driving Internal Combustion Engines', U.S. Patent No. 1.901.709, applied 15th March, 1930.
12. R.A. Erren & W.H. Campbell, *J. Inst. Fuel,* **6,** 277 (1933).
13. R.A. Erren, 'Internal Combustion Engine Using Hydrogen as a Fuel', U.S. Patent No. 2.183.674, Application 10 September 1936.
14. R.A. Erren, 'Method of Charging Internal Combustion Engines', U.S. Patent No. 2.164.234, Application 17 September 1938.

[15] R.O. King *et al.*, 'The Hydrogen Engine: Combustion Knock and Related Flame Velocity', Trans. E.I.C., Vol. 2, 1958.
[16] R.O. King, W.A. Wallace & B. Mahapatra, *Can. J. of Res.*, **26,** Sec. F, 264 (1948).
[17] R.E. Billings & F.E. Lynch, 'History of Hydrogen-Fueled Internal Combustion Engines', Publication No. 73001, Utah, 1973.
[18] R.J. Schoeppel, paper presented at the Gas Symposium of the Society of Petroleum Engineers of AIME, Omaha, Nebraska, 21-22 May 1970.
[19] R.G. Murray & R.J. Schoeppel, paper presented at the Intersociety Energy Conversion Engineering Proceedings, Boston, Massachusetts, 3-6 August 1971.
[20] R.J. Schoeppel & R.G. Murray, paper presented at the Frontiers of Power Technology Conference, Stillwater, Oklahoma, 28-29 October 1968.
[21] Perris Smogless Automobile Association, 'An Answer to the Air Pollution Problem: The Hydrogen and Oyxgen Fueling Systems for Standard Internal Combustion Engines', First Annual Report, Perris, California, 1971.
[22] M.R. Swain & R.R. Adt, Jr., paper presented at the 7th Intersociety Energy Conversion Engineering Conference, San Diego, California, 25-29 September 1972; paper presented at Intersociety Conference of Energy Conversion, University of Pennsylvania, 1973, August 13-16.
[23] L.O. Williams, *Cryogenics*, p. 693 (December 1973).
[24] P. Sporn, priv. comm., 1974.
[25] Figure supplied by C. Heath, 1971.
[26] W.F. Libby, *Science*, **171,** 499 (1971).
[27] P.M. Miller, *Scientific American*, **228,** No. 2, 78 (1973).
[28] J.I. Deen & R.J. Schoeppel, paper presented at the Frontiers of Power Technology Conference, Stillwater, Oklahoma, 15-16 October 1970.
[29] R.J. Schoeppel, priv. comm., August 1974.
[30] N.C. Hallett, 'Cost, Study and Systems Analysis of Liquid Hydrogen Production', Air Products and Chemicals Inc., Contract NAS 23894, June 1968.
[31] R.D. Witcofski, priv. comm., 27 February 1974.
[32] L. Jones, priv. comm., July 1974.
[33] J.E. Johnson, paper presented at the Cryogenic Engineering Conference, Atlanta, Georgia, 10 August 1973.
[34] J.W. Michel, 'Hydrogen and Exotic Fuels', Oak Ridge National Laboratory, Report ORNL-TM-4461, June 1973.
[35] L.O. Williams, Astronautics and Aeronautics, p. 42 (February 1972).
[36] R.D. Witcofski, paper presented at the American Chemical Society Symposium on Non-Fossil Chemical Fuels, Boston, Massachusetts, 9-14 April 1972.
[37] R.D. Witcofski, paper presented at the 7th Intersociety Energy Conversion Engineering Conference, San Diego, California, 25-29 September 1972.
[38] W.J. Small, D.E. Fetterman & T.F. Bonner, Jr., paper presented at the Intersociety Conference on Transportation, Denver, Colorado, 23-27 September 1973.
[39] F.S. Kirkham & C. Driver, paper presented at AIAA 5th Aircraft Design, Flight Test and Operations Meeting, St. Louis, Missouri, 6-8 August 1973.
[40] J.V. Becker, *Astronautics and Aeronautics*, **9,** No. 8, 32 (1971).
[41] J.V. Becker, paper presented at the 7th Congress of the International Council of the Aeronautical Sciences, Rome, Italy, 14-18 September 1970.
[42] D.E. Wilcox, C.L. Smith, H.C. Totter & N. C. Hallett, paper presented at the ASME Annual Aviation and Space Conference, 16-19 June 1968.
[43] W.J.D. Escher & G.D. Brewer, paper presented at the AIAA 12th Aerospace Sciences Meeting, Washington, D.C., 30 January-1 February 1974.
[44] Lewis Laboratory Staff, 'Hydrogen for Turbojet and Ramjet Powered Flight', NACA RM E57D23, 26 April 1957.
[45] W.J.D. Escher, 'Prospects for Liquid Hydrogen Fueled Commercial Aircraft', Report PR-37, Escher Technology Associates, St Johns, Michigan, September 1973.
[46] S. Weiss, 'The Use of Hydrogen for Aircraft Propulsion in View of the Fuel Crisis', Technical Paper (NASA TMX-68242), paper presented at NASA Research and Technology Advisory Committee on Aeronautical Operating Systems, Ames Research Centre, Moffett Field, California, 7-8 March 1973.
[47] H.C. Watson, E.E. Milkins & J.V. Deslandres, 'Efficiency and Emissions of a Hydrogen and Methane fuelled spark engine', Report T/8/74 Version 2, Mechanical Engineering, University of Melbourne, Australia (presented Fisita Conf., Paris, May 1974).

CHAPTER 16

Environmental

INTRODUCTION

ONE WOULD NOT be complete in a presentation of the various energy possibilities if the pollutional pros and cons were not considered. There can be no doubt, as far as atmospheric pollution is concerned, that the Hydrogen Economy is an absolutely satisfactory economy, for the long term. However, in considering the environmental effects of an economy, it is not sufficient to consider only its effect upon the quality of air and water. We must also ask about the consumption of resources, and other effects upon the environment, which the introduction of such a technology would involve.

This type of analysis has been sparse, but it has been done recently in a preliminary way by Plass[1], and it is his analysis which has been the main influence in the following considerations.

RESOURCES AND THE HYDROGEN ECONOMY[1]

There is a tendency in considering a Solar-Hydrogen Economy to regard it as using no resources, because it converts water to hydrogen and oyxgen, which then come back to water. However, materials would be used for the construction of the system, that is for collectors, electrolysis plants, pipelines, storage tanks and etc.. From a conservationist's viewpoint, as distinguished from the air and water pollutional viewpoint, a Hydrogen Economy based upon fossil or nuclear fuels should be compared with a conventional economy based on such fuels.

In doing this, Plass uses the following equation:

$$\frac{R}{T} = P \times \frac{H}{PT} \times \frac{R}{H} \qquad (16.1)$$

where R = quantity of the resource in question (e.g. tons for minerals, or Btu for energy resources)

T = time (years)

P = population

H = quantity of hydrogen consumed (Btu).

The second factor H/PT is the per capita demand for hydrogen per year. The third factor, R/H is the resource cost per unit of hydrogen. These are both important. When conversion efficiencies from heat, etc., are high, H/PT is low. For low efficiencies, the reverse is true. If the hydrogen manufacturing design and distributing facility is good, R/H

is small, whereas poor design will cause an increase in R/H. It is necessary in applying equation 16.1 to make reasonable estimates about the factors, H/PT and R/H.

The resources consumed in the production, transport and storage of hydrogen are the fuel (except for the solar case), iron, aluminum and copper, land occupied by collector farms, pipelines and storage facilities.

Plass[1] studied a coal-based economy; a coal-based synthetic fuel economy; a nuclear-based hydrogen economy; and a nuclear-based electric economy. The results obtained are summarised in Table 16.1.[1]

Thus, in spite of the environmental advantage of a Hydrogen Economy, it would consume more energy than other economies in the setting up stage. This would not be so, however, were the economy to be based upon an inexhaustible source, such as the solar source (or, conceivably, fusion). However, even with these inexhaustible sources, caution must be taken that the introduction of hydrogen does not run away with too high a proportion of mineral resources, all of which must be regarded as limited in amount.[1]

ENVIRONMENTAL EFFECTS

The added resource costs which would come from hydrogen used along with coal and fission might be thought to be overcome by the environmental advantage of forming only water back from water. A simple equation which can be used for looking at the environmental effects of hydrogen is that given by Commoner[2]:

$$\frac{D}{T} = P \times \frac{H}{PT} \times \frac{D}{H} \quad , \qquad (16.2)$$

where P = population
H = quantity of hydrogen consumed (Btu)
T = time (years)
D = environmental damage (dollars).

The first two factors have been discussed. The third factor D/H is the amount of environmental damage per unit of hydrogen produced. Its value depends upon the technological process, for example, D/H for hydrogen produced by coal would be greater than that produced were solar energy the origin of the hydrogen.

Plass[1] considers alternatives for utilising coal, at the same time taking into account land destruction by strip-mining. Synthetic fuel plants, uranium mines, solar farms and other means, are also considered.

To make a conclusion of relative pollutional effects, Plass[1] utilises a classification of environmental effects suggested by Brubaker.[3] This is an environmental matrix which classifies environmental problems, causative factors, character of the environmental insult, problem threshold, area affected and other allied considerations.

The hydrogen situations are more advantageous environmentally than any other. However, in converting coal to hydrogen, more coal is needed to provide the necessary energy for the nation's economy, if hydrogen is used as a medium, and a correction has to be made for this. *A solar-based Hydrogen Economy is only about 40% as polluting as the corresponding nuclear-based hydrogen system.*

TABLE 16.1[1]

ENERGY APPETITES OF FOUR SYSTEMS, TWO OF WHICH ARE HYDROGEN ECONOMIES

System	User requirements (1970)			Production ratios					
	Elec-tricity E/T	Syn-thetic fuel F/T	Hyd-rogen H/T	R coal / E	R coal / F	R coal / H	R nucl / E	R nucl / H	Total re-source demand R/T
I (coal-hydrogen)	7.98	–	40.57	3.09	–	3.92	–	–	183.7
I* (coal-synthetic fuels)	7.98	40.57	–	2.92	1.52	–	–	–	84.9
II (nuclear hydrogen)	7.98	–	40.57	–	–	–	3.56	4.52	211.8
II* (nuclear all electric)	38.30	–	–	–	–	–	3.96	–	151.5

All of above $\times 10^{15}$ Btu/yr. Btu resource/Btu of E, F, or $H \times 10^{15}$ Btu/yr.

Present energy economy, at 1970 levels, requires 64.6×10^{15} Btu/yr of raw energy from various sources.

TABLE 16.2[1]

ENVIRONMENTAL DAMAGE COMPARISONS

System	Environmental Damage Index (1970 levels)	Ratio of Damage to that of Present System
I (coal-hydrogen)	670	1.64
I* (coal-synthetic fuel)	554	1.36
II (nuclear-hydrogen)	483	1.18
II* (nuclear-electrical)	461	1.13
III (solar-hydrogen)	259	0.63
III* (solar-electric)	173	0.42
Present mixed system (calculation omitted)	408	1.00

If environmental damage alone were the criterion, then the optimal system is the solar-electric alternative. The final results are shown in Table 16.2.[1]

Thus, the advantages of the solar systems come out well, and the solar-hydrogen systems excellently, particularly if utilised with wind power. Wind utilising sea-borne stations would probably have a better index than even the solar-electrical economy.

GENERAL DISCUSSION[4]

Considering, then, the environmental contaminants produced by a Hydrogen Economy – rather than resource exhaustion – there could be only two pollutants, NO and water. The NO can be eliminated by using fuel cell conversion.

Contamination with water must be considered. Table 16.3[4] shows the relative production of water. Water production would be an advantage, not a difficulty, if it were condensed and then collected, because it would be produced free from other gases, and hence would be clean.

There is hydrogen peroxide production during internal combustion of hydrogen with air, and the amount is shown in Table 16.4.[5] This contaminant (0.02) of H_2O in *condensed* water from internal combustion is probably not of importance because of the rapid decomposition of H_2O_2 in normal vessels and in the presence of light.

An advantage of a Hydrogen Economy will be its capacity for energy storage. The absence of overhead transmission lines will be an environmental bonus.

Electricity and hydrogen are both clean burning, but, at the generating plant, hydrogen has the environmental advantage. If transmitted by pipe, it is environmentally more acceptable than the transport of electricity *via* a wire grid. Underground electricity transmission is many times more expensive than overground.

In respect to transport, the limitation of hydrogen is in the storage technique. A cylindrical tank, 10" × 60", would hold about 35 gallons of hydrogen, equivalent to 9 gallons of gasoline in energy. It would need special handling techniques, but these seem well justified in view of the environmental advantage over methane.

SUMMARY OF ENVIRONMENTAL EFFECTS

In respect to air and water pollution, hydrogen is the ideal fuel and its general use would lead to a radical reduction of air and water pollution. In respect to resource use-up, so long as one speaks of a Solar-Hydrogen Economy, the situation is excellent. A Solar-Electrochemical Economy would be, in respect to resource use-up, still better. Table 16.2[1] quantifies the added pollution which would arise from the use of coal.

TABLE 16.3[4]

WATER PRODUCTION FOR VARIOUS ENERGY SOURCES

Fuel	lbs water/Btu
Gasoline (C_8H_{18})	74×10^{-6}
Methane (CH_4)	105×10^{-6}
Hydrogen	174×10^{-6}

TABLE 16.4[5]

CONCENTRATION OF H_2O_2 IN THE EXHAUST OF AN ENGINE USING H_2 GAS AS A FUEL

Test No.	H_2O_2 in the exhaust (ppm)
1	230
2	220
3	220

REFERENCES

[1] H.J. Plass, paper presented at The Hydrogen Economy Miami Energy (THEME) Conference, University of Miami, Florida, 18-20 March 1974.

[2] B. Commoner, 'The Environmental Cost of Economic Growth', in *Energy, Economic Growth and the Environment*, ed., Schurr, John Hopkins University Press, 1972.

[3] S. Brubaker, *To Live on Earth,* Mentor Book, 1972.

[4] R.L. Savage, L. Blank, T. Cady, K. Cox, R. Murray & R.D. Williams, edd., 'A Hydrogen Energy Carrier', Vol. II – Systems Analysis, NASA-ASEE, September 1973.

[5] E.J. Griffith, *Nature*, **248,** 458 (1974).

CHAPTER 17

Alternative Economies

VARIOUS ASPECTS of a Solar-(Atomic) Hydrogen Economy have been presented. Its advantages would often seem significant, and, in some areas, essential. One cannot, however, present the Hydrogen Economy, associated with solar and atomic sources, without the presentation of the strengths of alternative and competitive energy bases.

The rivals to a Solar-Hydrogen Economy are time-dependent in strength. Thus, fossil fuels will run out: it is not controversial to rationally state that it would be better to change away from them as soon as economically and environmentally possible. Such a change will be forced by rising prices resulting from exhaustion; and encouraged by the constantly rising air pollution. But it should also be encouraged by the great desirability of having a supply of fossil fuels left for the synthesis of petrochemicals; which otherwise would have to take place increasingly after 2000 *via* hydrogen and atmosphere-derived CO_2.

THE DIVERSITY OF VIEWPOINT IN ENERGY AFFAIRS

The topic of energy is connected directly with the topic of the economy, and the Standard of Life. Finally, it reflects the relation of how much a person has to work for what he can buy. It produces much rhetoric. It may dismay the serious student of energy to see the great pre-occupation with the very near time, namely the frequency and intensity of discussions bearing upon, for example prices of natural gas in the next six months. Thus, the coal industry may be less interested in how to research and develop over decades to Abundant, Clean Energy, but is very interested in how to have regulations made so that it can get on with strip-mining next year. The financial institutions are interested in how moves in the energy field will affect the balance of payments for this year – and maybe for next year. Utilities are interested in the granting of licences to build atomic power stations as of *now*. The oil industry shows interest in taxes and import quotas more than in the degree to which their products pollute. The automobile industry is interested in the immediate reduction of standards concerning pollutants – far more than in the successful development of an alternative means of propulsion. The gas industry sees the matter in terms of the conditions set by regulatory agencies. Environmentalists are interested in what will happen to the situation ecologically, but pay little attention to whether their recommendations would allow transportation at practical cost.

Reading reports of energy matters in the '70s, there is an impression of a warring series of mandarins, each after the very near term advantage for his group, none looking beyond 10 to 20 years ahead (and very few looking 1-2 years ahead), and few preparing for the exhaustion of the fossil fuels. In contrast with this, the time at which oil and gas exhaust is 15 to 30 years and coal 30 to 60 (*if* it becomes available); and the time of building a new system of Abundant, Clean Energy: 25 to 50 years.

Pollution seems to have been erased from the mind of the vocal sources; and the impression is given that the spectre of exhaustion is temporary, will soon be replaced with that of plenty, and all will continue as it was when the capital in the fossil fuel account was being spent without knowledge that it was finite. The phrase 'The Energy Crisis' itself implies something which will last a year or so and be gone. The concept of the "winds of permanent change" is entirely lacking.

Those who *are* concerned and knowledgeable in the big and long term picture are not in a position directly to affect the situation. What will actually happen is determined by the short term and united advantages of the largest corporate groups. The most dangerous impression (and it is one which is still in the minds of many) is that there will be oil and natural gas for ever ('they always find some more'), because the most certain aspect of the present picture is that the world is in the last few decades of these sources, and also that during this time inflation will be increasingly rampant (partly because the price of oil and natural gas will rise continuously, forcing up the price of most other products). Correspondingly, advertisements are often published which spread the false impression that, although there may be difficulties with oil and natural gas, there stands an immense Mountain of Coal, which can be consumed for centuries to come (but see Chapter 3).

Here there is an attempt to put the prospects in perspective, and it is stressed again that they are all exceedingly time-dependent.

GASES AND LIQUIDS FROM COAL

The pros and cons of an economy based on fuels from coal are:

The price of the synthesis of natural gas or oil from coal in the U.S. was quoted in November 1973 (Lessing[1], Stroup[2]) as \$1 to \$1.50 for natural gas from coal per MBtu. However, the price is proportional to the price of coal (which has increased by *c.* 250% between June 1973 and August 1974). As the prices of oil and natural gas rise, the price of coal will rise, for reasons none other than the price of something is the highest amount of money the buyer can be made to pay, which depends on the price of the alternatives. But it is clear that methane from coal is likely to be cheaper than hydrogen from coal-based electricity, because more than 60% of the energy in converting coal to electricity is lost as heat in the Carnot loss of the engine which drives the generator.

Thus, only if the cost of the clean-up of methane to a purity of electrolytic hydrogen increased the price of the hydrocarbon by this percentage or more,* could the price of electrolytic hydrogen be competitive with that of methane from coal.

This does not mean, however, that hydrogen *from coal* may not

*Although see the thermal-electric method for producing hydrogen outlined in Chapter 9.

be competitive with methane from coal. It should be about the same price. Alternatively, new methods of the production of hydrogen from photosynthesis, or hydrogen from a photoelectrochemical technique, may be successful in producing hydrogen at lesser cost than CH_4 from coal. That is why research in these (and other) areas of cheap hydrogen production is an important base to an early and desirable change in the energy future.

Thus, if coal can be produced at a sufficient rate to supply our energy needs (but see Chapter 4), and before cheap nuclear electricity, or electricity from one of the solar alternatives, is available, the choice is likely to be either hydrogen from coal, or methane from coal.

Such a conclusion is not entirely rational. Systems analysis (Chapters 6 and 7) suggests that there are more permanent alternatives available in solar energy right now, such as in the tracking mirror method developed by Hildebrandt and his colleages; and in the Ocean Thermal Gradient (or Solar Sea Power) method developed by Anderson, Zener and others. Both these methods seem, *according to systems analysis*, to be buildable now, and to be able to produce electric power and hydrogen at acceptable prices, e.g. at less than $3 per MBtu. They could be a part of a permanent future, whereas a coal-based possibility is certainly very temporary. However, systems analysis is a less firm basis than reality. The South African coal gasification industry exists, and can make South Africa essentially independent of imported oil.* It is more likely, then, that the South African model will be the one which the U.S. will follow. There are subjective grounds for preferring a new way which is basically similar to the old way. Also, and, increasingly the coal is becoming owned by the oil companies, and these will become the future Energy Companies.

Consideration has to be given to the two alternatives, i.e. hydrogen or methane from coal, in terms of ease of transmission. Here, methane scores, for the energy per unit volume is 3.5 times that of hydrogen. Again methane scores in needing less modification to plant than one would have to carry out to use with hydrogen.

Conversely, were hydrogen to be produced from coal, rather than methane, an improvement in the pollutional situation would occur. CO_2 would only be produced at the base plant; and the other impurities are easier to remove at such a plant than from individual automotive power plants.

Two major negative points in a development using coal – which have been brought out in Chapter 4 – will be mentioned again here:

(1) If industry is expanded to develop synthetic natural gas, the economic situation of the amortisation of the plant might meet difficulties with greenhouse effects which could mean a retreat from coal-based energy soon after 2000. Could large scale synthetic natural gas plants be operating before 1980?

(2) Mining of sufficient coal within such a short time as would be necessary if coal were to be used as a main source of energy, has logistical difficulties (Chapter 4), and success in underground gassification has evaded fifty years of research.

---

* At present, coal-based gasoline supplies only a few per cent of South Africa's needs.

Balance of the likelihood of large scale development of synthetic natural gas

It is easier to see the long-term rather than the short-term probabilities in this case. It would be attractive to develop a synthetic natural gas economy based upon coal: and – depending on the greenhouse effect – its attractiveness might last into the new century.

However, another path exists. A gradual conversion to a Hydrogen Economy would be consistent with the undoubted fact that the final energy sources will be atomic and solar, when the large scale use of hydrogen will be very likely. Thus, hydrogen should be extracted from coal, or made with coal-based electricity, depending upon which gives the cheaper product, taking into account the facts that the products to be compared must both be equally clean and that thermal-electrochemical methods seem to offer much in the way of price reductions for electrochemically produced hydrogen. As the atomic and solar sources get built up, there would be no further disruption or change needed. The Hydrogen Economy would be in being (Chapter 18). There would be some advantages of this course in the longer term economics: there would be no difficulty in amortisation, for the plants producing hydrogen could have their normal life of 40 to 50 years. The difficulty which does still exist is logistical, the difficulty of getting out coal at a sufficient rate. The continuity of normal life past about 2000 seems dependent on this, i.e. upon the success of a new form of accelerated mining which seems to provide the only possibility.

Indeed, the essence of a very real Energy Crisis, which will be felt independently of the monopolistic price setting which is the origin of some of the economic difficulties of the 1970s, is that it will not be possible to build *in time* sufficient new coal mines, or sufficient atomic plants – or sufficient solar alternatives – to deal with the rate of exhaustion of oil and natural gas. The expansion rate of the Western Economies has brought on the exponential end-phase with oil and natural gas before the alternative Abundant, Clean Energy sources have been prepared (Fig. 17.1).

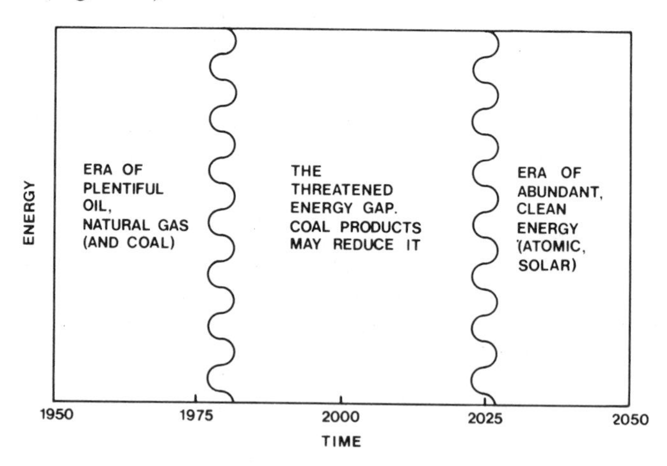

Fig. 17.1 The area of difficulty in the short-term energy future.

Thus, the path which seems to offer the best short-term option is to maximise the production of coal, reduce the pollutional problem by using it to generate electricity at central plants (where pollution from the stacks can be fairly well controlled), and start a Hydrogen Economy with this coal-based electricity. Alternatively, separate coal to hydrogen synthesis plants could be set-up. At the same time the build-up would be accelerated of a clean breeder technology to its utmost and research and development very greatly accelerated in solar and solar-gravitational conversion. Included would be the immediate development of wind power in high energy density wind-belt areas of the world (Chapter 5)[3], with the actual building of large scale plants envisaged in the early 80s.

Development of liquid fuels from coal

The picture is not attractive, in respect of price and pollution. The (1974) price of coal-based petroleum, as at the South African plant, is 75 cents per imperial gallon of petroleum. The pollutional problem would be as bad as pollution from oil, the cause of much of the present air-pollutional difficulties, to which less attention is now paid, because of the increasing prospect of an energy shortfall.

The underground gasification of coal and its effect on the coal availability problem

The prospect of a latter-day fossil fuel economy with synthetic methane has been shown in Chapter 4 to depend upon the solution of logistic difficulties. The possibility that these may be overcome by underground gasification will be briefly examined.

Firstly, the underground gasification of coal is an old topic and efforts to achieve it have been made for many years, particularly in the U.S.S.R., the U.S.A. and the U.K. For some time, in these countries, plants were built and underground gasification occurred, but, during the 1950s, they were all shut down because of competition from cheap natural gas. The point is that much is known about the underground gasification of coal.

There are several methods used. In all, there is a controlled burning of coal in the seam to produce an oxidising reaction which results in the formation of CO by the reduction of CO_2. Small quantities of hydrogen and hydrocarbons are released. A typical composition is shown in Table 17.1.[4]

The gas produced is a very low Btu gas, about 100 Btu SCF^{-1} (one-tenth that of natural gas). The other difficulty is the large amount of nitrogen present.

Many mechanisms of underground gasification have been tried. In one, underground galleries are prepared. Air is blasted in at an ignited face, and the product gas is removed and collected. In a different approach, boreholes are drilled from the surface. Air is forced down each and the head of the hole is ignited. Horizontal drills intersect the vertical ones and the gas is withdrawn from them.

The advantage of these approaches is that they can be tried out on poor quality coal for which gasification would not otherwise be possible. The difficulties are[5]:

(1) Large areas are required, e.g. 30 million tons to supply a 100 MW station for 25 years.

(2) The environmental consequences would be poor because of subsidences, etc.
(3) The quality of the gas is very poor.

Research can often improve a situation. However, underground gasification has received much research and does not have good prospects.

TABLE 17.1[4]

TYPICAL COMPOSITION OF UNDERGROUND GASIFICATION OF COAL

	Underground gas
CO_2	10.5 %
O_2	0.9 %
H_2	8.4 %
CO	10.7 %
CH_4	1.8 %
Other hydrocarbons	0.3 %
N_2	67.4 %
Btu/cu ft	86.0

A METHANOL ECONOMY

Methanol is a sufficiently important possible fuel for the phase between the end of oil and natural gas and the primary dependence on atomic and solar energy, to be considered. It is one of the options which must be kept open, and its synthesis from coal at a sufficiently cheap price is being researched by a number of oil companies.

A Methanol Economy would have several advantages:

(1) Natural gas has a low pollution index, but it is a gas and it would be more convenient (i.e. more applicable to the plant now owned) if the new fuel could be had in liquid form.

(2) Methanol can be produced from coal, but it can also be produced from vegetative materials and this makes it a possibility as a medium of solar energy. It could also be a medium of atomic energy in the sense that CO_2 could be extracted from the atmosphere and used with hydrogen from nuclear-based electrolysis.[6]

(3) Internal combustion engines may probably be run on methanol with few changes on present engines.

(4) Methanol would be less pollutive than gasoline.

(5) Storage could be more convenient than that of gaseous hydrogen; and cheaper than that of liquid hydrogen.

(6) Transmission could be a fraction cheaper than that of hydrogen (Fig. 17.2[7]).

(7) Safety would be better than that of hydrogen.

(8) Materials problems would be less than that of hydrogen.

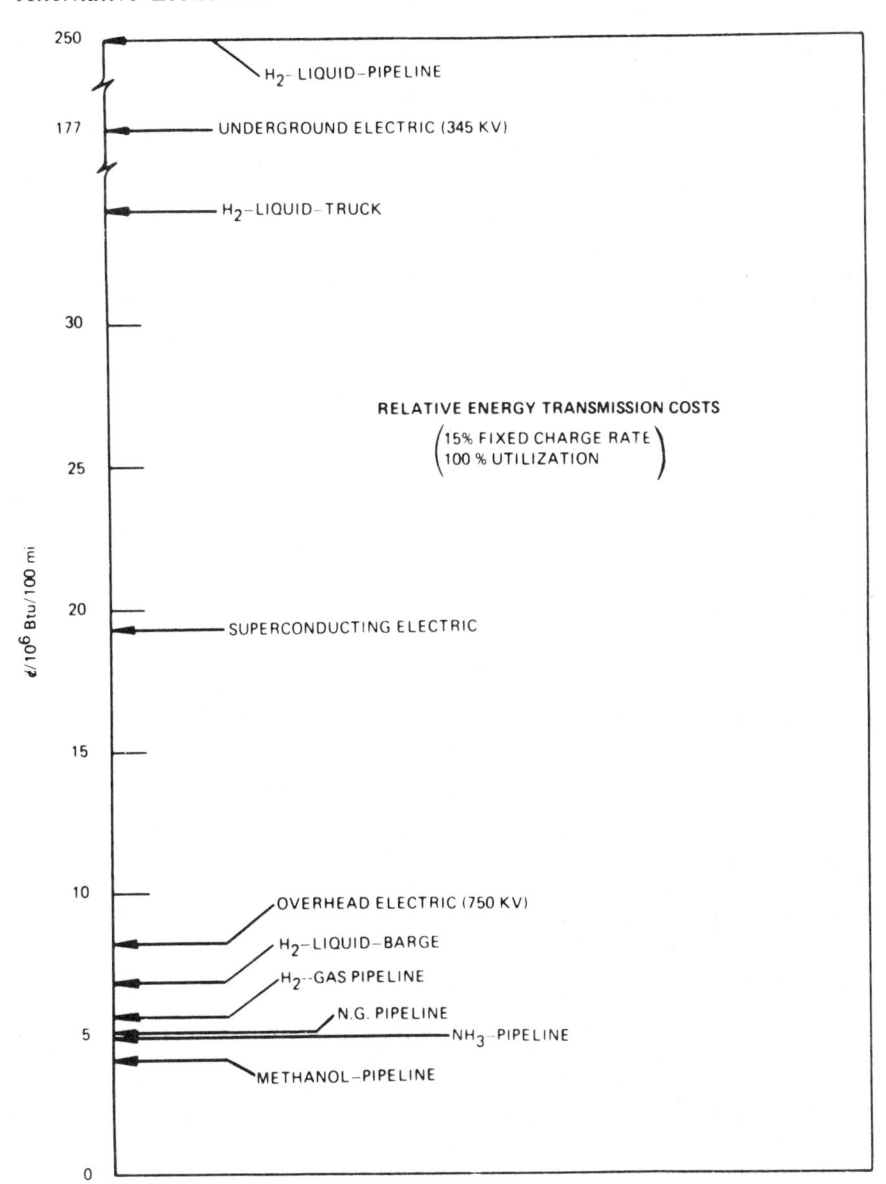

Fig. 17.2 Energy transmission costs.[7]

(9) It could leave the capital invested in the fossil fuel technology much less disturbed than movement to another technology.

The advantages of a Methanol Economy must be compared with the following negative points:

(1) The CO_2 difficulty after 2000 would still be present. Some alleviation would be obtained by using methanol in fuel cells. Thus, methanol trans-

duces well to electricity in fuel cells, whereas methane and hydrocarbon oils are not good fuels for such cells. Assuming, roughly, that the efficiency of energy conversion to electricity is twice times in a fuel cell what it is in the corresponding heat engine (an estimate which is low for transportation, but too high for central power plants), then the CO_2 production rate would be halved for the same energy production, if all fuel burning were methanol in fuel cells instead of gasoline and oil in chemical engines.

(2) The transduction possibilities with hydrogen are superior to those of methanol. Thus, hydrogen makes an excellent pollutionless fuel (negligible NO) in internal combustion engines. Methanol would still give CO_2 and fails to meet the NO index criterion.

(3) Hydrogen is inevitable for future aircraft and it would be advantageous to have the same fuels for land-based and air-based transportation.

(4) In a sense, methanol is simply a fuel, whereas hydrogen could be the basis of an economy. Thus, all the pollutional advantages of running factories and industry on hydrogen; the advantages of having cheap excess hydrogen in metallurgy and for recycling; the side effects of fresh water production; and the synthesis of proteins from hydrogen, nitrogen, CO_2 and enzymes; are lost if methanol, rather than hydrogen, were the general medium of energy.

(5) The prospective price of methanol in 1974 dollars is about $3.50 per MBtu, and the future price is predicted to be at $4.00 or above. Cheaper prices quoted involve production from natural gas or naphtha, clearly not to be considered for a future which would have to last for many decades to justify the investment of capital in plants.

Production of hydrogen by several methods should be cheaper than this. After coal is no longer sufficiently plentiful, methanol will have to come from (the necessarily cheaper) hydrogen, produced by one of the methods of Chapter 9. One question then, would be whether the cost of liquefying hydrogen could be less than the cost of making CH_3OH from hydrogen and air-extracted CO_2. It is difficult to see how pure methanol from hydrogen could be cheaper than gaseous hydrogen.

Methods of producing methanol

Three methods may be mentioned:

(1) The gasification of coal followed by the Fisher-Tropsch method with an iron catalyst (Chapter 4). Methanol is produced at present in small quantities by this method, but, by adjustment of conditions, the methanol content could be increased.

(2) Anaerobic digestion of vegetable matter giving rise to methane; the reforming of methane.[8]

(3) Electrolysis of water to hydrogen, dissociation of limestone to CO_2 (or CO_2 from air), and then the reduction of CO_2 with hydrogen to methanol.

The methods in which natural gas is converted to methanol and naphtha is steam reformed to this material, may be neglected, in considerations of a future energy base.

The future price of methanol in massive production

The price of methanol must be estimated, because no large-scale production exists. The estimates of Table 17.2[9] may be of value, the two columns referring roughly to mid-1973 and mid-1974.

H. Linden[10] predicts a methanol price of \$3.00 to \$4.00 per MBtu.

TABLE 17.2[9]

PROJECTED COSTS OF LARGE-SCALE METHANOL PRODUCTION
(cost per MBtu)

Year	Coal gasification to methanol	Anaerobic digestion of vegetable matter	Nuclear electrolysis to hydrogen, CO_2 + hydrogen, etc.
1973	2.25*	3.00	4.50†
1974	4.50**	3.45	5.17‡

* = coal at \$5.00 per ton
† = 1973 dollars, 10% return on investment
** = coal at \$10 per ton
= 1974 dollars, 15% return on investment

Methanol as a fuel for the internal combustion engine

There is a difference in the octane values between methanol and gasoline which would need minor alterations being made to engines intended to run on methanol.

Studies carried out by Pefley, Adelman and McCormack[11] on methanol-fueled internal combustion engines are very encouraging. The emissions

TABLE 17.3[11]

PURE METHANOL VEHICLE PERFORMANCE RECORD

Vehicle:	*Use:*
1972 Plymouth Valiant	Operated by Meter Readers
Owner:	*Service Time and Mileage:*
City of Santa Clara	30 months and 20,000 miles

Best Anti-pollution Performance

	HC Gr/Mi	CO Gr/Mi	NO_x Gr/mi
City Vehicle:			
Air Resources Board*	0.0	0.42	1.30
Santa Clara University (CVS-1)	0.026	1.42	0.67
1976 Standards:			
California State (CVS-2)	0.9	9.0	2.0
Federal (CVS-2)	0.41	3.4	2.0

Principal Problems:
Cold start, fuel system materials, sluggish performance when set at minimum pollution levels

* The test by the Air Resources Board was the 7-mode test with the results converted to CVS-1.

are much less than in gasoline-fueled vehicles. Part of the work was carried out on what was called 'dissociated methanol', i.e. hydrogen.

Cars were run on pure methanol for 20,000 miles with no unusual difficulties in performance. Some results are shown in Table 17.3.[11]

Such vehicles have been placed first in clean air tests. Emissions as a function of methanol are shown in Fig. 17.3.[11] NO is about the same as with gasoline (Table 17.3).

One of the difficulties is holding methanol and gasoline in solution in the presence of traces of water. Excess aldehyde production is another.[11]

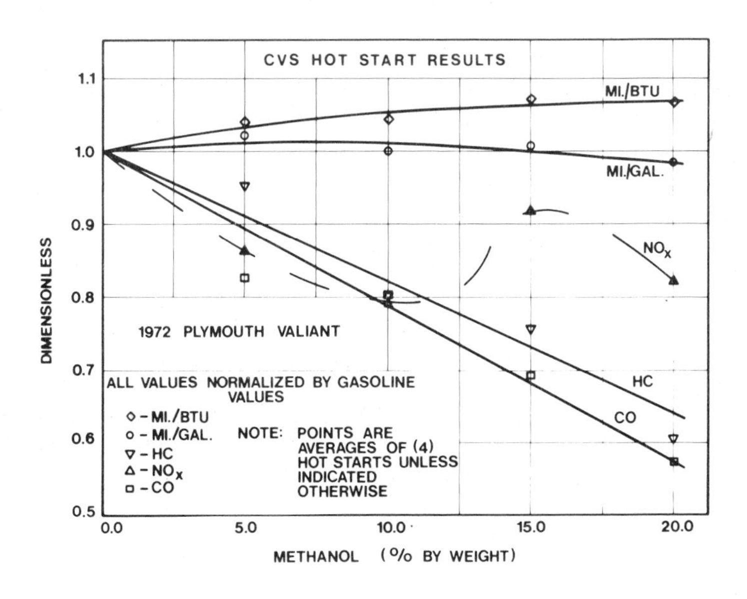

Fig. 17.3 CVS hot start results.[11]

Summary of a possible Methanol Economy

The most attractive aspect of a methanol economy is that it would give us a liquid fuel, at first producible from coal, later possibly producible from solar energy *via* photosynthesis, or atomic energy *via* atmospheric or limestone CO_2, which would give a direct substitute for oil. Further, methanol burns cleaner than oil.

An advantage of methanol, compared with synthetic natural gas, is that it burns electrochemically in a fuel cell[12] more readily (less loss in overpotential) than does oil or natural gas.

The major difficulties of using the methanol are the production costs, the remaining CO_2 pollution, and the lack of a common fuel for air and land transportation. There seems to be no prospect of reducing the price of methanol to compete with natural gas, or hydrogen, from coal. In an era of cheap nuclear electricity, it would become cheap, but hydrogen from water would then be cheaper still.

A NUCLEAR-ELECTROCHEMICAL ECONOMY

General

Classical thoughts about the post-fossil fuel energy situation have been that it would be nuclear-electric. This is tantamount to a nuclear-electrochemical economy, because the electricity has to be stored (Chapter 10) and the most convenient and general way by far is to store it in hydrogen. The electricity would be regenerated *via* fuel cells, used in industry and transportation, particularly in chemical and metallurgical technology. These would then have to convert to electrochemical technologies, i.e. large-scale technologies would have to convert from being based on thermal, chemical reactions, with waste heat, and often polluting gases, to ones electrochemically acting, with a waste product of electricity, or driven by electricity against a free energy gradient. This is the real competition for the Hydrogen Economy. It offers a possibility of a clean technology, so long as the electrochemical versions of present thermal processes can be developed.

Ideal development of a Nuclear-Electrochemical Economy

The timing of a Nuclear-Electrochemical Economy depends firstly upon the time of development of a satisfactory breeder, one which breeds and does not emit to the air significantly, and for which a satisfactory solution to the waste problem (possibly disposal beneath antartic ice?) is engineered. Correspondingly, it depends upon the development of fusion and the degree to which the threatened greenhouse effect proves actual, and upon what degree of funding is placed behind research in electrochemical technology, without which (and in the absence of a hydrogen base) a nuclear-based economy cannot function.

Electrochemical and Hydrogen Economies in various applications

Transmission: In an Electrochemical Economy, electricity would have to be transmitted over long distances to reach the site of use. A comparison of costs is seen in Fig. 17.4.[7] Here there is no doubt about the advantage of hydrogen as a medium for energy.

Storage: In this respect, the electrochemical path (batteries) offers little help compared with the excellent possibilities with hydrogen.

Transportation: The electric car is more attractive energy-wise, than the petroleum car, and Fig. 17.5[13] shows this. Tracing back the efficiencies of the use of energy to be taken from nuclear sources; fed into batteries to work electric motors, used to electrolyse water, and the hydrogen thus produced to be burnt chemically – then the nuclear-chemical (i.e. hydrogen) way is less than half as efficient as the nuclear-electrochemical way.[13]

The energy-efficiency consideration is important. However, the chemically fueled car gives a far better performance than the electric car. It makes air-conditioning and heating possible; these functions will be difficult with battery-driven cars, for the batteries which are now usable are still too heavy. The chemically-fueled car, however, has been polluting. Here, hydrogen changes the situation. The performance of the hydrogen-fueled car can equal that of the gasoline-driven car (Chapter 15).

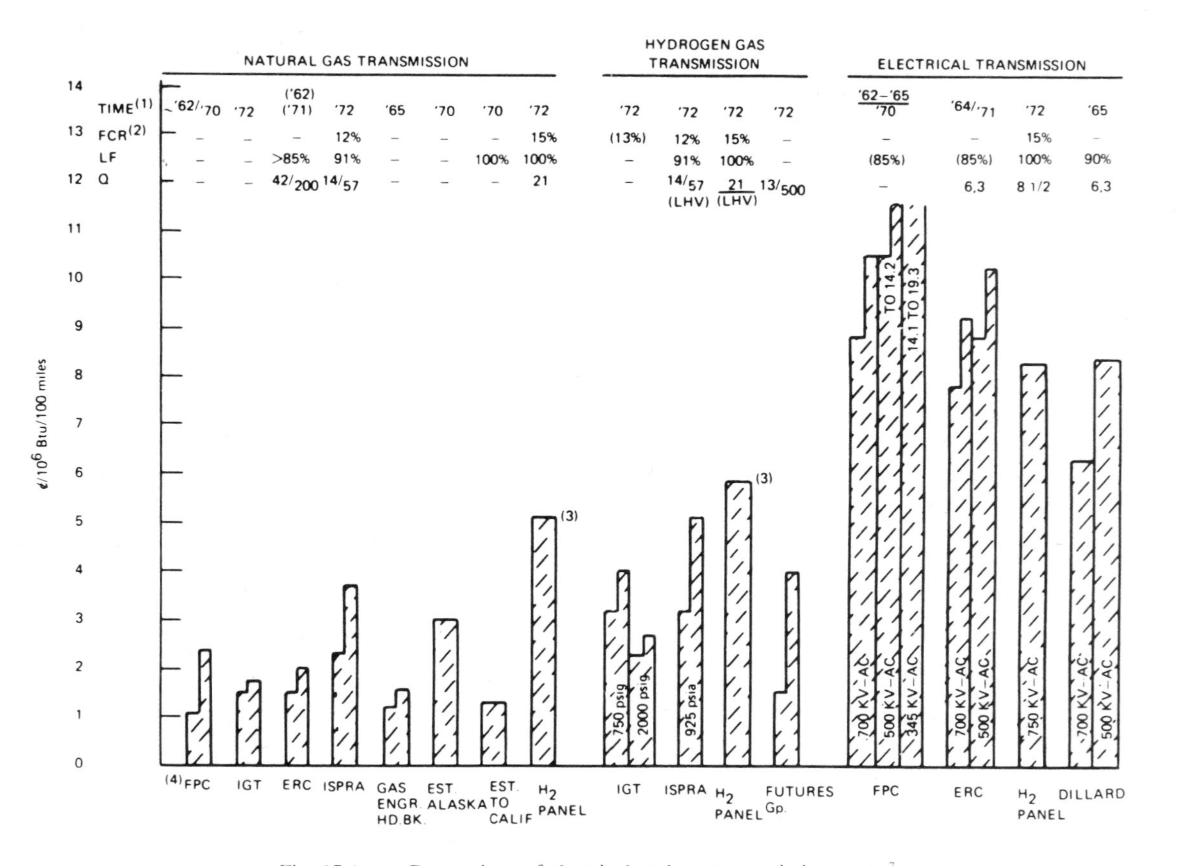

Fig. 17.4 Comparison of electrical and gas transmission costs.[7]

NOTES TO FIG. 17.4

(1) Year in which cost data was developed, where two or more dates are given, the top date represents the year(s) in which the data was developed and the bottom the year of the publication that contained the data.

(2) Abbreviations:

FCR = Fixed charge rate
LF = System load or use factor
Q = Design capacity of system, 10^9 Btu/hr

(3) This estimate applies to a "near—urban environment."

(4) References

FPC — Federal Power Commission 1970 National Power Survey, Parts 1 and 11, U.S. Govt. Printing Office, Washington, DC, 1971.

IGT — D. P. Gregory and J. Wurm, "A Hydrogen—Energy System," 7th IECEC, San Diego, Calif., Sept. 25—29, 1972.

ERC — Electric Utilities Industry Research and Development Goals Through the Year 2000, June 1971.

ISPRA — G. Beghi *et al.*, "Transport of Natural Gas and Hydrogen in Pipelines," Internal Report — EURATOM ISPRA — 1550, May 1972.

GAS ENGR. HANDBOOK, 1965, The Industrial Press, N. Y.

EST. — B. Foster, "Projected Costs of Alternate Sources of Gas," IGT, IIT, Chicago, 1970.

H PANEL — "Hydrogen and Other Synthetic Fuels," Ref. (3), Sept. 1972.

FUTURES Gp — E. Fein, "A Hydrogen Based Energy Economy," The Futures Group, Conn., Report 69—08—10, October 1972.

DILLARD — J. K. Dillard, "Transmission Above 700 kV Hits Economic Roadblock," Electric Light and Power, 43, pp. 44—49, February 1965.

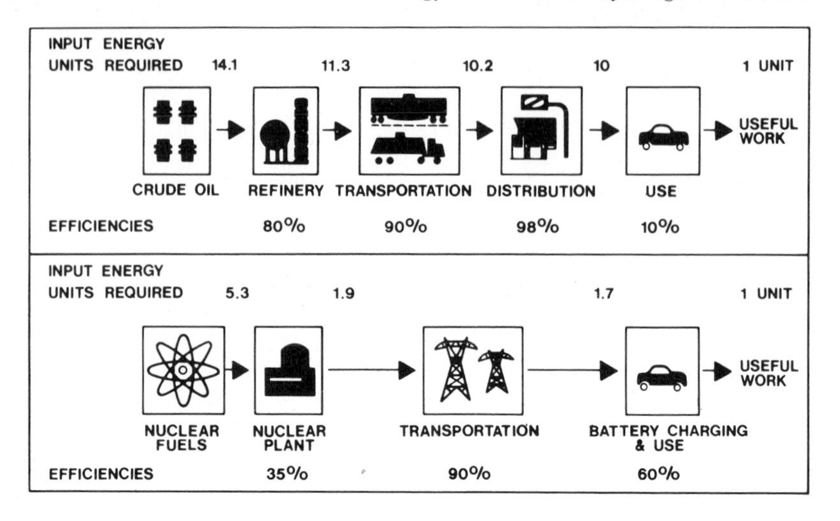

Fig. 17.5 Comparison of electric and gasoline-powered cars.[13]

High energy density batteries have excellent prospects, they await research funding in an environment amicable to the success of the research. By 2000, all U.S. cars could be run electrochemically or by hydrogen, from excess generating capacity available during the night.[14]

An option which to some extent compensates the poorer performance of the electric car is a fuel cell-battery hybrid. The fuel cell has excellent energy per unit weight characteristics and could be run on liquid hydrogen. The battery could be a cheap one (e.g. iron-air), and of low capacity, because it would be used in accelerative periods only. It would receive its recharge from the fuel cell.[15]

Rail transport: This is four times as energy-efficient as road transport, and seven times more energy-efficient than air transport.[13] Therefore, in the future, an increasing amount of use will be made of rail. The pros and cons between hydrogen and electrochemical drive in railway trains is less clear. British Rail has decided to go to an electrochemical drive (sodium sulphur) for branch lines, whilst taking electric pick-up drive (same locomotives) for main-lines.[16]

Thus, the electrochemical solution seems attractive for railways, but a drop in the cost of stored hydrogen towards $3.00 per MBtu (see analysis of its costs in Chapter 16) could change the prospect.

Shipping: There will be in the future an increasing amount of sea transportation because of its efficiency in terms of weight $\times$ distance $\times$ $(\text{cost})^{-1}$. There is a case for fuel cells here, i.e. hydrogen fuel, for there is no problem with regards to power per unit weight on a ship. Batteries would have to be too big for long journeys.

Residential energy: The energy used at present in U.S. residences is shown in Fig. 17.6.[13]

An important prospect in space-heating in houses is the universal introduction of the heat pump[17], to replace electric resistance space heating. The heat pump is a machine attempting to refrigerate the air outside

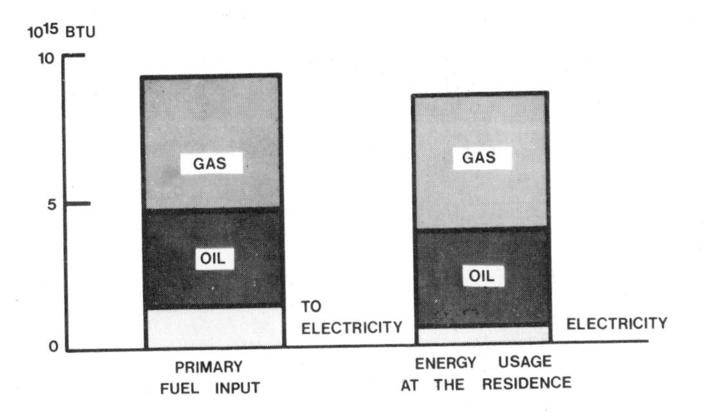

Fig. 17.6 1968 U.S. residential energy use.[13]

the house, and making a poor job of it. The inefficiency of the refrigeration of the external air shows up as efficiency in respect to heat produced by the refrigerator which is used to heat the house. Using gas or oil for heating, one obtains 50% efficiency. Using a heat pump, about 200% efficiency, the inefficiency of the refrigeration turns up as a great efficiency in converting the energy which works the refrigeration cycle into heat (which comes from that in the colder external air). Thus, electricity driving heat pumps would provide household energy satisfactorily, but even the heat pump, with its electrical drive, would have no advantage compared with the total energy concept of a hydrogen-driven fuel cell in each house. The cell produces the household electricity, and the waste heat of this electricity production is used to heat the house. Compared with a 200% efficient heat pump, the price of heating in a total energy concept does not have to include a component for the fuel, which has been paid for in the cost of the electricity.

Industrial sector: For most thermal processes in the present 'thermal' economy, there is an equivalent process which is electrochemical, and could be developed into a segment of an Electrochemical Economy. Some examples are given in Tables 17.4[18] and 17.5.

Hydrogen could substitute for processes in industry now fueled by natural gas (Chapter 14). Gas turbines could be run on hydrogen. Many processes could be run electrochemically without pollution; the decision would be a detailed one in each industry and product, with comparisons between various trade-offs. It depends much upon the degree and type of funding for research on electrochemical technology.

Safety: Here, all the advantages are with the Electrochemical Economy, compared with that of hydrogen.

Materials: The difficulties are not severe with either. The possible embrittlement difficulties of hydrogen-carrying pipes have to be traded off against the material exhaustion of lead for batteries (but iron-air batteries are possible).

TABLE 17.4[18]

EXAMPLES OF THE ALTERNATIVE THERMAL AND ELECTROCHEMICAL PATHS CHEMICAL HAPPENINGS

Phenomenon or process	Thermal	Electrochemical
The determination of free-energy changes and equilibrium constants in chemical reactions	Determine equilibrium constant and use $\Delta G^\circ = -RT\ ln\ K$	Determine thermodynamic cell potential and use $\Delta G_o = -nFE$
Synthesis, e.g. water from hydrogen and oyxgen	Occurs heterogeneously, presumably by non-charge transfer collisional processes $H_2 + \frac{1}{2}O_2 \rightarrow H_2O$	Occurs in electrochemical cell by reactions $H_2 \rightarrow 2H^+ + 2e$ $\frac{1}{2}O_2 + 2H^+ + 2e \rightarrow H_2O$ $H_2 + \frac{1}{2}O_2 \rightarrow H_2O$
Biochemical digestion	Series enzyme-catalysed chemical reactions	Some enzymatic reactions may act through electro-chemical mechanisms analogous to local cell theory of corrosion
Many so-called chemical reactions, e.g. chemical synthesis of Ti	$TiCl_4 + 2Mg \rightarrow Ti + 2MgCl_2$ (apparently a thermal collisional reaction)	$2Mg \rightarrow 2Mg^{++} + 4e$ $Ti^{4+} + 4e \rightarrow Ti$ $4Cl^- \rightarrow 4Cl^-$
Production of electrical energy	$H_2 + \frac{1}{2}O_2$ explodes, produces heat, expands gas, causes piston to move, and drives generator	H_2 and O_2 ionize on electrodes, as above in this column, and produce current
Storage of electrical energy	Electricity pumps water to a height and allows it to fall on demand to drive generator	Allow to cause some electrochemical change, e.g. $Cd^{++} + 2e \rightarrow Cd$, $Ni^{++} \rightarrow Ni^{4+} + 2e$, which will be reversed on demand
Syntheses of inorganic and organic material, e.g. Al, adiponitrile	$2Al_2O_3 + 3C \rightarrow 3CO_2 + 6Al$; Tetrahydrofuran → 1,4-dichlorobutane → adiponitrile	$Al^{3}+ + 3e \rightarrow Al$ $H_2C{=}HC{-}CN \xrightarrow[\text{coupling}]{\text{cathodic}} NC{-}(CH_2)_4{-}CN$
Spreading of cracks through metal	Amount of stress at the apex of crack per unit area is so high that crack is propagated into metal bulk	Bottom of crack dissolves anodically, obtaining current from local cell formed with surface (which is an electron donor, probably to O_2 from air)

TABLE 17.5

GROUPS OF INDUSTRIAL PROCESSES: CHEMICAL, ELECTROCHEMICAL AND HYDROGEN ORIENTED

st of energy. alue of the product[17]	Product	Chemical	Electrochemical	Hydrogen
0.3	Steel	Classical blast furnace: carbon reduction of oxides of iron.	Reduction of iron in solution to powdered iron. Zero pollution[29].	Direct reduction of iron oxide by hydrogen a known process for iron (see Chapter 14). Zero pollution.
0.4	Aluminum	$2Al_2O_3 + 3C \rightarrow 3CO_2 + 6Al$	Electrochemical deposition from Al_2O_3 in sodium fluoride. Cheaper processes plentifully available, e.g., $AlCl_3$ in KCl-LiCl, etc.	Reduction Al_2O_3, T > 2000°C. High yield per unit volume of plant. But not yet researched.
0.05	Copper	Burning of Cu_2S in air and electrochemical refining.	Several electrochemical processes exist for copper from molten salts.	Direct chemical reduction of sulphide followed by thermal decomposition of H_2S.
0.1	Magnesium	$2MgO + C \rightarrow CO_2 + 2Mg$ (outmoded).	Electrochemical reduction from molten salts.	Hydrogen reduction might be advantage in respect to space.
0.3	Glass	At present chemical.	Electrodiffusion of ions into glass gives increased toughness.	Research?
0.04	Organics	Chemical processes are the normal ones.	Some processes electrochemical, e.g. electrochemical polymerisation. Acrylonitrile to its dimer.	Many hydrogeneration processes known.
0.3	Paper	Present technology.	Research?	Research?
0.2	Inorganic chemicals	Many made by chemical methods, but pollution often the problem.	Many made with electrochemical methods, reduced pollution. Fuel cell mode of manufacturing compounds with by-product electricity little developed.	Ammonia.
0.5	Cement	All processes chemical.	Research?	Research?
0.1	Oils, fat.	Present technology uses hydrogen.	Research?	Present technology hydrogen-oriented.

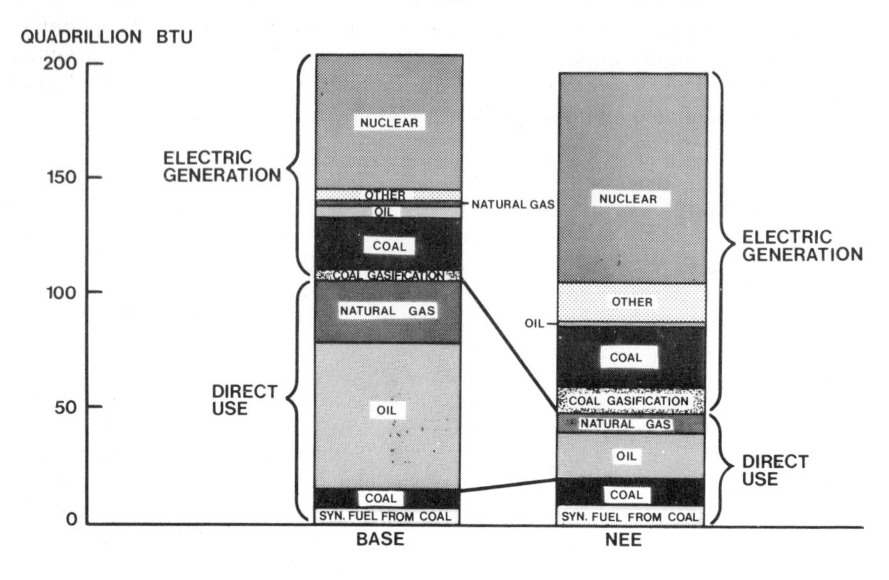

Fig. 17.7 Total energy economy – year 2000.[13]

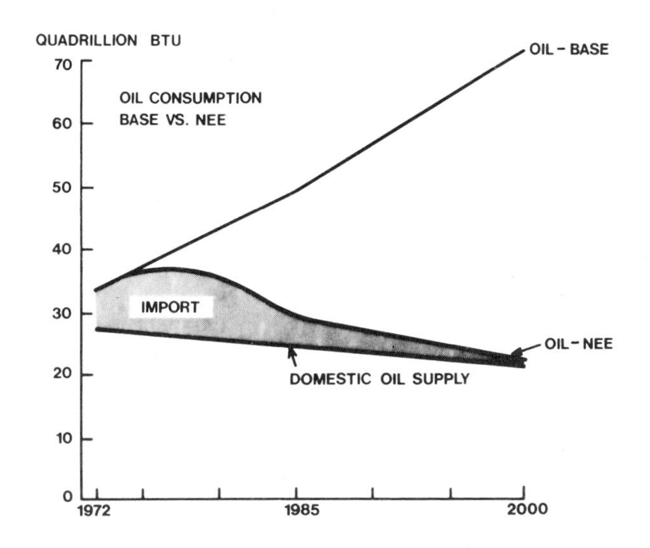

Fig. 17.8 Predicted consumption *versus* domestic U.S. oil supply.[13]

THE ADVANTAGES OF A NUCLEAR-ELECTROCHEMICAL ECONOMY

Compared with the Fossil Fuel Economy, there are numerous advantages of a Nuclear-Electrical Economy, e.g. see Fig. 17.7[13] and 17.8.[13] Compared with a Nuclear-Hydrogen Economy, however, the advantages of a Nuclear-Electrochemical Economy are less marked and may become negative. The distinction will depend on the distance between user and source of energy and the existence of situations where hydrogen may be essential, e.g. when storage is important and when the weight of the fuel is paramount, as in air transportation. The advantages of easily available, cheap hydrogen in technological chemistry and in metallurgy must be weighed.

A NUCLEAR-ALUMINUM ECONOMY

That an energy system could be based upon aluminum, rather than hydrogen, was proposed by Hyman[19], who founded his suggestions on the advocacy of aluminum energy systems put up by Zaromb.[20–22] The advantage of aluminum compared with hydrogen is in storage – it would have a lesser weight per unit of energy stored, because hydrogen has to be housed in tanks. Aluminum as a storage medium may be superior to others. There are practical routes for converting aluminum back into electricity. Aluminum is a more concentrated source of energy than hydrogen in terms of energy per unit volume (Table 17.6[19]). Hydrogen yields a greater energy per gram, but the difficulty is in the container and its weight. Beryllium and lithium have interesting properties, but are scarce. Aluminum, however, is widely available, and the possibility of its extraction from clay means that it will be easily available, and because it is one of the few metals for which steep price rises are not to be expected within the century. Magnesium is inferior in important respects to aluminum and sodium is harder to handle. Thus aluminum has many attractions as a storage medium.

TABLE 17.6[19]

FREE ENERGY OF SOME POSSIBLE FUEL OXIDATIONS AT 25°C

Reaction	K joules/mole	fuel k joules/g	non-volatile oxide	fuel k joules/cc	non-volatile oxide
$H_2 + \frac{1}{2}O_2 \rightarrow H_2O$	237.25	117.69		8.24*	
$C + O_2 \rightarrow CO_2$	394.48	32.87		64	
$CH_4 + 2O_2 \rightarrow CO_2 + 2H_2O$	818.17	51.01		21.17*	
$CH_8H_{18} + 25/2O_2 \rightarrow 8CO_2 + 9H_2O$	5308.38	46.47		32.64	
$2Li + O_2 \rightarrow Li_2O$	559.97	40.35	18.74	21.55	37.72
$2Na + O_2 \rightarrow Na_2O$	376.65	8.19	6.08	7.94	13.8
$Be + \frac{1}{2}O_2 \rightarrow BeO$	581.71	64.55	23.26	119.41	70.0
$Mg + \frac{1}{2}O_2 \rightarrow MgO$	569.70	23.43	14.13	40.77	50.6
$Zn + \frac{1}{2}O_2 \rightarrow ZnO$	318.27	4.87	3.91	34.76	21.9
$2Al + 3/2O_2 \rightarrow Al_2O_3$	1576.78	29.22	15.46	78.96	61.3

* As a liquid at cryogenic temperatures.

Recovery of energy depends on electrochemical cells working with aluminum.[20] According to Hyman[19] and to Zaromb[20-22], aluminum-powered vehicles may be competitive with the hydrogen-fueled vehicle.

One aspect of Aluminum Economy is that a more economic method of making aluminum than the present Hall-Herriault process may have been discovered. This is the method developed by Toth.[19] It could mean a reduction in price by 33%.

Thus, the advantages of aluminum is safety, light-weight (no container needed), and the clean properties of an entirely Electrochemical Economy. If the Nuclear-Electrochemical Economy develops rather than the Nuclear or Solar-Hydrogen Economy, aluminum may become an important part within it.

The aluminum alternative is not to be taken as a primary competitor of a Hydrogen Economy. Hydrogen does not feel competition from aluminum in respect to transmission, transduction (that of aluminum could only be electrochemical), or in chemistry and metallurgy. Pollution by Al_2O_3 dust could become a problem. Rather, if as a result of other pressures, nuclear plants near to towns become acceptable (perhaps underground), then storage in Al, and Al as a fuel for fuel cells in cars, should be considered.

TYPES OF ECONOMIES

Some of these are shown in Tables 17.7 and 17.8. It is likely that a mix of these energy bases will drive (and determine) economies of the various countries as a function of time. The evolution away from the fossil fuels and towards solar sources and perhaps fusion is certain. That hydrogen will be extensively a part of the energy system seems at present likely (with the reservation concerning the absence of cheap room-temperature super-conductors). The possible development is discussed in Chapter 18.

ON NEGATIVE ASPECTS OF HYDROGEN AS A MEDIUM OF ENERGY

Although acceptance of, and even enthusiasm for, the proposition that hydrogen should be the general medium of energy, has been widely shown by engineers and scientists, particularly those in fields such as energy conversion, aeronautics, pollution control, the gas industry, the atomic energy industry, and etc., there are those – mainly from the oil industry – who are negative to the concept. It is essential to examine the weight and validity of the criticisms.

Before they are listed, however, it is desirable to make several statements:

(1) Most of the considerations associated with a Hydrogen Economy depend on the time frame in which it is purported to play an important role in Energy Technology. Those who advocate the change to hydrogen view the present Fossil Fuel Economy as in its terminal stage: the time scale for the end of fossil fuels is 'near' (by which is meant short compared with the time needed to research and *build* a new energy system), and so they do not often consider a competition between a Hydrogen Economy and one based on oil or natural gas. Atomic and solar sources are those for which hydrogen is to be a medium. The cost estimates which are made are for projections, and these are projections both in the sense of time, and, often, technology.

On the other hand, those who criticise a Hydrogen Economy, usually see it as a direct and immediate competitor to natural gas and oil. For them, the time scale is 'now', the ending of the useful price range for fossil fuels is 'far off' (by which they mean beyond their career times), and, if coal is included, 'hundreds of years' away (compare the material of Chapters 3 and 4).

Obviously, the merits and defects of hydrogen look different if seen from this short-term point of view. Thus, if in a thought experiment, one changes over to a Hydrogen-run Economy, in 1970s technology and costs, then the main gains would be in the reduction of air pollution and the giving to advanced aircraft a greater payload. The first of these is a very great advantage, but it would be balanced by many disadvantages, above all the disruption of the present system, and (if obtained *via* electrolysis or nuclear heat) the greater cost of the fuel. The only supporters of the introduction of a Hydrogen Economy 'today' (were this conceivable) would be environmentalists, city planners, ecologists, etc. Economics would win.

TABLE 17.7

ENERGY BASES TO THE ECONOMY
1980 – 2000 – 2040

Energy-Base	Comment
'Synthetic natural gas', coal based	Many attractions and perhaps cheapest. But would not lead to coupling with energy bases which will inevitably develop. Amortisation? Rate of availability of coal? Pollution?
Methanol from coal	Attraction is liquid fuel, lesser pollution, fuels fuel cells. But price significantly greater than methane or finally hydrogen. No coupling with future energy bases. Temporary palliative.
Nuclear-electric	Already in being to small extent. But could only provide electricity. Pollution? Wastes?
Nuclear-electrochemical	Could couple with industry and transportation. Price as function of size? Thermal pollution? Distance source from sink? Air pollution? Wastes? Price if clean?
Nuclear-hydrogen	Capital intensive nature of nuclear power means storage in hydrogen. Long distance (pollution) source to sink needs hydrogen linkage. Suitable for industry for transport. Wastes? Price if clean?
Solar-electric	As residential energy source, seems very likely for countries latitude 30N to 30S. As main source, ecologically ideal. Cost prospective now favourable. But for countries outside latitude 30N to 30S problem is transmission energy from regions high insolation. Power relay satellites?
Solar-hydrogen	Prospects look economic for sources in 30N to 30S up to sinks of 4000 miles distant. Ecologically excellent, though would require more resources than solar-electrical.

TABLE 17.8

ENERGY BASE TO ECONOMY
Post 2040

Energy-base	Summarised Comment
Nuclear(fusion)-electrochemical	Reduced cost of electric power and reduction of pollutional hazards could mean viability of plants near cities. Feasibility of fusion process?
Nuclear(fusion)-hydrogen	Very large and therefore economic on sea or remote areas because of thermal pollution. Ecologically acceptable. Energy per capita to 20 kW: necessary for recycling in steady state world. Feasibility of fusion process?
Solar(land based)-hydrogen	North Africa, Saudi-Arabia, Carribean, Australian, tropical seas as source areas. Source to sink always < 4,000 miles.
Solar(space based)-electrochemical	Sufficient decrease cost of lift to space will make orbiting collectors possible, microwave to earth collector areas < 100 miles from cities. Electricity used directly; run transportation and industry electrochemically.
Solar(space based)-hydrogen	Space-based solar collectors very capital intensive, 'night' functioning. Hence, store of energy needed, best is H_2.

The reason for considering the Hydrogen Economy urgently *now* is that the time for a change to it is long compared with the time at which oil and gas will exhaust (15-30 years). And, as far as probable synthetics from coal are concerned, their lifetime will be such – and pollutional difficulties such – that conversion 'as soon as possible' to hydrogen becomes a rational goal.

(2) Another giant dividing line – not quite the same one as that discussed in (1) – is whether consideration should be given to coupling with nuclear and (the potential) solar sources, or not. If they are examined closely, and the time frame is 'now', *then* the claims of hydrogen are transformed. Then, indeed, even with present technology, they look very good and even essential. The pros and cons are laid out in this chapter. The nearest competitor is methanol. But methanol would have to come from hydrogen.

(3) Much depends on the extent to which the whole picture and the long-term are considered. If these are considered, hydrogen becomes a very likely medium with electricity the alternative, dependent for its competitive power on the development of electrochemical technology. But it is quite possible to take individual points – important points – in which hydrogen offers a disadvantage compared with other possibilities. If these points alone are examined, within the present time frame (and particularly if there is an appeal to 'commonsense' and a few emotive descriptions*), it is possible for an anti-hydrogen case to be presented, using facts and without making any false statements.

* For example, L. W. Russum[23] of the Amoco Oil Company, describes hydrogen as a 'fearsomely expensive, last-resort sort of fuel to fall back on only when fossil fuel supplies are exhausted'. The same author refers to Pratt and Whitney's diffused power concept, now in practical test in houses using natural gas – as a 'fanciful suggestion'.

The main criticisms of the use of hydrogen are as follows:

(1) *Making hydrogen is too expensive*
This objection is entirely true in 1974. Natural gas in the U.S.A. is \$0.44 per MBtu (a controlled artificially low price). If one takes present technology, hydrogen would be over \$5.00 per MBtu. The 'fearsomely expensive' fuel of Russum becomes clearly visible.

However, and somewhat obviously, the considerations which should be made are those of the situation at least ten, perhaps twenty and most probably thirty and forty years (and more), ahead. The 1974 price of Mid-East oil was \$11 per barrel, and the equivalent ex-refinery cost of gasoline would be in the region of \$4 to \$5 per MBtu. In view of the price-availability relation for oil products, it is manifest that hydrogen will become cheaper than petroleum within years, not decades (see Chapter 15). The price of hydrogen in constant dollars will decrease due to technological advances in its production (Chapter 9) and these will tend to compensate inflational rises in coal and electricity prices. The price of coal and electricity will increase at a slower rate than those of oil. Nuclear and solar power will bring cheap electricity and hydrogen when breeders can work in low-use regions of the diurnal cycle; or if one of the cheaper possibilities for solar power now only systems-analysed (see Chapter 7) becomes proven technology.

(2) *'It is better to produce methane from coal, not hydrogen'*
This is a true criticism and, whilst CH_4 is cheaper at the arrival point than H_2 (as may be the case whilst both come from coal), hydrogen will not become the general fuel, although it is likely to be used for special purposes, such as aircraft. But (cf. the time scale possibilities seen in Chapter 18), with a beginning of development in this decade, a Hydrogen Economy would still be in practice in experimental cities in 2005. Cheap nuclear power is expected by 2020. Oil and natural gas will long be gone as general fuels. To propose CH_4 as a rival to H_2 results from thinking in the wrong time scale.

(3) *The double Carnot difficulty*
Whilst fossil fuels can be taken out of the ground (ready made) and used without serious pollutive difficulty, this objection is to be weighed very seriously in automotive and similar applications. It does not apply to many of the proposed future methods of making hydrogen (Chapter 9). However, the criticism is again in the wrong time scale. Double Carnot simply ends up as a price per ton mile in transportation. The correct question is: Would the price per ton mile using hydrogen be less than that using methanol, or using electricity; and what are the environmental relative weights? These questions are considered in Chapters 15, 16 and 17.

(4) *Energy can be transmitted over long distances more cheaply in hydrogen than in wires. But the electricity at the other end will be more expensive*
Whether this criticism is true depends on the distance from source to sink; and the efficiency of the reconverter fuel cells. (To a smaller extent, it depends on the transmission voltage.) Up to 800 miles for

700 kilovolts transmission, with a fuel cell efficiency of 60%, the criticism is true. Fuel cell efficiencies for hydrogen as a fuel up to 66% seem likely to be attained within a decade and perhaps up to 75% within two decades. But the most important variable is the distance. Transmission lines of several thousand miles are to be expected, particularly with solar sources.

In any case, the criticism is weak because it implicitly assumes that most of the hydrogen will be converted again to electricity. This is unlikely. The ratio of use of hydrogen fuel as electricity at the site of use, or as a fuel, is difficult to assess. At present, electricity is about 10 to 20% of the whole. If such a ratio remains, the predominant position in respect to the conditions for the transmission of energy is as given in Fig. 17.4[7].

(5) *'Hydrogen is terribly dangerous'*
Hydrogen is a more dangerous fuel than natural gas. The situation is weighed in Chapter 11.

(6) *Difficulties of introducing hydrogen as a fuel in transportation would be very great*
This criticism does not seem true, even at this time. The situation has been presented in Chapter 15. In this instance, there does seem a considerable case for a changeover in fuel to hydrogen before economics force it. Thus, the automotive industry can either obey government regulations and produce an expensive car with poor performance; have the regulations changed and build a car with an acceptable performance which pollutes with fuel costs which are already beginning to limit its use; or turn to hydrogen from coal and be able to build high performance cars which don't pollute and have running costs which need not rise with the cost of oil. The disadvantages are the need for a larger fuel tank and precautions to avoid venting in an improperly ventilated closed space. The reader will make his own conclusions, and of course they will be time dependent because pollution and petrol will continue to increase in degree and cost respectively.

SOME NEAR-TERM (PRE-2000) RESEARCH GOALS

There are many changes in stress in research funding which should arise from the present perceptions. Some radical changes ought to be made, too; for example, the vast energies and expenditure in trying to keep the doomed oil-based economy alive till the last decade (e.g. a transfer of support should be made away from trying to make gasoline engines clean to developing them to run on hydrogen).

A few of the R&D goals are:

(1) New methods for the rapid mining (or gasification in situ) of coal.

(2) The engineering of breeded reactors to function economically, though built underground, and with permanently satisfactory waste disposal.[24]

(3) Reliable predictions of the feedback effects associated with increase in the temperature caused by CO_2 and which so much affect the possible time during which one could continue to use fossil fuels.[25]

A few of the R&D goals are:

(1) New methods for the rapid mining (or gasification in situ) of coal.
(2) The engineering of breeded reactors to function economically, though built underground, and with permanently satisfactory waste disposal.[24]
(3) Reliable predictions of the feedback effects associated with increase in the temperature caused by CO_2 and which so much affect the possible time during which one could continue to use fossil fuels.[25]
(4) New methods for the production of hydrogen: in particular, photo-oriented methods and thermal-assisted electrolysis.
(5) Cheaper (i.e., longer-lasting) fuel cells.[12]
(6) The development of electrochemical technology, particularly electrogenerative reactors.[26]
(7) The economic and storm-proof construction of very large ocean-borne platforms.
(8) Economic solar energy collection, particularly the engineering of ocean-thermal gradient collectors and large-scale wind collectors for use in high wind-belt regions.
(9) Reduction of costs in hydrogen liquefaction and storage.
(10) Hydrogen-driven vehicles; fuel cell-battery hybrid-driven vehicles.[12, 15]
(11) Recycling processes[27], particularly those using electricity.
(12) The electrochemical route for food from atmospheric CO_2 and air N_2.[28]

REFERENCES

[1] L. Lessing, *Fortune*, p. 138 (November 1972).
[2] R. Stroup, Westinghouse Company, priv. comm., 1974.
[3] cf. General Survey, *Science*, **184**, 247 (1974).
[4] P.N. Thompson, 'Appraisal of Underground Gasification — Suggested Procedures', U.K. National Coal Board, Operational Research Executive, May 1974.
[5] P.N. Thompson, priv. comm., 10 September 1974.
[6] M. Steinberg, priv. comm., 1974.
[7] J.W. Michel, 'Hydrogen and Exotic Fuels', ORNL-TM-4461, Oak Ridge National Laboratory, Oak Ridge, Tennessee, June 1973.
[8] D.L. Klass, *Chem. Tech.*, p. 161 (March 1974).
[9] Table constructed with the aid of information provided by C. Heath, 30 May 1974.
[10] H. Linden, *Weekly Energy Report*, 17 June 1974.
[11] R.K. Pefley, H.G. Adelman & M.C. McCormack, 'Methanol-Gasoline Blends – University Viewpoint', paper presented at the Methanol Conference, Henniker, New Hampshire, June 1974.
[12] J.O'M. Bockris & S. Srinivasan, *Fuel Cells: Their Electrochemistry*, McGraw-Hill, New York, 1970.
[13] P.N. Ross, 'Development of the Nuclear-Electric Energy Economy', Westinghouse Electric Corporation, East Pittsburg, Pennsylvania.
[14] E.H. Hietbrink, J. McBreen, S.M. Selis, S.B. Tricklebank & R.R. Witherspoon, in *The Electrochemistry of Cleaner Environments*, ed. J.O'M. Bockris, Plenum Press, New York, 1972.
[15] K. Kordesch, in *Modern Aspects of Electrochemistry*, Vol. 10, edd. J.O'M. Bockris & B.E. Conway, Plenum Press, New York, 1974.
[16] J.L. Sudworth & M. Hames, *Power Sources*, **3**, 227 (1973).
[17] 'Technology of Efficient Energy Utilisation', Report of a NATO Science Committee Conference, Les Arcs, France, 8-12 October 1973.

[18] J.O'M. Bockris & A.K.N. Reddy, *Modern Electrochemistry*, Rosetta Edition, Plenum Press, New York, 1974.

[19] H. Hyman, paper presented at the Third World Congress of Engineers and Architects, Tel Aviv, Israel, 17-21 December 1973.

[20] S. Zaromb, 'Aluminum Fuel Cell for Electric Vehicles', Symposium on Power Systems for Electric Vehicles, Columbia University, New York, April 1967; Public Health Service Publication No. 999-AP-37, pp. 255-267.

[21] S. Zaromb, paper presented at the 4th Intersociety Energy Conversion Engineering Conference, American Institute of Chemical Engineers, New York, 1969, pp. 904-910.

[22] S. Zaromb, *J. Electrochem. Soc.*, **109**, 1125-1130 (1962).

[23] L.W. Russum, 'Engineering Perspective of a Hydrogen Economy', paper presented at the National Meeting of the American Chemical Society, Division of Petroleum Chemistry, Atlantic City, New Jersey, 9 September 1974.

[24] E.J. Sternglass, *Low Intensity Radiation*, Earth Island Press, 1973.

[25] G.N. Plass, in *The Electrochemistry of Cleaner Environments*, ed. J.O'M. Bockris, Plenum Press, New York, 1972.

[26] J.O'M. Bockris, ed., *The Electrochemistry of Cleaner Environments*, Plenum Press, New York, 1972.

[27] A.T. Kuhn, in *The Electrochemistry of Cleaner Environments*, ed. J.O'M. Bockris, Plenum Press, New York, 1972.

[28] H.G. Schlegel & R.M. Lafferty, in *Advances in Biochemical Engineering*, Vol. 1, Springer-Verlag, Germany, 1971.

[29] J.O'M. Bockris, in *The Electrochemistry of Cleaner Environments*, ed. J.O'M. Bockris, Plenum Press, New York, 1972.

CHAPTER 18

Stages in the Development of a Solar-Hydrogen Economy

INTRODUCTION

IT IS OF INTEREST to suggest some possible steps towards the setting up of some of the Abundant, Clean Energy schemes discussed in this book. These may be expected to be pushed forward not only by government planners, but also by an increasing number of companies who will find opportunity in the energy shortage situation.[1]

DATE OF COMMENCEMENT AND IMPLEMENTATION

Development will be over several decades. If hydrogen is the energy medium in relying on the use of coal as an intermediate energy source, the pollutional problems associated with coal can be reduced; and the advantage arises that the system can be coupled to the emerging nuclear, and the potential solar sources, thus reducing the uncertainties in the lifetime and hence ammortization considerations.

STAGE I

Stage I would be connected with the gasification of coal. Coal gasification plants should be built for the synthane process of hydrogen from coal in a coal-rich region (Joliet, Illinois, has been suggested by Savage *et al.*).[2]

In any experimental beginning, there should be a substantial length of pipeline, e.g. 100 to 200 miles, between the producer plant and the user site. There, one would erect a fuel cell and alternator. CO_2 would be no problem because of the reduced time aspect, and injection of other impurities into the atmosphere could be avoided by appropriate purifying processes not practical to carry on a vehicle.

An economic incentive must be offered. It has been suggested[2] that a 10 to 20 year contract be given to power plants for the purchase of gaseous fuel at a constant price, an attractive condition in an inflational economy.

The contractor who builds the plant has to obtain his amortisation costs, service and maintenance, construction and materials, and then come up with a price which will cover these, and allow sufficient profit to motivate him. If the price of the electricity so attained comes out above the price of the time, then there should be initially, something similar to a Ford Foundation or government subsidy[2] to bring it down to the average price of electricity obtained from the more polluting sources.

If the necessary initial organisational work can be done expeditiously, it would seem reasonable to expect completion of such a first trial by the early '80s.

STAGE II

The second stage would involve the coupling of hydrogen with an off-shore fission plant supplying a smoggy area. The plant could supply electricity only and could be implemented by, say, some 10 years after the first order had gone in, so that by the late '80s, Los Angeles, for instance, could have its first hydrogen from nuclear electricity. Household energy could be supplied to some districts and information on the total energy concept in respect to heating and electricity supply could be gathered. A trial use in transportation could begin with a decade of municipal transports run on hydrogen. If this were successful, permanent smog removal from Los Angeles could be on the way by the 1990s.

STAGE III

This stage would attempt to implement a solar source backing a hydrogen generation plant with the hydrogen fueling local transportation.

It would be suitable to choose a small town, one of some 10,000 people. The town might be a 'government town'. Los Alamos might be a possibility. The solar collecting station could involve the mirror concentrator type of solar collector as discussed by Hildebrandt *et al.*.[3] Aerogenerators in the advantageous mountain regions might be used to add to the supply. Fresh water would be needed. It might have to be brought a significant distance from one of New Mexico's lakes.

If this project could be under way building by the early '80s, completion in 5 years would be a reasonable goal. A solar collecting station, with concentration onto boilers (Chapter 6 and 7) should be an easier piece of engineering than that of building a breeder atomic reactor.

STAGE IV

In an advanced trial stage, one would run at first one, and then all, of the Hawaiian Islands on a Hydrogen Economy.

Here, the original source could be the large aerogenerators, suggested by Mullett (Chapter 5): the western Hawaiian Islands receive substantial winds for some 10 months of the year. A sea solar-power plant could be used in the Hawaiian area. A water-borne nuclear reactor could compete, experimentally, with the other two sources.

In the Hawaiian stage, all energy should be *via* the medium of hydrogen: the first full test of a Hydrogen Economy. There is sufficient industry in the State for the experiment to be representative. Subsidies would have to be given for trial conversions and some new hydrogen-oriented technology developed.

The first major trial of an Abundant, Clean Energy Concept would be attractive to Hawaii, with its image as a tourist centre. The attractivity would increase as the pollution of the rest of the U.S. increases. As by 2000 much more coal will be being burned in the U.S., escape to a clean Hydrogen Economy on the Islands would be particularly attractive.

If the project could be begun by 1985, one might expect a significant degree of conversion before 1995 and a largely converted Economy around 2000.

These plans would give sufficient data for an assessment of the Hydrogen Economy to be made, and the degree to which it could be coupled to the solar and atomic sources. At the same time, if massive gasification of coal becomes practised (cf. Chapter 4), this could already have given rise to a substantial use of hydrogen. By the early '80s, it will be time to start passing laws – e.g. all equipment built after a certain date should be built so that it could be converted to operate in a Hydrogen Economy.

Correspondingly, final safety codes would be formulated and disseminated.

There could be one last main large scale experiment before the degree to which hydrogen should be official policy is decided – this would be the running of a large city on the mainland on hydrogen. An opportunity could be found in Miami, excellent for its proximity to the Gulf Stream for the use of sea-solar plants and a floating nuclear station. The aim could be the conversion of the city to these sources by 2010.

HYDROGEN AS A FUEL FOR TAKE-OFF AND LANDING OF AIRCRAFT

Present jet aircraft could be fitted as of now with wing tanks which would allow planes to take off and land on hydrogen, thus eliminating pollution near air-fields.

THE FIRST INDUSTRIAL PROCESSES

The first chemical process associated with a Hydrogen Economy would be an ammonia plant which could be built near a block of steel mills, so that the oxygen produced could be used. The sale of the oxygen product would help to reduce early hydrogen costings.

HOUSEHOLDS, AND THE SOLAR-HYDROGEN ECONOMY

The running of solar houses, i.e. houses in which the entire energy load (including electricity) is collected from solar energy on the roof, is exemplified by the *Solar I*, the house designed by Boer at the Institute of Solar Energy, University of Delaware.

This house stores its roof-collected electricity in lead-acid batteries. Cadmium sulphide panels are the origin of the energy. Storage in hydrogen (perhaps in hydrides) would have the advantage that each house could be the source of the energy for an electric or hydrogen fueled vehicle.

TRUCKS AND BUSES

Heavy trucks (with their long journeys and great power needs) present a poor prospect for battery drive, but a good one for stored hydrogen. Fuel costs exceed depreciation in running trucks, so that there would be marked cost lowering. An early start on hydrogen-fuelled buses would be advantage in stressing high conversion efficiency, and this would favour fuel cells. An early start on hydrogen-fuelled buses would be advantageous as a lead to the development of hydrogen-based power plant for transportation.

BUILDING COMPLEXES RUNNING ON HYDROGEN

In city areas in the Northern Hemisphere, piping of hydrogen (rather than roof-top collection of solar energy) would be the path for energy, in a way similar to that commenced on a pilot plant scale with natural-gas driven fuel cells by the Pratt & Whitney Company.

Direct solar energy fueling of houses would be practical between latitudes *c.* 30N and 30S. Consider the Flinders University of South Australia, a campus of 350 acres on high ground overlooking Adelaide at latitude 35°. The kilowatt hours used on this campus in 1973 was 5.8 million, and the cost $150,000. The average hours of sun over Adelaide, South Australia, is 2,500 per year. Hence, the area for solar collectors at 5% efficiency would be about 10^5 sq m – equivalent to a square patch of side about 300 metres.

Heating and cooling would be carried out with heat pumps working with fuel cell derived electricity. The campus could produce the hydrogen fuel for its motor transports.

The capital cost of converting this campus to a Solar-Hydrogen Economy depends principally upon the cost of the photovoltaic panels assumed as the sources to be used. Let this be $500 per kW (see Chapter 7). Only the actual hours of sunlight have been used so that no allowance for diurnal variation is necessary. Allowing a factor of 2 for diffuse light, the figures suggest that a 5 MW source would be necessary. The cost for the solar converters alone would be about $2.5M. It would take about 16 years to break even, *if* the price of electricity remained constant. Obviously, it will rise, and the time at which the investment in the CdS equipment would be paid for in savings may reasonably be estimated as occurring within a decade. Such rough computations neglect several elements, e.g. the fuel cells ($250 per kW), the liquid

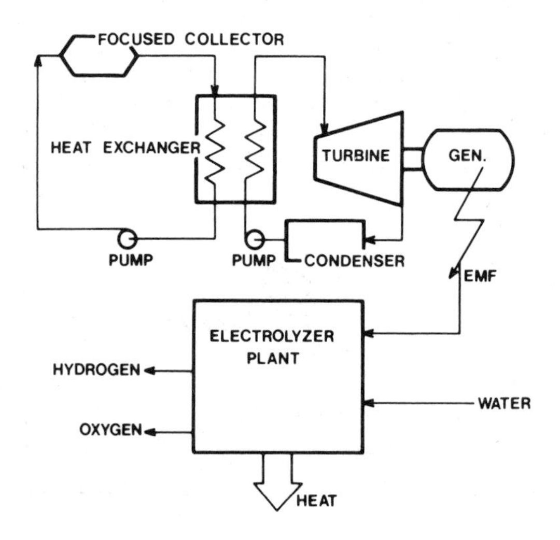

Fig. 18.1 Hydrogen fuel *via* electrolysis of water.[4]

hydrogen storage plant ($150 per kW), and the heat pumps ($100 per kW). They neglect maintenance and insurance costs. The extra costs of $500 per kW, and the ancillary costs, will roughly double the cost estimate made, and the time of breaking even. There is only tenuous evidence (Chapter 7) which would give a 20 year life to the CdS photovoltaics. This extremely rough estimate suggests that land-based solar collectors, at a price about double that now estimated for the panel material, if costed and ordered in 1976 at the estimated price given, would have paid for themselves in two to three decades. The principal uncertainty is in the life of the photovoltaic panels and these might be replacable by silicon, although not as early as 1976 for economic reasons.

Once a scheme such as that suggested had been completed, the Campus's cars could be supplied with locally produced hydrogen fuel.

A question of where the solar collection area would be found arises. Zero land costs have been allowed. The collector would be in three areas on the University's site, two covering land which at present is parkland and the other covering roofs of the laboratory and administration buildings. Sun tracking solar panels and collectors could be laid out on the flat roofs of the buildings. The campus buildings are far enough apart to make successive layers of panels, built upwards, effective without shadow effects.

ECONOMIC COMPARISONS OF TWO SOLAR-HYDROGEN CONCEPTS

McCulloch, Pope and Lee[4] have worked on a cost estimate of two Solar-Hydrogen plants.

System 1 is shown in Fig. 18.1.[4] The collector system for the solar energy would consist of cylindrical parabolic collectors.

Fig. 18.2[4] shows a cross-section of a single parabolic collector. A collector system of this description is efficient at about 400° F. The collectors (the calculations were done for Albuquerque, New Mexico) were tilted in a north-south plane 35° from the horizontal towards the south, and the spacings between collectors were equal to their dimensions to prevent the shadowing of one by another. The system is connected to a Rankine cycle engine. The fluid temperature was assumed to be 400° F, turbine inlet 360°, outlet 180°, and condenser at 100°. Efficiency would be 18%. The turbine output drives an electrolysis vessel, assumed to be 75% efficient.

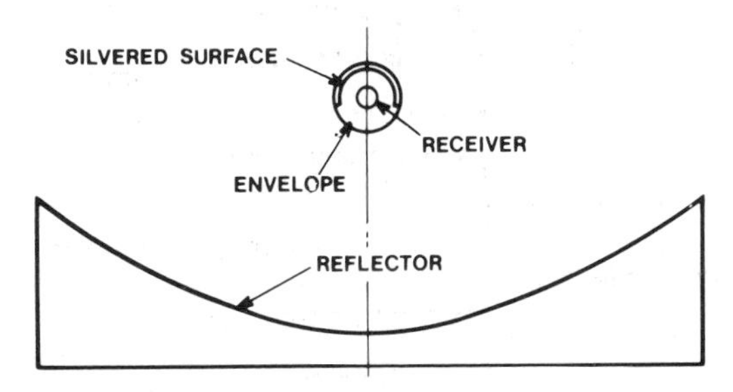

Fig. 18.2 Cross-sectional view of focussed collector.[4]

TABLE 18.1[4]

SUMMARY OF ASSUMED COSTS IN THE COST ESTIMATE OF McCULLOCH, POPE AND LEE

Capital Costs	
Collector array	$1.26 ft^{-2}
Land $0.011/ft^2	
Collectors 1.00/ft^2	
Fluid processing 0.25/ft^2	
Generating plant	$200 kW^{-1} (output)
Electrolyser plant	$75 kw^{-1} (input)
Chemical dissociation plant	$30 ft^{-3}—hr (output)
Operational Costs	
Labour and maintenance	$0.25 yr^{-1}—ft^{-3} —hr

In System II, see Fig. 18.3[4], the collector system was assumed to be the same as that for System I. However, the thermal energy used in a chemical process, e.g. that of de Beni and Marchetti.[5]

In Table 18.1[4], some of the costs assumed are given. The fluid processing equipment costs were taken as $0.25 per square foot of occupied area.

The electrolyser plant costs at $75 per kW is low.[6,7] The operational costs for labour and maintenance were taken at about $1.65 per year per kW capacity.[8] The operational costs for both systems are about $0.25 per year per standard cubic feet hourly capacity. No allowance has been made for sale of the oxygen.

The performances do not indicate a preference for either of the two systems. The electrolysis one is easier to build, and uses established technology. Capital cost figures assumed are shown in Table 18.2.[4] The plant cost for the energy equivalent of 1 kW is about $1,000 and this is about five times that for present conventional plants using fossil fuels. The final production costs are shown in Table 18.3.[4]

The costs of electricity by this process would be about 5 cents per kWh. This is not an encouraging figure, about five times more than the present one. However, the present one involves fossil fuels, and

TABLE 18.2[4]

CAPITAL COSTS FOR SYSTEMS I AND II (McCULLOCH, POPE AND LEE)

	System I	System II
Collector array	$35,000,000	$35,000,000
Generating plant	$10,500,000	
Electrolyser plant	$ 3,900,000	
Chemical dissociation plant		$18,600,000
TOTAL	$49,400,000	$53,600,000

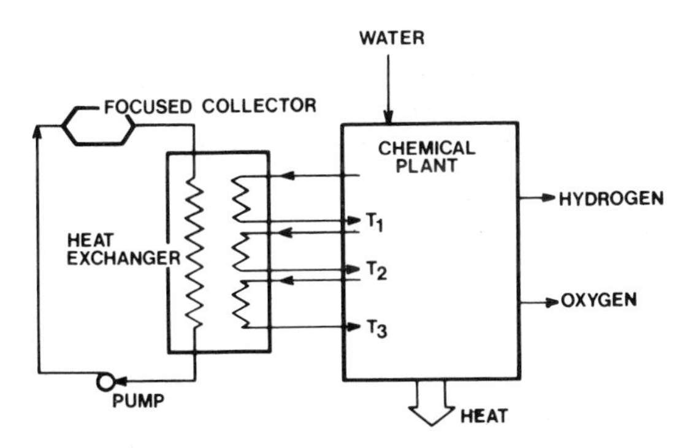

Fig. 18.3 Hydrogen fuel *via* thermochemical reactions.[4]

TABLE 18.3[4]

HYDROGEN PRODUCTION COST (McCULLOCH, POPE AND LEE)

Input data	
Plant cost	\$50,000,000
Capitalisation rate	8%/yr
Operation cost	\$160,000/yr
Consumables cost	
Output	1.02×10^9 ft^3/yr
Results	
Unit cost	\$0.0041/ft^3
ENERGY EQUIVALENT	\$15/10^6 Btu \$0.052/kW–hr

price rises of several hundreds of per cent in the next 2 to 3 decades seem likely. The competition is more nearly that of *breeder reactors* where 1974 cost estimates approach \$1,000/kW (initial electricity costs, 1.2-1.5 cents kWh^{-1}).

The principal cost element in the system is the collector array, i.e. the collectors should receive the largest share of the solar research and development money, when this is oriented to household electricity.

Alternatively, take the Tyco process for the production of semi-conductor Si at \$250 per kW for the actual Si, and \$300 per kW of collector machinery, i.e. \$550(kW)$^{-1}$ for the overall plant, including storage. Let us take this to produce about 0.5kW (latitude, weather) for one-third of the time. Then, for a cost of \$55 per year (10% money), one obtains $365 \times 24 \times \frac{1}{6}$ kWh, i.e. 3.7 cents per kWh at site, or about 4 cents at 1,000 miles distance from the source. Converted back to electricity at 66% efficiency, the electricity would cost 6 cents kWh^{-1}.

Much lower cost estimates come from sea-solar power plants (Chapters

6 and 7). There, the systems analysis cost of electricity at source is 0.6 cents kWh^{-1}, or about 1.1 cents at 1,000 miles.

The costs of wind power electricity or hydrogen are more difficult to establish, because there is a wide spread in estimates of the cost per kW of a wind generator (Chapter 5). In the constant high velocity wind belts, one could make a very rough order calculation by starting with the cost of material for a generator with an average production of 6 MW. The cost of metal (aluminum) for this generator would be some $30,000, i.e. $5 per kW. If we allow an order of magnitude increase for labour and erection costs ($50), and $100 per kW for the electrolysis plant, one gets $150 per kW. This is equivalent (10% money) to about 0.2 cents per kWh at source; about 0.7 cents per kWh at 1,000 miles.

Other estimates of aerogenerator costs are up to five times higher. These would give prices of electricity about half those resulting from the analysis made by McCulloch, Pope and Lee.[4]

These 1,000 mile costs of electricity from three solar schemes can be converted into the cost of hydrogen by use of the equation:

$$\text{Cost of 1 MBtu (cents)} = 229Ec + 40 \tag{9.32}$$

Thus, the costs of hydrogen would be (with a cell voltage of 1.5):

Photo-voltaic solar collectors:	$12.97
Ocean thermal gradients:	4.17
Wind:	2.80 to 5.55

These costs ranges stress the attractions of the newer methods proposed for converting solar energy, rather than those generally supposed to be in the forefront of arrangements for conversion. Although the cost projections for photovoltaics have dropped so precipitously since 1972, they still need a cost lowering of several hundred per cent to break even. It is quite conceivable that photovoltaic silicon at half its present estimated cost could be attained. Still lower costs would soon run up against the costs of energy in making the substance. It may be that the use of photovoltaics will be restricted to the roofs of single family units in the 30N to 30S zones. Much depends upon the actual cost of realised sea-solar and wind energy at source.

REFERENCES

[1] General Survey, *Science*, **184**, 247 (1974).
[2] R.L. Savage, L. Blank, T. Cady, K. Cox, R. Murray & R.D. Williams, edd., 'A Hydrogen Energy Carrier', Vol. II – Systems Analysis, NASA-ASEE, September 1973, p. 149.
[3] A.F. Hildebrandt, G.M. Haas, W.R. Jenkins & J.P. Colaco, *E. & S.*, **53,** No. 7 (July 1972).
[4] W.H. McCulloch, R.B. Pope & D.O. Lee, 'Economic Comparison of Two Solar/Hydrogen Concepts', Sandia Laboratories, October 1973.
[5] G. de Beni & C. Marchetti, *Euro-Spectra*, **IX,** No. 2, 46 (June 1970).
[6] W.E. Heronemus, 'The United States Energy Crisis: Some Proposed Gentle Solutions', presented to a joint meeting of the local sections of ASME and IEEE, West Springfield,

Massachusetts, 12 January 1972, printed in the Congressional Record of a Hearing before the Special Subcommittee on the National Science Foundation of the Senate Committee on Labour and Public Welfare, 4 May 1972.

[7] 'Hydrogen Fuel Use Calls for New Source', *Chemical and Engineering News*, **16**, (3 July 1972).

[8] T. Baumeister & L.S. Marks, edd., *Standard Handbook for Mechanical Engineers*, 7th edn., McGraw-Hill, New York, 1967.

CHAPTER 19

Prospects

POLITICAL CHARACTER OF FIELD

ANY TOPIC WHICH concerns the supply of energy per capita to the population must be political.

The reasons may be spelt out:

(1) The income per capita is linear with the energy per capita as shown in Fig. 3.3, Chapter 3.

(2) The West has got into an era of high inflation. The inflation is associated partly with the rising cost of energy.* Thus, the price of energy could have an influence upon the stability of the Western political system.

THE MULTI-NATIONAL CORPORATIONS

The directors of profit-making organisations are appointed with the primary task of making a profit, and replaced if they do not make one. The profit is measured upon a *yearly* basis. Long-term considerations, even five-year ones, must be neglected by the directing board if they interfere with the maximisation of short term profit. Investment for research in things which would take more than five years to yield a profit is usually declined.

In former times, national governments in Western countries had a greater degree of control over the direction of the primary financial trends in their countries than at present. To some extent, small in Western countries, there was longer range planning, the limited degree of which has added to the present situation, and which has now become a necessary basis to the solutions of the future.

At present the degree of the silent influence in Western countries of multi-national corporations on the economic situation of a country is comparable with the influence of the government. Planning with long-term goals has been reduced and seems likely to continue to diminish.

Thus, the influence of the directors of multi-nationals on the energy situation lacks the long-range planning necessary to meet the continuous rise in the price of energy expected, whilst the sources are oil, natural gas and (to a lesser extent) coal. Their influence must be towards maintaining, beyond the point which broader considerations would dictate, the present technology, in which the capital of their corporations is invested.

---

* A 400% increase in the energy price would devalue a buyer's currency by *c.* 40%.

Conversely, there is an advantageous side to the strong presence of the profit-oriented corporations within so-called democratic states. Their frankly hierarchical organisation brings an efficiency not possible in democratic or socialistic organisations. If it is possible to couple progress towards Abundant, Clean Energy with the effectiveness of competing corporate bodies, the solution to the Energy Problem may be attained within the framework of our present political structures. What seems to be needed are government organisations of the NASA type, in which a heavily funded central administration with long-term plans, effects its aims by giving contracts for the attainment of specific new technologies to profit-making corporations.

VESTED CAPITAL AS A BRAKE TO MAKING THE CHANGES IN TIME

Capital, invested in a certain technology, needs this technology to continue so that the capital can be amortised. Disastrous is the change in technology which brings *unplanned* obsolescence.

This influence would apply in socialist systems, as well as capitalist ones. However, the diffused (and academic) nature of ownership in the socialist ones may make the weight of the difficulty in the path of nimble changes in technology less frought with career disaster for leaders and hence less resisted.

THE CHAIRMANSHIP OF RESEARCH FUNDING COMMITTEES

Increase in living standards is preceded by the spread of new energy-using technology; the new technology depends upon development work; and the development work in new areas depends upon research. The research cannot be carried out *independently* without substantial financial grants from governments, to organisations which are not involved in technologies which would be driven to obsolescence if progress were made by the research funded.

Hence, the man who steers the support of free research in a field to university teams, research institutes, or government organisations, exerts a strong influence upon the direction of technological change and such men are important determiners of technological influences on the standard of life. During the last few decades, most of the *technological* change has originated in the U.S. Thus, those who are appointed chairmen of U.S. research funding committees have an unusual power to accelerate, or reduce, progress in a given direction. One thing is certain: progress in that direction will not occur if there is no research funding. There are no more back-room inventors. In energy development, such chairmen clearly have a long-range influence upon the kind of lives which citizens will live a few decades later.

The necessity here, then, is that the chairmen of committees for decisions for funding must be free of affiliations with organisations invested in older technology. A U.S. Government Committee, considering the funding of research into electrochemical energy storage, is less likely to come to positive recommendations for effective funding if its members are employees of oil or automotive corporations. By such a statement, there is no doubt implied as to the integrity of the individuals, but recognition of the consequences of a certain mode of thinking and associated assumptions of what is acceptable. Thus, such corporations would be exposed to financial adversity, were sufficiently cheap and long-lived high energy density batteries for cars developed: for the technology which represents their capital would become obsolete. There are no second-hand values for an obsolescent technology.

Another matter to be watched, but one more difficult to distinguish without help of independent expert evaluation, is 'weak funding'. This is a sort of 'fobbing off' technique, whereby money may be put into a field at an intensity and for a time too small to give rise to significant progress. After, for instance six years with six men, no communicable progress has been made. Thus, research progress is not linear with funding: below some critical amount, zero progress is likely. Support can then be withdrawn, and the threatening development successfully put off for a number of years. Suggestions about re-funding can be met with the evidence that 'this has been tried'.

The intelligent layman can get an idea of whether support is being seriously (and not merely defensively) given by asking about levels in competing areas. In monetary terms relevant to the 1970s, support of research and development into big nationally important projects (e.g. breeder reactor research and gasification of coal) will run in the U.S.A. at a level of hundreds of millions of dollars per year. There are different degrees of development in fields – a field which has reached pilot plant building level will be about an order of magnitude more expensive in respect to needed research funds than one which is at an earlier level of advancement. However, if gas from coal is being supported in the hundreds of millions of dollars per year, whereas *tens* of millions of dollars are being spent on the development of solar energy (which is also at the pilot plant stage in, say, sea-solar power), it is synthetic natural gas from coal, not hydrogen from solar energy, which will be produced. The reason given will be that the solar energy is 'too expensive' and the reason for that – which may not be stated – is that the necessary R. & D. was not funded at a sufficient level at a sufficiently early time.

The *source* of the support, not only its magnitude, is of great importance. \$100M research support for solar energy from an oil and coal vending corporation, particularly if it purchases a blooming research company 'to develop the invention', has not the same degree of positive significance in respect to commercialisation as the same sum given by a government to an organisation which would profit if the work were *successful*. The development of cheap solar energy would hardly be advantageous to a corporation building giant new coal mines.

THE INEVITABILITY OF LIMITS TO GROWTH

During the last few years there has been an increasing consciousness of a threat of limits to economic growth. Part of this threat is associated with the production of energy, although it may be connected with the lack of recycling of material resources, and particularly of air pollution from fossil fuels. In this book, suggestions have been brought together, which would, if their pursuit resulted in successful technology, make it possible for a ten billion people world to have an energy budget per person equivalent to that enjoyed by Americans of the early 1970s.

Such a state of affairs – very far from achievement at this time – would then lead eventually to another limit. In fact, if 'perfect recycling' and 'Abundant, Clean Energy' were both achieved before, say, 2050, the next limit would perhaps be the heat produced by raising the energy per capita to $>$ 20 kW per person, which would lead to the beginning of a significant increase in world temperatures.

MISUNDERSTANDINGS OF THE CONCEPTS OF A HYDROGEN ECONOMY

Some of the interest shown in the Hydrogen Economy in the early '70s is associated with several misunderstandings. Thus, it is implied that it is proposed to introduce a Hydrogen Economy immediately, in view of its undeniable ecological advantages.

The principal consideration, of course, is the cost of hydrogen fuel *versus* that of other fuels. The time for liquid hydrogen from low-use period electric power and modern electrolysers, to become cheaper than gasoline, seems to be a shorter one than it would take to build such an economy. It depends upon the development of emerging electrolysis technology and the practical availability at rational rates of the low-use period electric power. The cost of liquid hydrogen from coal would seem to be competitive with that of other liquid fuels. Hydrogen from nuclear and solar sources, *if no new technological advances in hydrogen production occur*, will have to wait till gassoline is about 50% above its 1974 price. In the relevant time scale at which a Hydrogen Economy could be built, there will be no competition from oil or natural gas.

Competition is to be seen from an electrochemical economy, or, for a few decades, one based on CH_4 or CH_3OH from coal. The pros and cons have been presented in Chapter 16.

One misunderstanding is that comparisons of present prices of fossil fuels and price estimates for hydrogen are relevant. Such comparisons turn out to be not unfavourable for hydrogen. What is much more important is that hydrogen will be cheaper than present liquid or gaseous fuels at the pertinent time, in the mid '80s.

Another misunderstanding of the Hydrogen Economy is to see it independently of the Energy Source. Hydrogen itself is not a fuel: it is a *medium* of energy derived from coal, nuclear or solar sources. Hydrogen, as a medium in which to transport energy over long distances, would make available to the great energy-using sites of the world the solar energy from the highly insolated deserts, and from distantly sited very large breeders.

A SOLAR-HYDROGEN ECONOMY AND A LEVELLING OF LIVING STANDARDS

The determining factors for the living standards of the populace in earlier times contained elements concerning organisation, but were much dependent on the energy and initiative (indeed, endurance and physical strength) of the individual members of the society. The presence of fuel and raw materials to couple with this energy and initiative gave rise to the prosperous country, e.g. 19th century England.

This classical picture of the origins of affluence became less true as machines, and then automated factories, replaced individual workers.

Such technology can be bought, but the prosperous country has still to have the cheap indigenous fuel supply. Such was the mid-20th century U.S.A.

As cheap fossil fuels exhaust, machines and automation increase, and the *nuclear* sources develop, there is the possibility of purchase of such complex technology from only three or four vendor countries. A certain neo-dependence can be seen in the less technologically developed countries existing on the leash of corporations who vend the equipment and service it. In such a situation, a levelling of living standards between those of the vendors and buyers would be less rapid.

A *Solar*-Hydrogen Economy would have a different effect. Thus, the technology of solar collectors, and the associated technology of energy generation and transmission (e.g. through sea solar power or aerogenerators in the high intensity wind belts), looks as though it would be far simpler than that of breeder reactors, or fusion. Safety requirements would certainly be simpler to achieve. It is easier to conceive the building up of the machinery of solar-energy generation in many of the developing countries by means of the indigenous engineering capability. A levelling in economic standards would tend to occur more quickly.

Finally, in a Solar-Hydrogen Economy, those countries which receive a particularly high degree of insolation (North African States, Saudi Arabia, Australia, South Africa and parts of India) will be able to collect and export solar energy in hydrogen, or by means of power relay satellites, to the industrial producer countries. This will take wealth from the highly industrialised areas and give it to the countries in the insolated areas. Conversely, the availability of hydrogen gas lines through other countries, and of solar energy generation in those of latitudes 30N to 30S, and its transfer to distant places, will increase the energy per capita of the solar-rich and thus decrease the imbalance of wealth among nations.

Such a levelling would tend to decrease the probability of some kinds of wars.

INTERNATIONAL POLITICAL IMPLICATIONS

There are already strains felt among the Western countries as a result of the increasing cost of energy. It will probably be necessary to raise large tax levies in order to build the New Energy System. This would be tantamount to reducing living standards and could tend towards a decrease in long-term political stability, more in affluent countries where it will be worse.

During the gap between the exhaustion of indigenous supplies of fossil fuels and the availability of massive amounts of nuclear and solar power, the U.S. will have to import gigantic quantities of oil from the OPEC countries. Such imports will be tantamount at 1985 to the arrival of a 250,000 ton tanker every two hours.[1] It would be easy to cut off this stream of tankers by naval interdiction.

Competition between the capitalist countries, particularly Japan and the U.S.A., for oil may bring ugly conflicts to the fore. The Soviet countries, with their surplus of natural gas, may be able to exert undesirable pressures on the Western economies.

One aspect which looks positive is the possibilities of export. This refers to the solar energy rich countries exporting hydrogen or beaming power, to the Northern Hemisphere countries, but also to the possibility of fusion or breeder electricity being beamed from the U.S. to distant countries. In this way, it is possible that technologically advanced countries will be able to win back some of the losses which they began to suffer in payments to the oil possessors after November 1973.

SPECULATIVE POSSIBILITIES IN ALTERNATIVE SCENARIOS

Some speculations are recorded in Table 19.1. Speculative political consequences are included in the *scenarios* because of the likelihood that political changes would follow gross variations in energy price, and hence in living standards.

SOME LEADING CONCLUSIONS OF THIS BOOK

(1) Fossil fuels will be exhausted (i.e. the price will be higher than those of alternatives), in less than three decades in respect to oil and natural gas; and in some six decades with respect to the coal known to exist in the ground. These statements are based upon the assumption of a continued growth of the world economies. Were this to halt, or even to go backwards, i.e. suffer long-term depression, the estimates would be modified upwards. As the growth of population seems not to be doubred, a reversal of growth in energy needs would be tantamount to a fall in living standards.

(2) The availability of a large fraction of the coal in the ground for use as a fuel is, however, doubtful, because a much greater rate of production would be necessary within 15 to 20 years than at present, if coal is to take over substantially from natural gas and oil. Expansion of the coal industry within this, the remaining time, to the degree which would be necessary appears an impractical goal.

(3) A new coal-based energy source; sufficient breeder reactors and practical solar energy will not be ready to the degree needed by 2000. Indeed, a large (perhaps 50%) contribution of energy from these new sources will be needed even by 1985, when the United States would be importing oil in huge quantities. There does seem, therefore to be the threat of a serious and even disastrous energy shortage, between the time when the prospective shortage of natural gas and oil drives their prices up to an unusable height, and the built-up of alternative and cheaper resources. The gap would fall between the middle 1980s, and the early tens of the next century in the U.S. It is at present unclear how this period will be bridged. The underground gasification of coal and the rapid development of massive sources of wind power are possibilities: the former often tested unsuccessfully, the latter little developed, but seeming (along with sea-solar power) to be the most hopeful prospect for very rapid (e.g., 10-20 years) development.

(4) Coal, seen as an interim source, does not exist in 'hundreds of years' worth, as has often been stated. Such estimates do not allow for the effects of population growth, and the increase of energy consumption per person, which follow long-going trends. The extrapolations which have given rise to the over-optimistic estimates of the availability of coal have been based upon its use for electricity production, and this

TABLE 19.1

SPECULATIVE SCENARIOS OF ENERGY TO 2050

Pessimistic	Probable	Optimistic
Assumptions Western countries cannot keep up payments for oil to OPEC countries. Underground gasification fails. Logistical fears for coal confirmed. Clean breeders cannot be developed before exhaustion of available fossil fuels and uranium. Solar energy is only residential.	*Assumptions* U.S. oil and natural gas only ½ needs by 1985. Rest mainly OPEC at this time. Military confrontation concerned with independence of Arab States. Climax of difficulties (i.e. maximum price energy) 1990s. CH_4 and H_2 supply from coal significant from mid-1980s. Residential solar plentiful 1985. Breeders increasingly significant 2000. Sea solar power a significant part of energy economy, 1990. Greenhouse effect, and nuclear-solar-hydrogen economy visible 2000.	*Assumptions* Coal based CH_4 and H_2 available 1980s. Sea solar power pilot plant running 1980. Massive use of high intensity wind belts (Southern Hemisphere): pilot plants for these in early 80s. Breeders satisfactory, clean, and commercial 2000. Pilot fusion reactor works, 1995. Greenhouse effect not visible 2010s.
Results Energy too expensive. kW's per capita begin to fall in 1970s. Gross unemployment, social upheaval. Situation ameliorated by massive national purchase Soviet natural gas and oil at price agreed to on condition of break-up of large international corporations.	*Results* Europe falls economically and politically to Soviets as price energy supply from OPEC too great. Rapid build-up massive solar begins mid-80s, significant 2000, thereafter greatly increased. Solar-Hydrogen Economies begin to spread after 2005. Fusion? Australia exports solar energy produced hydrogen to Japan 2020s. Great difficulties with resources. Recycling degree insufficient. Maximum difficulties of this kind 2030s. Fusion experimental 2010s, commercial 2040s.	*Result* Hydrogen Economy, 2000. Coal products still important 2000, but then fade. Living standards rise towards 20 kW per capita by 2025. ('Cheap Nuclear Power'.) Energy supply at 2010 mix breeder; sea solar power, wind and residual use solar. Corporate-giant capitalism blooms, intertwined in Soviet-based countries is base of their technology U.S. beams power to Africa and South America. Solar energy transferred up to 6,000 km by pipeline; further by power relay satellite. Resource exhaustion met by attainment negative population growth and availability abundant, clean energy for extra deep mining and recycling.

is only a small fraction of our energy use. If it were used massively to replace oil and natural gas, the date at which the world's coal reserves will exhaust, is reduced to a few decades. If the logistic difficulties to using even these limited coal resources were overcome, massive use may cause climatic problems after 2000.

(5) Abundant, Clean Energy sources are likely to be solar (eventually perhaps fusion). The position of breeder reactors is uncertain – the cost of making them and their waste products clean brings this uncertainty. Systems-analysis of solar energy by tracking mirror concentrators and by sea solar power, makes this energy source look economically feasible (and much more ecologically sound), in respect to development prospects. Research and development in these solar directions promise more for ecology and a permanent Abundant, Clean Energy source, than does the development of atomic energy, with the exception of fusion. But the solar methods offer a path to Abundant, Clean Energy which seems far more easily attainable than that of fusion.

(6) Wind energy, previously regarded as a trivial source, could be used as a massive source of energy, if aerogenerators could be located in the high intensity wind belts of the world, where the mean annual wind is more than 15 mph. Hydrogen could be made from the electricity generated there and transported to distant points.

(7) Solar energy does not look like a difficult and expensive resource, as has often been stated. Technological break-throughs occurred in 1972 and 1973 which reduce the projected cost of photovoltaic solar energy by two orders of magnitude. However, the newer methods mentioned in (5) and (6) give better cost projections than do the photovoltaic prospects, and promise to yield electricity at source more cheaply than that at which it was available from coal and oil in early 1973. However, it must be stressed that all matters with solar energy are in terms of *projections.* Hard facts cannot be to hand until pilot plants have been built.

(8) The first three areas for collection of solar energy in the world are the North African States, Saudi Arabia and Australia. The criterion of goodness is the solar intensity multiplied by the desert land area available. The export potential from these countries is enormous. The solar energy which could be collected from the Australian deserts could support the entire energy needs of Japan and the U.S.

(9) Economic projections for massive solar energy now look favourable. The main difference from the older views came with the use of tracking mirror concentrators; of ocean thermal gradients (solar sea power); and the possibilities of transmission in hydrogen, or by power relay satellites. The cost projections for photovoltaics have been reduced by the use of the edge-defined-film-growth.

(10) There is difficulty in seeing cryogenic low resistance conduction as the method for the transport of electricity over long distances. The possibilities seem to be: transport in hydrogen and transport *via* power relay satellites. Hydrogen seems to be cheaper up to about 4,000 miles. This estimate is based upon transfer in pipes on land; a sea-borne passage in pipes is an unstudied area.

(11) *Chemical* methods for the production of hydrogen involve Carnot-like efficiency factors and these added to the cost factors connected

with the maintenance of high temperature equipment, would seem likely to put the costs of the hydrogen thus produced greater than that of it produced from the heat → electricity → hydrogen route.

(12) The price of electricity is the main controlling parameter in the production of hydrogen electrochemically. There is confusion between two types of 'off-peak power'. Literal 'off-peak' power is unlikely to be available. However, cheaper, low-use time power (the use of generators not needed during the night) *should* be available. Looked at rationally, for 1974 in the U.S., the costs of such power should be about 5 mils and be available at central points for about 6 hours per day. Such power could be the beginning of a Hydrogen Economy. Liquid hydrogen produced from this electricity and with present electrolysis devices, would be only 50% more than 1975 gasoline.

(13) Two new processes for the production of hydrogen look outstanding:

(a) Thermal assisted electrolysis in which up to 50% of the electricity which produces hydrogen arises from an outside heat source, converting without a Carnot loss to hydrogen.
(b) Photo-oriented methods.

The first of these methods can be cost analysed, and promises hydrogen at some $2 per MBtu (U.S. gasoline *c.* $2.50). Cost analysis of the photo-oriented methods has as yet not much to go on, but if single crystals do not have to be used, they should give hydrogen at less than $2 per MBtu.

(14) Storage of hydrogen in solid hydrides is expensive and limited in use. Massive storage in underground cavities is feasible. The pipeline in a lengthy hydrogen transmission system will have a significant storage capacity. Liquid hydrogen is best for small quantity storage and transports. The cost of liquid hydrogen should become less than that of gasoline during the 1980s, i.e. before a Hydrogen Economy could be implemented.

(15) Hydrogen is a more difficult fuel to handle than methane or gasoline, because of the larger inflammability limits and lower ignition temperatures in contact with air. But these difficulties are reduced by the lower energy per unit volume in a potential explosion; and the fact that the gas rises, instead of dangerously spreading over the ground in the circumstances of an accident, is an improvement compared with the behaviour of the heavier hydrocarbon vapour. Safe transportation of liquid hydrogen over road and rail has been a fact for many years.

(16) There may be a problem in getting steel pipes to carry high pressure hydrogen for long periods without embrittlement. However, low alloy steels should be able to last for more than 50 years with a two-cycle per day pressure variation.

(17) Transduction of hydrogen could be chemical or electrochemical. If power per unit weight is the main thing, the transduction should be chemical; if energy (and hence economy) per unit weight is the main point, the transduction should be electrochemical. The two means will doubtless be developed in competition. Development of the fuel cell would be the most helpful single method by which we could conserve the energy of the remaining fossil fuels.

(18) The availability of massive quantities of hydrogen would give

a permanently clean metallurgical and chemical industry. In residences, a total energy concept would be available. Eventually, food production using hydrogen with atmospheric nitrogen and enzymes could be possible.

(19) Transportation would lose its difficulties of pollution and the rising price of its fuel if it operated on hydrogen originating chemically or electrochemically at first from coal as the energy source. Internal combustion engines need little modification to run satisfactorily on hydrogen. Jet air transports would have a longer range and higher payload if fueled with hydrogen. Hypersonic transports must use hydrogen.

(20) Environmentally, a Solar-Electrochemical Economy or a Solar-Hydrogen Economy, are the most favourable options in sight.

(21) The optimum fuel-based energy system is time-dependent. Formerly, it was thought that the main advantage of a Hydrogen Economy would be in respect to the removal of pollution. However, within a decade, liquid hydrogen will be the fuel of advantage on the basis of price alone. Its only competitor would be methanol, but this would have eventually to be made from hydrogen. The basic energy source would be, at first coal, thereafter nuclear, and then solar.

(22) The implementation of a Hydrogen Economy may start with a small electricity plant running on hydrogen from coal. Hawaii would seem to be a major test area for a full-scale rehearsal of a Hydrogen Economy, using inexhaustible, clean energy sources.

(23) The energy per capita is equivalent to the living standard per capita, and lowering it is a potential cause of depression. There must be confrontation between the influence of vested capital and its tendency to retain the present energy economy in terms of oil and natural gas based fuels (thus leading to a decline in average living standards), and more rational views which would bring the end of the use of fossil fuels before their exhaustion (because of their need in petrochemicals) and hence guide the economy dynamically towards reliance on nuclear and solar sources. The degree of advancement in this direction depends upon government leadership in the setting up of NASA-like Research and development organisations, particularly if the funding is oriented towards the simpler (i.e. the solar) possibilities.

The way is still open for an entirely satisfactory outcome with only minor and temporary decreases in living standard. However, the present situation in respect to the energy supply could result (as a consequence of the increasing cost of being able to do massive research and development) in Fundamental Difficulties of a long-term character.

REFERENCES

[1] H.C. Hottel, 'Challenges in Production of Fossil Fuels', Chemical Engineering Progress, June 1973.

APPENDIX I

A Note on Price Estimates

Many aspects of the present book describe alternative prospects in the technology and energy production, storage, transmission, transduction and useage. Often, the discussions are based upon extrapolation, with the danger that changes of condition at future times may decrease the validity of the estimate.

A major factor which affects the validity of these estimates is that the 1970s have been characterized by an inflation rate in the region of 10% in many countries. Thus, an estimate of, say, $2.50 for a MBTU of hydrogen made at present is not expected to be applicable literally in future years, for all prices will clearly be affected by inflation. However, *relative* prices should remain much less changed, except for those connected with exhausting resources (in particular, oil, and then coal). These will be likely to undergo extra-inflationary price increases. Conversely, it must be remembered that, for a few years yet, the price of OPEC oil will be artificially high, and, in the face of a threat of loss of markets to a new method of energy production, could be instantly lowered to destroy the competitor.

Another aspect of prices quoted is that they may involve differences for the price of the same quantity within the book. If, for example, there is a difference of 15-25% in the price quoted for the liquefaction of a certain amount of hydrogen, it is because this price is not a firm one at this time, and various estimates are available, depending on place, technology used, price of energy, and other factors.

A distinction throughout must be made between the price which can be estimated on technological grounds (cost of materials, money, labour, energy, manufacture, etc.) and the selling price. The latter will include the price of delivery, the mark-up of the manufacturer and the retailer, the State and Federal taxes, etc. For example, on September 23rd, 1974, the Oil and Gas Journal published the following prices for a U.S. gallon of gasoline:

Average price at pump	Average price ex tax	Ex-refinery (depending on location)
51¢ (1973, 39¢)	40¢ (1973, 27¢)	25-36¢

The ex-refinery price is the one with which comparison should most properly be made with the cost estimates for hydrogen, and other fuels, by technology not yet commercialized. Of course (as stressed by the above figures), the price of gasoline, the standard, is liable to rise, as is hydrogen produced from coal or atomic energy, although the price from the latter source will be affected principally by inflation and technological changes, and less by exhaustion. In respect to hydrogen from coal, however, there would be expected to be an increase parallel to the increase of the price of oil, because the price of a commodity will not be much less than that for the nearest competitor, and it may be expected that coal prices will rise during the exhaustion phase of oil, due to an increase of what profit can be demanded without fear of competition.

Lastly, it should be noted that, although competitive *price* is usually considered to be determinative in the evolution of some product, this is only so if the various products have equal quality. For example, if hydrogen is to be produced from coal before the advent of cheap nuclear or solar electricity, it may be cheaper from coal than by electrolysis. However, if the coal-born hydrogen contains sulphur dioxide in significant quantities it may not be competitive with hydrogen from electrolysis, a higher price. The ecological factor is a new element in economic thinking. Correspondingly, in considering the estimates given for methods of producing hydrogen, only very large (e.g. an order of magnitude) differences in the predicted prices among the various alternatives considered should change the priorities of further research. For example, in Table 9.6, estimates are given for these various methods but, within a decade, these estimates will not only be changed by inflation, but also by technological innovations, which will not be uniform among the methods, and by changes in price of the components in the manufacture, so that the relative prices in Table 9.6 (apart from their inflational equivalents) may be significantly changed.

APPENDIX II

On the Weight of Price Comparisons with the Cost of Fossil Fuels

Price considerations are frequently made in this book because, normally, engineering arguments as to the preferability of a technology, the end results of which are the same, are decided on comparative costs. However, in respect to the comparisons among fuels, the price arguments are only partial. It is not only that, in many of its aspects, hydrogen is a superior fuel, e.g. in respect to its ecological implications. The other consideration is that the present price of fuels from OPEC countries (the main source of future liquid hydrocarbons) is an arbitrary price, made possible by monopoly, and is many times greater than the sum of the real cost of removal from the ground, purification, etc.

Were hydrogen to become available at a price below that of gasoline, the price of the oil from which it is derived could be lowered. However, there is still weight to the quotation of comparative prices, estimates and projections. Without them, one loses one foot from the ground. Further, the development of a Hydrogen Economy (see Chapter 18) will not be realizable on a large scale until, at the best, around the century's end. Before that, a real exhaustion of liquid and gaseous hydrocarbons is likely, and the reality of exhaustion will have forced the price of hydrocarbon fuels, except those from coal, to values which will greatly limit their use, and which will not, then, be lowerable to meet competition.

Estimates and projections of the price of hydrogen (always related to dollars of a certain date), in-so-far as they are in the range of gasoline prices whilst this fuel remains usable, serve to indicate that the economics of hydrogen will not be prohibitive.

Index

This book is Royal 8vo format
Set in 10pt
Times on an eleven point body
Printed on
Matt Coated Art 90 GSM

9625